Nutrient Metabolism Studies in Companion Animals

Nutrient Metabolism Studies in Companion Animals

Editors

Lian Li
Baichuan Deng
Giulia Pignataro

MDPI

Basel • Beijing • Wuhan • Barcelona • Belgrade • Novi Sad • Cluj • Manchester

Editors
Lian Li
Nanjing Agricultural University
Nanjing
China

Baichuan Deng
South China Agricultural University
Guangzhou
China

Giulia Pignataro
University of Teramo
Teramo
Italy

Editorial Office
MDPI AG
Grosspeteranlage 5
4052 Basel, Switzerland

This is a reprint of articles from the Special Issue published online in the open access journal *Metabolites* (ISSN 2218-1989) (available at: https://www.mdpi.com/journal/metabolites/special_issues/FCSXWCIYU1).

For citation purposes, cite each article independently as indicated on the article page online and as indicated below:

Lastname, A.A.; Lastname, B.B. Article Title. *Journal Name* **Year**, *Volume Number*, Page Range.

ISBN 978-3-7258-2633-9 (Hbk)
ISBN 978-3-7258-2634-6 (PDF)
doi.org/10.3390/books978-3-7258-2634-6

Contents

metabolites

MDPI

Article

Effects of Softening Dry Food with Water on Stress Response, Intestinal Microbiome, and Metabolic Profile in Beagle Dogs

Limeng Zhang [1,2,†], Kang Yang [1,†], Shiyan Jian [1], Zhongquan Xin [1], Chaoyu Wen [1], Lingna Zhang [1], Jian Huang [3,*], Baichuan Deng [1,*] and Jinping Deng [1,*]

1 Maoming Branch, Guangdong Laboratory for Lingnan Modern Agriculture, Guangdong Provincial Key Laboratory of Animal Nutrition Control, National Engineering Research Center for Breeding Swine Industry, College of Animal Science, South China Agricultural University, Guangzhou 510642, China
2 Research Center of Pet Nutrition, Guangzhou Qingke Biotechnology Co., Ltd., Guangzhou 510642, China
3 Institute for Quality & Safety and Standards of Agricultural Products Research, Jiangxi Academy of Agricultural Sciences, Nanchang 330200, China
* Correspondence: hjian661105@126.com (J.H.); dengbaichuan@scau.edu.cn (B.D.); dengjinping@scau.deu.cn (J.D.)
† These authors contributed equally to this work.

Abstract: Softening dry food with water is believed to be more beneficial to the intestinal health and nutrients absorption of dogs by some owners, but there appears to be little scientific basis for this belief. Thus, this study aimed to compare feeding dry food (DF) and water-softened dry food (SDF) on stress response, intestinal microbiome, and metabolic profile in dogs. Twenty healthy 5-month-old beagle dogs were selected and divided into two groups according to their gender and body weight using a completely randomized block design. Both groups were fed the same basal diet, with one group fed DF and the other fed SDF. The trial lasted for 21 days. The apparent total tract digestibility (ATTD) of nutrients, inflammatory cytokines, stress hormones, heat shock protein-70 (HSP-70), fecal microbiota, short-chain fatty acids (SCFAs), branch-chain fatty acids (BCFAs), and metabolomics were measured. Results showed that there was no significant difference in body weight, ATTD, and SCFAs between the DF and SDF groups ($p > 0.05$), whereas feeding with SDF caused a significant increase in serum cortisol level ($p < 0.05$) and tended to have higher interleukin-2 ($p = 0.062$) and HSP-70 ($p = 0.097$) levels. Fecal 16S rRNA gene sequencing found that the SDF group had higher alpha diversity indices ($p < 0.05$). Furthermore, the SDF group had higher levels of *Streptococcus*, *Enterococcus*, and *Escherichia_Shigella*, and lower levels of *Faecalibacterium* ($p < 0.05$). Serum and fecal metabolomics further showed that feeding with SDF significantly influenced the purine metabolism, riboflavin metabolism, and arginine and proline metabolism ($p < 0.05$). Overall, feeding with SDF caused higher cortisol level and generated effects of higher intestinal microbial diversity in dogs, but it caused an increase in some pathogenic bacteria, which may result in intestinal microbiome disturbance and metabolic disorder in dogs. In conclusion, feeding with SDF did not provide digestive benefits but caused some stress and posed a potential threat to the intestinal health of dogs. Thus, SDF is not recommended in the feeding of dogs.

Keywords: water-softened dry food; pet food; beagle dog; inflammatory response; fecal microbiota; metabolomics

Citation: Zhang, L.; Yang, K.; Jian, S.; Xin, Z.; Wen, C.; Zhang, L.; Huang, J.; Deng, B.; Deng, J. Effects of Softening Dry Food with Water on Stress Response, Intestinal Microbiome, and Metabolic Profile in Beagle Dogs. *Metabolites* **2022**, *12*, 1124. https://doi.org/10.3390/metabo12111124

Academic Editor: Saulo L. Silva

Received: 18 October 2022
Accepted: 12 November 2022
Published: 16 November 2022

1. Introduction

Dogs serve as great companions, possibly due to their remarkable social cognitive abilities [1]. Owning a dog can provide emotional benefits and improve one's physical and mental health [2–4]. Dog owners care deeply for their pets and treat them as family members, which is evidenced by the increased pet-related expenditure on things such as pet food, veterinary service, training, and pet-sitting [5]. Therefore, maintaining health and expanding the lifespan of dogs could be of significant interest to owners.

The gut is now considered the largest endocrine organ in mammals, and intestinal health can be a top priority in pet health. Gastrointestinal (GI) health not only refers to the aspects of digestion and absorption of nutrients, but also includes the intestinal microbiota, fermentation of product, and metabolite composition [6]. The gut microbiota is a complex ecosystem that has an impact on many areas of the host's health, including physiology, behavior, and fitness [7]. GI bacteria can influence host physiology and metabolism through direct contact, and also indirectly by way of microbiota-derived metabolites [8,9]. For instance, the GI microbiota aids food digestion and the production of metabolites, including vitamins and short-chain fatty acids (SCFAs) [10]. When there is an imbalance in the gut microbiota, it either helps or hinders intestinal immunity [11]. On the other hand, gut microbiota and metabolites can be affected by different diets in healthy dogs and cats [12–15]. Therefore, the composition of the microbiota community and fecal metabolites are critical measures for better evaluation and understanding of the gastrointestinal health of dogs.

In the case of young animals, some pet food companies suggest that puffed foods should be softened before use in feeding. However, these statements lack rigorous experiment-based evidence. There has been some research in recent years on the effects of various diets, including fresh and raw meat diets on the apparent total tract digestibility (ATTD), microbiota, and metabolome of the canine gastrointestinal tract [16–18]. However, as we know, no study has been carried out on the effect of water-softened dry food (SDF) on dog's health. Thus, the objective of this study was to evaluate whether feeding with SDF is beneficial to dogs. The effects of a three-week consumption of water-SDF on healthy dogs were examined in the current study by determining the ATTD of macronutrients, immune response, stress hormones, heat shock protein-70 (HSP-70), and fecal characteristics, microbiota, and metabolites.

2. Materials and Methods

2.1. Animals and Housing

Protocols for all experiments were approved by the Experimental Animal Ethics Committee of South China Agricultural University (protocol code 2021e028). This study included twenty beagle dogs, and their information was shown in Table 1. Subjects were all housed individually kennels measured 1.2 × 1.0 × 1.1 m in an enclosed room with constant temperature (24 ± 1 °C), relative humidity (70 ± 5%) and light control conditions (12 h dark–light cycle, light: 06:00–18:00) in the building of Experimental Animal Centre of South China Agricultural University. All dogs can interact with other dogs and staff daily and were provided opportunities to go out and play at least once a week.

Table 1. Detailed information of dogs in this study.

Group [1]	N (Male: Female)	Age (M)	Weight (kg)
DF	10 (4:6)	5.38 ± 0.35	7.33 ± 0.12
SDF	10 (4:6)	5.62 ± 0.29	7.25 ± 0.18

[1] DF: dry food, SDF: water-softened dry food.

2.2. Diets and Feeding

The basal diet used in this experiment was normal commercial dog food (Ramical, Guangzhou, China). The experimental dogs were fed 200 g every day (twice—at 09:00 and 17:00) to meet their energy requirements, which is calculated approximately according to the standard of the National Research Council (NRC, 2000). During the experiment, dogs had free access to clean water at any time through the water nipple. Table 2 showed the nutrient composition of the diet used in this experiment.

Table 2. The chemical composition of the basal diet.

Items [1]	Basal diet [2]
DM (%)	91.19
OM (% DM)	91.57
CP (% DM)	27.69
AHF (% DM)	11.21
TDF (% DM)	3.63

[1] DM: dry matter; OM: organic matter; CP: crude protein; AHF: acid-hydrolyzed fat; TDF: total dietary fiber. [2] Extruded diet: corn flour, flour, sweet potato, chicken powder, meat and bone meal, soybean meal, beef meal, calcium hydrophosphate, calcium chloride, lysine, methionine, vitamin A, vitamin D, vitamin E, copper sulfate, ferrous sulfate, zinc sulfate, and manganese sulfate.

2.3. Experimental Design

In this study, a total of 20 beagle dogs were chosen. All animals were physically examined at the vet clinic before the start of the trial; vaccination and deworming were completed two months before the experiment. Following an acclimation period of one month when all dogs were fed the basal diet, they were divided into two groups according to their gender and body weight (BW) using a completely randomized block design. One group continually fed the same DF as in the acclimation period (DF, n = 10, male:female = 4:6), and the other group fed SDF was made by the mixture of 1.5 mL purified water with 1 g of dry food (SDF, n = 10, male:female = 4:6). The water used in this test was filtered through a filter to ensure quality. Additionally, the water/dry food ratio was determined to ensure the pet food was neither too tough nor too soft. The whole trial lasted for 21 days. BW and body condition score (BCS) were recorded every ten days, and fecal score (FS) was recorded every day and evaluated as described previously [19]. Fresh fecal samples and blood samples were collected on the last day of the experiment period, and the total fecal collection was performed in the last 3 days.

2.4. Chemical Analysis and Digestibility Measurements

Digestibility was measured by the total fecal collection method. A mass of 200 g dry food of basal diet was collected every 10 days, and all feces was collected and 10% HCl was added to the nitrogen fixation for the last three consecutive days in this study. All fecal samples were oven-dried for 48 h (60 °C), then diet and dry fecal samples were ground in a grinder with a 1 mm screen for further chemical analysis. Dry matter (DM) and organic matter (OM) of the diet and feces were determined according to Association of Official Analytical Chemists (AOAC) (2000; method 950.46 and 942.05). A fatty analyzer (FT640, Grand Analytical Instrument Co., Ltd., Guangzhou, China) was used to determine the acid-hydrolyzed fat level (AHF) of the diet and feces (AOAC, 2000; method 954.01). An automatic fiber analyzer (FIBRETHERM FT12, C. Gerhardt GmbH & Co. KG, Königswinter, Germany) was used to analyze the total dietary fiber (TDF) content of the diet and feces (AOAC, 2000; method 962.09). Crude protein (CP) was determined using AOAC with the semi-automatic Kjeldahl apparatus (VAPODEST 200, C. Gerhardt GmbH & Co. KG, Germany). Besides, ATTD values were calculated using the equation as follows: [nutrient intake (g/d, DM basis) − fecal output (g/d, DM basis)]/nutrient intake (g/d, DM basis) × 100.

2.5. Blood Sample Collection and Analysis

On day 21, after overnight fasting, 5 mL of blood was taken via forelimb venipuncture. The samples were then clotted for 30 min and centrifugated for 15 min to separate the supernatant for further analyses. Commercial canine enzyme-linked immunosorbent assay (ELISA) kits (MEIMIAN, Jiangsu Meimian Industrial Co., Ltd., Yancheng, China) were used to measure the serum cortisol (COR), adrenocorticotropin hormone (ACTH), glucocorticoids (GC), HSP-70, immunoglobulin G (IgG), tumor necrosis factor-alpha (TNF-α), interferon-γ (INF-γ), interleukin-6 (IL-6), interleukin-4 (IL-4), and interleukin-2 (IL-2).

2.6. Microbial Analyses

2.6.1. DNA Extraction, Amplification, and Sequencing

Total genomic DNA was isolated from fresh fecal samples using the cetyltrimethylammonium bromide method. DNA concentration and purity were monitored on a 1% agarose gel. The DNA was diluted to 1 ng/μL with sterile water based on the concentration. The 16S rRNA genes for 16S V3-V4 were amplified using primers 341F (5′-CCTAYGGGRBGCASCAG-3′) and 806R (5′-GGACTACNNGGGTATCTAAT-3′) with barcodes. For all PCR reactions, a volume of 15 μL Phusion High Fidelity PCR Master Mix (New England Biolabs) was used, along with 2 μM of forward and reverse primers and nearly 10 ng of template DNA. The thermal cycling process was set as follows: initial denaturation for 1 min (98 °C), denaturation for 10 s (98 °C), annealing for 30 s (50 °C), and extension twice for 30 s and 5 min, respectively (72 °C). The PCR products were added with the same volume of (1X) loading buffer (including SYB Green) in an equal density ratio which was subsequently analyzed by electrophoretic manipulation on a 2% agarose gel. Using the Qiagen Gel Extraction Kit, the PCR product combination was then purified (Qiagen, Hilden, Germany). The TruSeq DNA PCR-Free Sample Preparation Kit (Illumina, San Diego, CA, USA) was used to build sequencing libraries, and index codes were then added. A Qubit@ 2.0 fluorometer (Thermo Scientific, Waltham, MA, USA) and an Agilent Bioanalyzer 2100 system were used to evaluate the library's quality. Finally, the Illumina NovaSeq technology was used to generate final paired-end reads of 250 bp.

2.6.2. Bioinformatics Analysis

Pairs of reads were combined using FLASH (V1.2.7, http://ccb.jhu.edu/software/FLASH/, accessed on 10 June 2022) [20], and the spliced sequence was referred to as the original label. The raw labels were quality filtered according to the quality control process of QIIME (V1.9.1, http://qiime.org/scripts/split_libraries_fastq.html, accessed on 10 June 2022) [21], resulting in high-quality clean labels [22]. Chimeric sequences were detected by comparing the tags with a reference database (Silva database, https://www.arb-silva.de/, accessed on 10 June 2022) using the UCHIME algorithm (UCHIME http://www.drive5.com/usearch/manual/uchime_algo.html, accessed on 10 June 2022) [23]. Then, we removed the chimeric sequences [24] and finally obtained the valid tags.

We used Uparse software (Uparse v7.0.1001, http://drive5.com/uparse/, accessed on 10 June 2022) (Edgar, 2013) to perform sequence analysis. Same OTU were defined as those sequences with ≥97% similarity. For each representative sequence, the classification information was annotated according to the Mothur algorithm and using the Silva database (http://www.arb-silva.de/, accessed on 10 June 2022) [25]. We used MUSCLE software Multiple to performing sequence comparisons (version 3.8.31, http://www.drive5.com/muscle/, accessed on 10 June 2022) [26] to investigate the phylogenetic relationships of different OTUs. Abundance information of OTUs was normalized using sequence number criteria corresponding to the least sequenced sample for further analyses of both alpha-diversity and beta-diversity. Alpha-diversity was measured by six indices, including Observed_species, Chao1, Shannon, Simpson, Ace, and Goods_coverage. All these indices in our samples were calculated using QIIME (version 1.7.0) and displayed using R software (version 2.15.3).

2.7. Fecal Short-Chain Fatty Acids and Branch-Chain Fatty Acids Analyses

The gas chromatograph (Shimadzu, Kyoto, Japan) coupled to a flame ionization detector and chromatographic separation capillary column (DB-FFAP, 30 m × 0.25 mm × 0.25 μm) was used in this study to analyze the concentrations of fecal SCFAs (acetic acid, butyric acid, and propionic acid) and branch-chain fatty acids (BCFAs) (isovaleric acid, isobutyric acid, and valeric acid). The instrument parameters and sample processing procedures of extraction of fecal volatiles referred to our previous study [27].

2.8. Untargeted Fecal and Serum Metabolomics Analysis

The sample processing procedures of extraction of fecal and serum metabolites referred to our previous study [27]. UPLC-Orbitrap-MS/MS analysis method was used as in a previous study [28], only with slight alterations. Thermo Fisher Scientific's compound Discover 2.1 data analysis tool was used to scan the mzCloud and mzVault libraries for metabolites. Performing multivariate analysis, MetaboAnalyst 5.0 (https://www.metaboanalyst.ca, accessed on 10 June 2022) was used. In this study, principal component analysis (PCA) and pathway enrichment analyses were performed on MetaboAnalyst 5.0.

2.9. Statistical Analysis

SPSS 26.0 was used for statistical analysis, and GraphPad Prism 8.0.3 software was used for graphical presentation. Student's t tests were employed to compare two groups. The mean ± standard error (SE) was used to express all data. Significance was set at $p < 0.05$, a trend was defined by $0.05 < p < 0.10$. Selecting the metabolites whose adjusted p value < 0.05 (calculated by Student's t test) and the variable importance in the projection (VIP) >2 to obtain the differential metabolites.

3. Results

3.1. Effects of SDF on Body Weight, Body Condition Score, and Fecal Score

As shown in Table 3, there was no difference in BW or BCS between the DF and SDF groups throughout the trial period ($p > 0.05$). The total fecal score (TFS) of the SDF group was markedly higher than that of the DF group ($p < 0.05$). The total soft feces ratio (TFSR) of the DF and SDF groups were both 0.48%. The total diarrhea ratio (TDR) of the SDF group was 0.95%, which was 2.38% higher than TDR of the DF group (3.33%).

Table 3. Body weight, body condition score, and fecal score in dogs fed DF and SDF on days 11 and 21.

Item [1]	DF [2]	SDF [3]	p Value
BW (kg)	8.01 ± 0.14	7.87 ± 0.18	0.540
BCS	5.05 ± 0.17	4.65 ± 0.11	0.660
TFS	2.55 ± 0.14	2.62 ± 0.25	0.028
TSFR (%)	0.48	0.48	-
TDR (%)	0.95	3.33	-

[1] BW: body weight; BCS: body condition score; all the test dogs were weighed and BCS were recorded with a nine-point scale on days 1, 11, and 21 before morning feeding. TFS: total fecal score. All samples were scored in 0.5 units increments using the following scale: 1 = hard, dry pellets, small hard masses; 2 = hard, stool, remains firm and soft; 3 = soft, formed, and moist stool, retains shape; 4 = soft, unformed stool, assumes shape of container; 5 = watery, liquid that can be poured; in which, $1 \leq FS < 2$ constipation, $2 \leq FS \leq 3$ normal, $3 < FS < 4$ soft stool, $4 \leq FS \leq 5$ diarrhea. TSFR: total soft feces ratio. Ratio of the number of soft feces to the total number of fecal scores. TDR: total diarrhea ratio. Ratio of the number of diarrhea to the total number of fecal scores. [2] DF: dry food. [3] SDF: water-softened dry food.

3.2. Effects of Soften Dry Food on Apparent Total Tract Macronutrient Digestibility

As shown in Table 4, fecal output, ATTD of DM, OM, CP, AHF, and TDF all showed no significant difference between the two groups ($p > 0.05$).

3.3. Effects of SDF on Serum Hormones, Heat Shock Protein-70, IgG, and Inflammatory Factors

As shown in Table 5, the SDF group tended to have higher IL-2 ($p = 0.062$) and HSP-70 ($p = 0.097$) levels than the DF group. The SDF group also had a higher level of COR compared with the DF group ($p < 0.05$). Besides, there was no difference shown in IgG, IFN-γ, IL-6, IL-4, TNF-α, GC, and ACTH of the two groups ($p > 0.05$).

Table 4. Food intake, fecal characteristics, and apparent total tract macronutrient digestibility in dogs fed DF and SDF.

Item	DF [1]	SDF [2]	*p* Value
Food intake [3], g/d (DM basis)	184.00	184.00	-
Fecal output, g/d (as is)	91.07 ± 4.78	89.00 ± 1.81	0.693
Fecal output, g/d (DM basis)	24.90 ± 1.27	23.82 ± 1.73	0.628
Digestibility, %			
Dry matter	88.55 ± 0.01	89.04 ± 0.01	0.628
%DM basis			
Organic matter	81.28 ± 0.26	82.39 ± 0.10	0.693
Crude protein	83.88 ± 0.01	85.52 ± 0.01	0.301
Acid hydrolyzation fat	94.83 ± 0.01	95.84 ± 0.01	0.104
Total dietary fiber	70.73 ± 0.02	66.82 ± 0.01	0.202

[1] DF: dry food. [2] SDF: water-softened dry food. [3] All dogs were fed a restricted diet during the trial and all dogs had no leftovers at each meal.

Table 5. Serum IgG, inflammatory factors, hormones, and HSP-70 in dogs fed DF and SDF on day 21.

Item [1]	DF [2]	SDF [3]	*p* Value
IgG	258.88 ± 16.04	227.38 ± 14.90	0.167
IFN-γ	42.10 ± 1.39	45.23 ± 1.23	0.108
IL-6	541.77 ± 20.15	567.00 ± 21.69	0.405
IL-4	163.72 ± 14.20	146.66 ± 9.65	0.334
IL-2	254.86 ± 11.54	287.23 ± 11.42	0.062
TNF-α	77.08 ± 4.15	77.12 ± 4.69	0.995
GC	321.91 ± 13.13	318.80 ± 9.23	0.887
ACTH	67.23 ± 3.31	68.10 ± 2.45	0.836
COR	394.24 ± 13.13	438.50 ± 9.23	0.014
HSP-70	334.50 ± 18.05	376.07 ± 15.42	0.097

[1] IgG = immunoglobulin G; IFN-γ = interleukin-γ; IL-6 = interleukin-6; IL-4 = interleukin-4; IL-2 = interleukin-2; ACTH = adreno-cortico-tropic-hormone; TNF-α = tumor necrosis factor; GC = glucocorticoid; COR = cortisol; HSP-70 = heat stress protein 70. [2] DF: dry food. [3] SDF: water-softened dry food.

3.4. Effects of SDF on Fecal Microbiota

As shown in Figure 1A, alpha diversity analyses of all samples were performed. Chao1 and Ace represent species richness, while the Shannon and Simpson indices are indicative of a diverse microbial population. We can see that the fecal microbial communities of the SDF group had more Observed_species, higher Ace and Chao1 indices levels than that of the DF group, whereas the fecal microbial of the SDF group had lower Goods_coverage over that of the DF group ($p < 0.05$). Besides, fecal samples were evaluated for beta diversity to analyze the structural difference of microbial community (Figure 1B). There was no separation between the DF and SDF groups in PCoA of weighted UniFrac distances ($p > 0.05$). However, PCoA of unweighted UniFrac distances revealed distinct separation between two groups ($p < 0.05$), which showed that SDF could influence the gut microbiota composition and structure in dogs.

The top 6 abundant phyla included Firmicutes, Fusobacteria, Bacteroidetes, Actinobacteria, Actinobacteriota, and Proteobacteria (Figure 1C). Additionally, the relative abundances at the family level indicated that the top 8 abundant families included Lactobacillaceae, Peptostreptococcaceae, Erysipelotrichaceae, Fusobacteriaceae, Ruminococcaceae, Streptococcaceae, Lachnospiraceae, and Muribaculaceae (Figure 1C). Finally, the top 10 abundant genera were Lactobacillus, Peptoclostridium, Fusobacterium, Allobaculum, Cetobacterium, Faecalibacterium, Streptoboccaceae, Blautia, Romboutsia, and Bifidobacterium (Figure 1C).

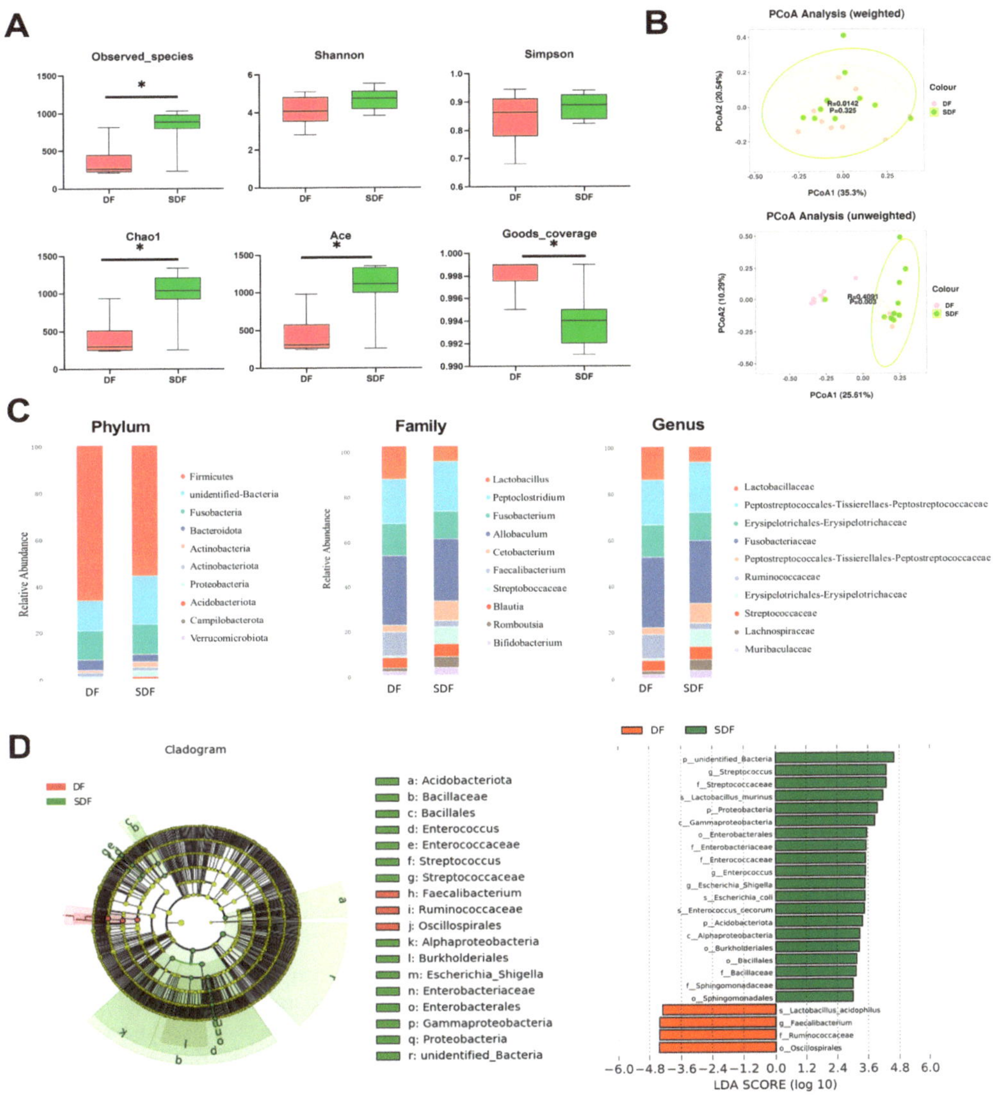

Figure 1. Alpha diversity indices (Observed_species, Shannon, Simpson, Chao1, Ace and Goods_coverage) of fecal microbial communities of the DF and SDF groups on day 21 (**A**). The symbol (*) indicates statistically significant differences between two groups. Principal Coordinate Analysis (PCoA) based on weighted and unweighted UniFrac distances of fecal microbial communities of the DF and SDF groups on day 21 (**B**). Predominant fecal microbial communities and different bacteria at the phylum, family, and genus levels of the DF and SDF groups on day 21 (**C**). LEfSe analysis identified gut bacterial biomarkers of the DF and SDF groups on day 21 (**D**). The histogram of the LDA scores presents species (biomarker) whose abundance showed significant differences between different groups. The LDA score represents the effect size. In the cladogram, circles radiating from the inner side to outer side represent taxonomic level from phylum to genus (species). Each circle's diameter is proportional to the taxon's relative abundance. Yellow nodes refer to the bacteria that contributed a lot in the DF group and blue nodes refer to the bacteria dominant in the SDF group. DF = dry food; SDF = softened dry food.

To select the differential bacteria, differential abundant taxa were confirmed by LEfSe analysis (LDA > 3.0). Proteobacteria and Acidobacteriota were enriched in the SDF group at the phylum level. At the family level, Proteobacteria, Enterobacteriaceae, Enterococcaceae, and Sphingomonadaies were enriched in the SDF group, and Ruminococcaceae was enriched in the DF group. Furthermore, at a genus level, Streptococcus, Enterococcus, and Escherichia_Shigella were enriched in the SDF group, and Faecalibacterium was enriched in the DF group (Figure 1D).

3.5. *Effects of SDF on Fecal Short-Chain Fatty Acids and Branch-Chain Fatty Acids*

As shown in Table 6, no significant difference occurred in the concentrations of fecal total SCFAs, acetic acid, propionic acid, butyric acid, total BCFAs, valeric acid, isovaleric acid, and isobutyric acid ($p > 0.05$) between the DF and SDF groups.

Table 6. Fecal SCFAs and BCFAs concentrations in dogs fed DF and SDF.

Item (μg/g)	DF [1]	SDF [2]	*p* Value
Acetic acid	1476.31 ± 35.88	1479.08 ± 54.89	0.968
Propionic acid	1079.94 ± 23.41	1063.59 ± 43.24	0.751
Butyric acid	442.42 ± 19.53	443.47 ± 13.93	0.891
Total SCFAs [3]	2998.66 ± 72.02	2986.13 ± 103.37	0.924
Isobutyric acid	92.21 ± 7.12	91.06 ± 9.95	0.955
Isovaleric acid	142.52 ± 12.76	141.41 ± 14.25	0.550
Valeric acid	19.70 ± 3.26	17.35 ± 2.20	0.891
Total BCFAs [4]	254.43 ± 20.20	249.82 ± 25.68	0.922

[1] DF: dry food. [2] SDF: softened dry food. [3] Total SCFAs: total short-chain fatty acids = acetic acid + propionic acid + butyric acid. [4] Total BCFAs: total branch-chain fatty acids = isobutyric acid + isovaleric acid + valeric acid.

3.6. *Effects of SDF on Fecal Metabolome*

To investigate the metabolic regulation in the gut of SDF, the fecal metabolites were analyzed using UPLC-Orbitrap-MS/MS. Metabolomics analysis from all fecal samples identified 249 metabolites. PCA of the fecal metabolomics shows that SDF influenced the fecal metabolites in the gut of dogs (Figure 2A). Furthermore, the Orthogonal Partial Least-Squares-Discriminant Analysis (ortho PLS-DA) for fecal metabolomics in two groups shows significant separation (Figure 2B). This result confirms the effects of SDF on fecal metabolites.

Twenty-four fecal metabolites were significantly different between the DF and SDF groups (Table 7). Dogs fed SDF had higher contents of isoquinoline, N-Acetylmuramic acid, riboflavin, guanosine, glutamylglutamine, tigilc acid, 5-Aminopentanoic acid, taurine, L-Glutamine, guanine and deoxyadenosine, and lower contents of 16-Hydroxy hexadecenoic acid, stearoylethanolamide, tryptophanol, N, N-Dimethylsphingosine, 4-Dodecylbenzenesulfonic acid Na salt, oleamide, anandamide, sebacic aced, 2-Piperidinone, carnosine, palmitic amide, adenosine, and fluoxymesteron. The enrichment analysis platform of Metaboanalyst 5.0 was adopted to explore the differential metabolic pathways (Figure 2C). Among these, purine metabolism (nucleotide metabolism) and riboflavin metabolism (metabolism of cofactors and vitamins) were significantly influenced. The specific fecal metabolites enriched in the differential pathways were showed in the bar plot (Figure 2D).

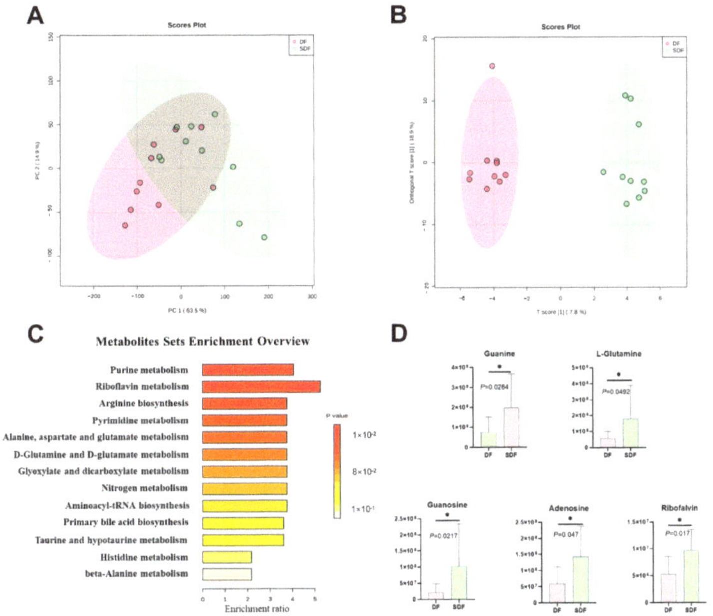

Figure 2. Principal components analysis (PCA) for fecal metabolomics data of dogs fed DF and SDF on day 21 (**A**). Orthogonal Partial Least-Squares-Discriminant Analysis (orthoPLS-DA) for fecal metabolomics data of dogs fed DF and SDF on day 21 (**B**). Enrichment bar plot based on the fecal differential metabolites of dogs fed DF and SDF on day 21 (**C**). Bar plot based on the fecal differential metabolites which were enriched in the differential metabolite pathways between two groups (**D**). The symbol (*) indicates the difference between the dry food (DF) and softened dry food (SDF) groups.

Table 7. Fecal differential metabolites were identified according to the standard of VIP > 1.5 and $p < 0.05$ between the DF and SDF groups.

Metabolites	*p* Value	VIP [1]	KEGG [2]	Trend (SDF [3] vs. DF [4])
16-Hydroxy hexadecenoic acid	0.000	2.03	-	Down
Stearoylethanolamide	0.000	1.96	-	Down
Isoquinoline	0.002	2.01	C06323	Up
Tryptophanol	0.002	2.26	C00955	Down
N, N-Dimethylsphingosine	0.002	2.17	C13914	Down
N-Acetylmuramic acid	0.004	1.58	C02713	Up
Riboflavin	0.004	2.28	C00255	Up
4-Dodecylbenzenesulfonic acid Na salt	0.004	2.24	-	Down
Oleamide	0.005	2.03	C19670	Down
Anandamide	0.005	1.63	C11695	Down
Guanosine	0.006	1.98	C00387	Up
Sebacic acid	0.006	2.15	C08277	Down
Glutamylglutamine	0.012	1.92	-	Up
2-Piperidinone	0.020	2.12	-	Down
Carnosine	0.027	2.08	C00386	Down
Tiglic acid	0.027	1.73	C08279	Up
Palmitic amide	0.023	1.82	-	Down
5-Aminopentanoic acid	0.028	1.72	C00431	Up
Taurine	0.029	1.61	C00245	Up
Adenosine	0.034	1.51	C00212	Down
L-Glutamine	0.035	1.56	C00064	Up
Fluoxymesterone	0.036	1.57	-	Down
Guanine	0.039	1.89	C00242	Up
Deoxyadenosine	0.040	1.51	C00559	Up

[1] VIP: variable importance in the projection. [2] KEGG: Kyoto Encyclopedia of Genes and Genomes. [3] SDF: water-softened dry food. [4] DF: dry food.

3.7. Effects of SDF on Serum Metabolome

To further investigate the effects of an SDF diet on metabolic regulation in dogs, we used UPLC-Orbitrap-MS/MS to analyze the serum metabolites. Metabolomics analysis from all serum samples identified 126 metabolites. PCA of the serum metabolomics showed no significant separation between two groups (Figure 3A). However, the ortho PLS-DA for serum metabolomics in the two groups shows significant separation (Figure 3B). This result shows that the SDF influenced the serum metabolites in dogs.

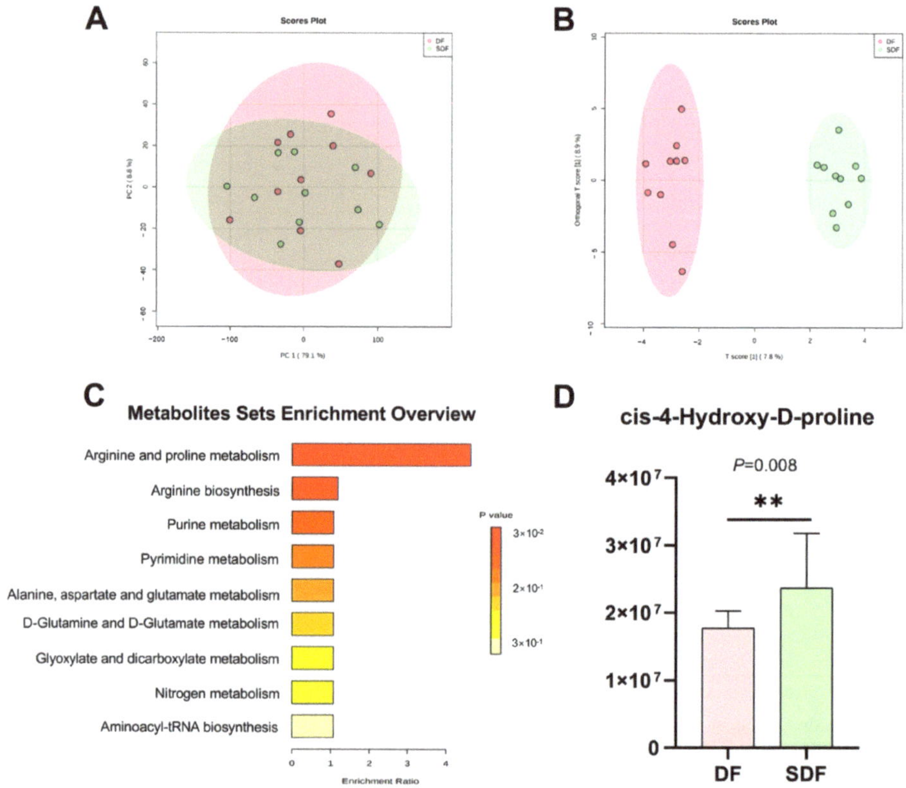

Figure 3. Principal components analysis (PCA) for serum metabolomics data of dogs fed dry food (DF) and softened dry food (SDF) on day 21 (**A**). Orthogonal Partial Least-Squares-Discriminant Analysis (orthoPLS-DA) for serum metabolomics data of dogs fed dry food (DF) and softened dry food (SDF) on day 21 (**B**). Enrichment bar plot based on the serum differential metabolites of dogs fed dry food (DF) and softened dry food (SDF) on day 21 (**C**). Bar plot based on the serum differential metabolites which were enriched in the differential metabolite pathways between two groups (**D**). The symbol (**) indicates the difference between the dry food (DF) and softened dry food (SDF) groups.

We selected the differential metabolites with VIP > 1.5 and $p < 0.05$, and 13 serum metabolites were significantly different between the DF and SDF groups (Table 8). Dogs fed SDF had higher contents of 3-Hydroxycapric acid, 3-indolebutyric acid, cis-4-hydroxy-D-proline, isoquinoline, alpha-hydroxyisobutyric acid, and L-glutamine, lower contents of N-acetylornithine, fluoren-9-one, 3-cresotinic acid, 3,4-dihydroxyhydrocinnamic acid, N-acetyl-L-phenylalanine, and indole-3-propionic acid. The enrichment analysis platform of Metaboanalyst 5.0 was adopted to explore the differential metabolic pathways (Figure 3C). Among these, arginine and proline metabolism were significantly influenced. Furthermore, the specific serum metabolites enriched in the differential pathways were showed in the bar plot (Figure 3D).

Table 8. The serum differential metabolites were identified according to the standard of VIP > 1.5 and $p < 0.05$ between the DF and SDF groups.

Metabolites	p Value	VIP [1]	KEGG [2]	Trend (SDF [3] vs. DF [4])
3-Hydroxycapric acid	0.001	2.49	-	Up
N-Acetylornithine	0.004	2.13	C00437	Down
3-Indolebutyric acid	0.006	2.16	C11284	Up
cis-4-Hydroxy-D-proline	0.008	2.02	C03440	Up
3-Hydroxybutyric acid	0.011	1.94	C01089	Up
Isoquinoline	0.011	2.05	C06323	Up
Fluoren-9-one	0.012	1.87	C06712	Down
3-Cresotinic acid	0.022	1.89	C14088	Down
3,4-Dihydroxyhydrocinnamic acid	0.024	1.84	C10447	Down
N-Acetyl-L-phenylalanine	0.026	1.85	C03519	Down
Alpha-Hydroxyisobutyric acid	0.027	1.90	-	Up
L-Glutamine	0.030	1.76	C00064	Up
Indole-3-propionic acid	0.042	1.83	-	Down

[1] VIP: variable importance in the projection. [2] KEGG: Kyoto Encyclopedia of Genes and Genomes. [3] SDF: water-softened dry food. [4] DF: dry food.

3.8. Correlation between Differential Metabolites and Fecal Bacteria at the Genus Level

Correlation between differential metabolites (fecal and serum) and fecal bacteria (at the genus level) is displayed in the heatmap (Figure 4). As we can see that Aeromounas were negatively associated with fluoxymesterone and fluoren-9-one, Dubosiella were negatively associated with fluxoymesterone, 4-Dodecylbenzenesulfonic acid Na salt, palmitic amide, 3-cresotinic acid and N-acetyo-L-phenylalanine. Besides, Collinsella had a positive association with isoquinoline, and Escherichia-Shigella had a positive association with serum cortisol, 3-indolebutyric acid, and isoquinoline.

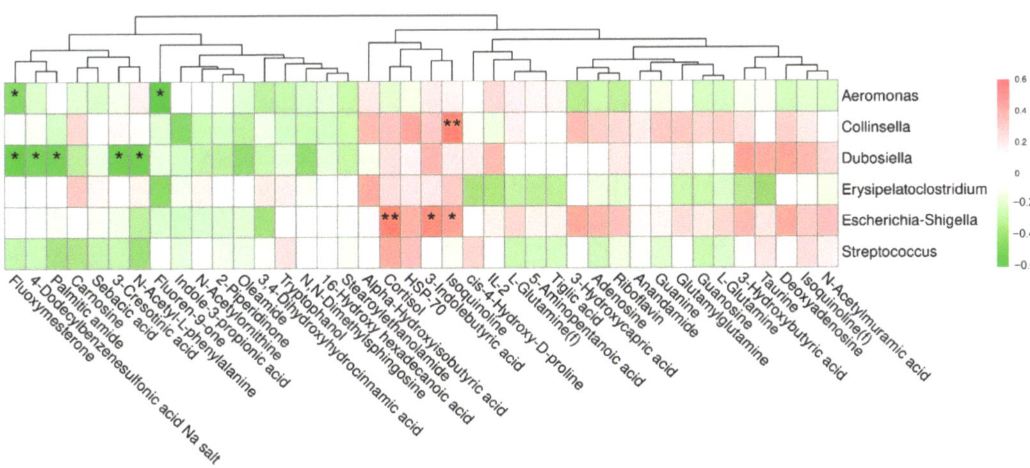

Figure 4. Correlation heatmap between differential metabolites and differential genera in the dogs on day 21. The symbol (*) indicates a significant correlation between serum metabolites and fecal bacteria (* $p < 0.05$, ** $p < 0.01$). Pink color indicates a positive correlation, and green color indicates a negative correlation.

4. Discussion

With scientific and healthy concepts being introduced more in pet feeding strategies, pet owners are paying extra attention to the diet quality and intestinal health of their pets. Although some pet food companies recommend that feeding pets with SDF is better for their health, there is no rigorous experiment-based evidence to suggest this. Therefore, we investigated the effects of SDF on ATTD, immune response, stress indicators, and fecal characteristics, microbiota, and metabolites in dogs.

Nutrient digestibility is an important measurement for evaluating the quality of a diet for dogs [29]. Many factors, including dietary ingredients, processing mode, and animal's physiology state can affect the nutrient digestibility, while fecal output may also be influenced by food intake, the chemical composition of the diet, physiological state, and nutrient digestibility of animals [30]. Higher nutrient digestibility usually results in lower fecal output. This can correspond to the results of the current study. Under the premise of daily intake of the same amounts of foods, there were no significant differences in fecal output and ATTD. Additionally, SDF had no effect on nutrient digestibility in dogs. In addition, nutrient digestibility, fiber content, DM intake, and fat tolerance can affect the fecal quality [31,32]. Even though the basal diet was the same for both groups in the present study, SDF resulted in softer stools and a higher diarrhea ratio than DF. This may be caused by the disturbance of the intestinal bacteria.

In this study, serum COR of the SDF group was significantly higher than that of the DF group on day 21. An increase in COR is commonly observed in stress response where internal and external stimulus activates the hypothalamic-pituitary-adrenal axis and triggers a cascade of hormonal release [33,34]. Besides, we found that serum COR level was positively associated with *Escherichia-Shigella*, indicating that the increase in *Escherichia-Shigella* may cause stress to dogs fed with SDF, thereby inducing an increase in th serum COR level. Additionally, the SDF group tended to have higher serum IL-2 and HSP-70 levels than the DF group. IL-2 is the major regulatory hormone of the immune system [35]. Heat shock proteins are induced by a wide variety of stressors and have broad cytoprotective functions; in particular, HSP-70 plays a vital role in cellular protection [36–38]. This may suggest that feeding with SDF may exert stress and have a tendency to cause inflammation in dogs.

In addition to the signs of gastrointestinal health (e.g., fecal state, diarrhea ratio, and volume) that are noticeable to pet owners, the abundance and activity of gut microbiota are critical to the long-term health [39–41]. Different diets can regulate the gut microbiota [12]. In our study, the results of alpha diversity in fecal microbial communities revealed significant differences between the two groups on day 21. Changes in the microbial communities occurred in the SDF group, which had more Observed_species, higher Ace and Chao1 indices, and lower Goods_coverage than the DF group. Moreover, a significant separation occurred between two groups in PCoA analysis based on unweighted UniFrac distances. In short, feeding with SDF could increase the species richness, the diversity of the gut microbiota and change the intestinal microbiome structure in dogs. The most abundant phyla found in the two groups were Firmicutes, Fusobacteria, Bacteroidetes, Actinobacteria, and Proteobacteria, which are similar to the previous studies [16,17,42]. The results of LEfSe showed that feeding dogs with SDF significantly changed with phyla Proteobacteria and Acidobacteriota, families Streptococcaceae, Enterococcaceae, Bacillaceae, Sphingomonadaceae, and Ruminococcaceae. Besides, genus *Streptococcus*, *Collinsella*, *Dubosiella*, *Escherichia_Shigella*, and *Enterococcus* were proved to be significantly enriched in the SDF group. Among these bacteria, *Streptococcus*, *Escherichia_Shigella*, and *Enterococcus* are known as pathogen bacterial associated with idiopathic small intestinal irritable bowel disease [40,43], and a previous study showed that *Escherichia_Shigella* had a strong negative correlation with *Faecalibacterium* [44], indicating that SDF caused the growth of some pathogenic bacteria. Similar to this study, our study results also showed that *Escherichia_Shigella* was enriched in the SDF group and *Faecalibacterium* was enriched in the DF group. *Collinsella*, the major taxon of the Coriobacteriaceae family, has a close correlation to insulin circulation [45]. *Dubosiella* has been recently identified as a novel member of the family Erysipeotrichaceae which thrives in such conditions, and a study showed it was enriched in the gut of obese LKO mice fed a high-fat diet [46,47]. The exact functions of *Dubosiella* are not clear, but a study showed the abundances of *Dubosiella* increased in DSS-induced mice [48]. Thus, we can speculate that the increase in fecal *Collinsella* and

Dubosiella in the SDF group may result in some potential inflammation in the present study. In conclusion, changes in bacterial phylogeny observed are not always indicative of functional changes. However, according to the results of previous studies, most of the increased bacteria genera in the SDF group are potential pathogenic bacteria from previous study; that is, SDF may lead to an increase to the intestinal pathogens. Moisture content is an important factor in the growth of bacteria in pet food, and SDF may increase the risk.

An abundance of gut microbial-derived metabolites exists in the digestive tract [49]. In this study, there was no difference in fecal SCFAs and BCFAs between the SDF and DF groups. However, fecal metabolites derived from microbiome activity are usually secreted into the intestine and transferred to the circulatory system through the intestinal barrier, serving as very important regulators of host metabolism [9,50]. Metabolomics helps to identify the key metabolites in different diets. Our study used UPLC-Orbitrap-MS/MS to analyze and investigate the changes in fecal and serum metabolites in dogs with 21 days of SDF feeding.

The results of fecal metabolites analyses showed that SDF significantly influenced the purine metabolism and riboflavin metabolism in dogs, and that there were five differential metabolites (guanine, L-glutamine, guanosine, adenosine, and ribofalvin) enriched in these two differential pathways. A study indicated that the release of guanine increases following hypoxia/hypoglycemia and permits astrocytes to exert neurotrophic effects [51]. Guanosine may have a function in stress response, biofilm development, and cellular damage protection [52–54]. A study reported that L-glutamine is an abundant nitrogen source in host serum and cells and serves as an environmental indicator and inducer of virulent gene expression [55]. Adenosine is an inhibitory modulator of neuronal activity, with potent inhibitory effects on synaptic activity. In a study, extracellular adenosine levels increase dramatically in the brain via hydrolysis of adenosine triphosphate during conditions of high energy demand [56]. Another study demonstrated that adenosine interferes with the production and release of various inflammatory mediators, and it activates cellular antioxidant defense systems, thus providing protective effects at multiple levels in the pathogenesis of ischemia and reperfusion [57]. Except for the above metabolites, some other differential metabolites are also worth noting. Taurine has been linked to antioxidant functions and defense against oxygen free radicals [58], as well as apoptosis, inflammation, cell death, and increasing NO production in endothelial cells [59,60]. Taurine can alleviate oxidative stress either directly by converting superoxides to taurine chloramine [61], or indirectly through a variety of processes, including the renin-angiotensin system [62]. In conclusion, the higher contents of guanosine and taurine in the SDF group also seem to indicate oxidative stress. In both in vitro and in vivo studies, oleamide has been shown to have anti-inflammatory and anti-allergenic properties [63–65]. Conversely, a study showed that oleamide and palmitic amide may significantly and positively correlate with activities circulating inflammatory markers, including TNF-α, hypersensitive-c-reactive-protein, and Lipoprotein-associated phospholipaseA2 activities [66]. While in this study, the lower contents of oleamide and palmitic amide may increase inflammation in dogs fed with SDF.

Further serum metabolites analyses showed that SDF significantly influenced the arginine and proline metabolism, and that cis-4-Hydroxy-D-proline was the differential metabolite in the differential pathway. However, the function and mechanism of cis-4-Hydroxy-D-proline in the serum are still unknown. Besides, high levels of 3-indolebutyric acid (IBA) and isoquinoline were observed in the SDF group. IBA, the naturally occurring metabolite of tryptophan, a study shows that it was the metabolic product of Clostridia species [67]. In another study, IBA was proven to co-occur with the incidence of inflammatory bowel syndromes in schizophrenic patients [68], and a potential mechanism for IBA to control inflammation is competitive inhibition of phospholipase A2 [69]. Isoquinoline was a type of cardiac glycosides and potent apoptosis inducers. Its derivatives are widely presented in many plants and foodstuffs, and readily cross the blood–brain barrier. In addition, isoquinoline derivatives are oxidized by monoamine oxidases to produce isoquinolinium cations with the concomitant generation of reactive oxygen species [70]. According to correlation analysis, we found that 3-indolebutyric acid and isoquinoline were positively associated with *Escherichia-Shigella*. Therefore, we suggest that 3-indolebutyric acid and isoquinoline could serve as potential biomarkers of the SDF stress response. However, the relationship between intestinal microbiome and metabolites needs further investigation.

These results could adequately explain the result of the increased diarrhea ratio and FS in dogs, and the increase in the pathogens caused some stress to dogs, thereby leading to the elevated serum IL-2, COR, and HSP-70. Based on the changes of specific bacteria and different metabolites, we speculated that bacteria may breed in the process of water-SDF, which leads to the increase in TFS and TDR, as well as abnormal fecal shape. However, although we observed that the contents of some bacteria and metabolites were changed and had some correlation, the exact biological functions and the underlying mechanisms in dogs still need further exploration.

5. Conclusions

We found that feeding with SDF did not influence the ATTD, fecal SCFAs and BCFAs in dogs. However, this feeding method increased the diarrhea ratio, serum stress indicators levels, and risk of inflammation. Although feeding with SDF increased the diversity of the intestinal microbiotas in dogs, it caused a significant increase in some pathogenic bacteria. Meanwhile, fecal metabolomics further revealed that feeding with SDF can induce metabolic disorders in dogs. In conclusion, feeding with SDF did not provide digestive benefits, but posed some stress and a potential threat to the intestinal health of dogs. Thus, SDF is not recommended in dogs.

Author Contributions: Conceptualization, L.Z. (Limeng Zhang) and K.Y.; methodology, L.Z. (Limeng Zhang); software, Z.X. and C.W.; validation, B.D., J.D. and L.Z. (Lingna Zhang); formal analysis, L.Z. (Limeng Zhang); data curation, L.Z. (Limeng Zhang); writing—original draft preparation, L.Z. (Limeng Zhang); writing—review and editing, K.Y.; visualization, S.J.; supervision, B.D. and J.H.; project administration, J.D., L.Z. (Lingna Zhang) and B.D. All authors have read and agreed to the published version of the manuscript.

Funding: This research was funded by National Key R&D Program of China (Grant No. 2021YFD1300400), National Natural Science Foundation of China (Grant Nos. 31790411, 32002186), Natural Science Foundation of Guangdong Province (Grant No. 2020A1515010322), Guangzhou Basic and Applied Basic Research Foundation (Grant No. 202102020850), and Start-up Research Project of Maoming Laboratory (Grant No. 2021TDQD002). The study meets with the approval of the university's review board.

Institutional Review Board Statement: Protocols for all experiments were approved by the Experimental Animal Ethics Committee of South China Agricultural University (protocol code 2021e028).

Informed Consent Statement: Not applicable.

Data Availability Statement: The datasets presented in this study can be found in online repositories. The names of repository/repositories and accession number(s) can be found below: https://www.ncbi.nlm.nih.gov/sra/PRJNA898496 (accessed on 5 November 2022).

Conflicts of Interest: L.Z. is employed by Guangzhou Qingke Biotechnology Co., Ltd. The remaining authors declare that the research was conducted in the absence of any commercial or financial relationships that could be construed a s a potential conflict of interest.

References

1. Udell, M.A.; Brubaker, L. Are dogs social generalists? Canine social cognition, attachment, and the dog-human bond. *Curr. Dir. Psychol. Sci.* **2016**, *25*, 327–333. [CrossRef]
2. Christian, H.E.; Westgarth, C.; Bauman, A.; Richards, E.A.; Rhodes, R.E.; Evenson, K.R.; Mayer, J.A.; Thorpe, R.J. Dog ownership and physical activity: A review of the evidence. *J. Phy. Act. Health* **2013**, *10*, 750–759. [CrossRef] [PubMed]
3. Islam, A.; Towell, T. Cat and dog companionship and well-being: A systematic review. *Int. J. Appl. Psychol.* **2013**, *3*, 149–155.
4. Carr, E.C.; Norris, J.M.; Alix Hayden, K.; Pater, R.; Wallace, J.E. A scoping review of the health and social benefits of dog ownership for people who have chronic pain. *Anthrozoös* **2020**, *33*, 207–224. [CrossRef]
5. Dotson, M.J.; Hyatt, E.M. Understanding dog–human companionship. *J. Bus. Res.* **2008**, *61*, 457–466. [CrossRef]

6. Beloshapka, A.N.; Wolff, A.K.; Swanson, K.S. Effects of feeding polydextrose on faecal characteristics, microbiota and fermentative end products in healthy adult dogs. *Br. J. Nutr.* **2012**, *108*, 638–644. [CrossRef]
7. Bordenstein, S.R.; Theis, K.R. Host biology in light of the microbiome: Ten principles of holobionts and hologenomes. *PLoS Biol.* **2015**, *13*, e1002226. [CrossRef]
8. Suchodolski, J.S. Companion animals symposium: Microbes and gastrointestinal health of dogs and cats. *J. Anim. Sci.* **2011**, *89*, 1520–1530. [CrossRef]
9. Canfora, E.E.; Meex, R.C.; Venema, K.; Blaak, E.E. Gut microbial metabolites in obesity, nafld and t2dm. *Nat. Rev. Endocrinol.* **2019**, *15*, 261–273. [CrossRef]
10. D'Argenio, V.; Salvatore, F. The role of the gut microbiome in the healthy adult status. *Clin. Chim. Acta.* **2015**, *451*, 97–102. [CrossRef]
11. Pereira, A.M.; Clemente, A. Dogs' microbiome from tip to toe. *Top. Companion Anim. Med.* **2021**, *45*, 100584. [CrossRef] [PubMed]
12. Handl, S.; Dowd, S.E.; Garcia-Mazcorro, J.F.; Steiner, J.M.; Suchodolski, J.S. Massive parallel 16s rrna gene pyrosequencing reveals highly diverse fecal bacterial and fungal communities in healthy dogs and cats. *FEMS Microbiol. Ecol.* **2011**, *76*, 301–310. [CrossRef] [PubMed]
13. Swanson, K.S.; Dowd, S.E.; Suchodolski, J.S.; Middelbos, I.S.; Vester, B.M.; Barry, K.A.; Nelson, K.E.; Torralba, M.; Henrissat, B.; Coutinho, P.M.; et al. Phylogenetic and gene-centric metagenomics of the canine intestinal microbiome reveals similarities with humans and mice. *ISME J.* **2011**, *5*, 639–649. [CrossRef] [PubMed]
14. Garcia-Mazcorro, J.F.; Dowd, S.E.; Poulsen, J.; Steiner, J.M.; Suchodolski, J.S. Abundance and short-term temporal variability of fecal microbiota in healthy dogs. *Microbiol. Open.* **2012**, *1*, 340–347. [CrossRef] [PubMed]
15. Hand, D.; Wallis, C.; Colyer, A.; Penn, C.W. Pyrosequencing the canine faecal microbiota: Breadth and depth of biodiversity. *PLoS ONE* **2013**, *8*, e53115. [CrossRef]
16. Beloshapka, A.N.; Dowd, S.E.; Suchodolski, J.S.; Steiner, J.M.; Duclos, L.; Swanson, K.S. Fecal microbial communities of healthy adult dogs fed raw meat-based diets with or without inulin or yeast cell wall extracts as assessed by 454 pyrosequencing. *FEMS Microbiol. Ecol.* **2013**, *84*, 532–541. [CrossRef]
17. Sandri, M.; Dal Monego, S.; Conte, G.; Sgorlon, S.; Stefanon, B. Raw meat based diet influences faecal microbiome and end products of fermentation in healthy dogs. *BMC Vet. Res.* **2016**, *13*, 65. [CrossRef]
18. Tanprasertsuk, J.; Perry, L.A.M.; Tate, D.E.; Honaker, R.W.; Shmalberg, J. Apparent total tract nutrient digestibility and metabolizable energy estimation in commercial fresh and extruded dry kibble dog foods. *Transl. Anim. Sci.* **2021**, *5*, txabo71. [CrossRef]
19. Middelbos, I.S.; Fastinger, N.D.; Fahey, G.C. Evaluation of fermentable oligosaccharides in diets fed to dogs in comparison to fiber standards. *J. Anim. Sci.* **2007**, *85*, 3033–3044. [CrossRef]
20. Magoc, T.; Salzberg, S.L. Flash: Fast length adjustment of short reads to improve genome assemblies. *Bioinformatics* **2011**, *27*, 2957–2963. [CrossRef]
21. Caporaso, J.G.; Kuczynski, J.; Stombaugh, J.; Bittinger, K.; Bushman, F.D.; Costello, E.K.; Fierer, N.; Pena, A.G.; Goodrich, J.K.; Gordon, J.I.; et al. Qiime allows analysis of high-throughput community sequencing data. *Nat. Methods* **2010**, *7*, 335–336. [CrossRef] [PubMed]
22. Bokulich, N.A.; Subramanian, S.; Faith, J.J.; Gevers, D.; Gordon, J.I.; Knight, R.; Mills, D.A.; Caporaso, J.G. Quality-filtering vastly improves diversity estimates from illumina amplicon sequencing. *Nat. Methods* **2013**, *10*, 57–59. [CrossRef] [PubMed]
23. Edgar, R.C.; Haas, B.J.; Clemente, J.C.; Quince, C.; Knight, R. Uchime improves sensitivity and speed of chimera detection. *Bioinformatics* **2011**, *27*, 2194–2200. [CrossRef] [PubMed]
24. Haas, B.J.; Gevers, D.; Earl, A.M.; Feldgarden, M.; Ward, D.V.; Giannoukos, G.; Ciulla, D.; Tabbaa, D.; Highlander, S.K.; Sodergren, E. Chimeric 16s rrna sequence formation and detection in sanger and 454-pyrosequenced pcr amplicons. *Genome Res.* **2011**, *21*, 494–504. [CrossRef] [PubMed]
25. Quast, C.; Pruesse, E.; Yilmaz, P.; Gerken, J.; Schweer, T.; Yarza, P.; Peplies, J.; Glöckner, F.O. The silva ribosomal rna gene database project: Improved data processing and web-based tools. *Nucleic Acids Res.* **2012**, *41*, D590–D596. [CrossRef]
26. Edgar, R.C. Muscle: Multiple sequence alignment with high accuracy and high throughput. *Nucleic Acids Res.* **2004**, *32*, 1792–1797. [CrossRef]
27. Yang, K.; Deng, X.; Jian, S.; Zhang, M.; Wen, C.; Xin, Z.; Zhang, L.; Tong, A.; Ye, S.; Deng, B. Gallic acid alleviates gut dysfunction and boosts immune and antioxidant activity in puppies under environmental stress based on microbiome-metabolomics analysis. *Front. Immunol.* **2022**, *5848*, 813890. [CrossRef]
28. Nuli, R.; Azhati, J.; Cai, J.X.; Kadeer, A.; Zhang, B.; Mohemaiti, P. Metagenomics and faecal metabolomics integrative analysis towards the impaired glucose regulation and type 2 diabetes in uyghur-related omics. *J. Diabetes Res.* **2019**, *2019*, 2893041. [CrossRef]
29. Furniss, G. The influence of nutrition on puppy and kitten growth and development. *Ir. Vet. J.* **2008**, *61*, 191–194.
30. Do, S.; Phungviwatnikul, T.; de Godoy, M.R.C.; Swanson, K.S. Nutrient digestibility and fecal characteristics, microbiota, and metabolites in dogs fed human-grade foods. *J. Anim. Sci.* **2021**, *99*, skab028. [CrossRef]
31. Brito, C.B.M.D.; Lima, D.C.D.; Souza, C.M.M.; Vasconcellos, R.S.; Oliveira, S.G.D.; Félix, A.P. Evaluation of dried apple pomace on digestibility and palatability of diets for cats. *Braz. J. Anim. Sci.* **2020**, *49*, 1. [CrossRef]

32. Wakshlag, J.J.; Simpson, K.W.; Struble, A.M.; Dowd, S.E. Negative fecal characteristics are associated with ph and fecal flora alterations during dietary change in dogs. *Int. J. Appl. Res. Vet. M* **2011**, *9*, 278–283.
33. McEwen, B.S. Stress, adaptation, and disease: Allostasis and allostatic load. *Ann. N. Y. Acad. Sci.* **1998**, *840*, 33–44. [CrossRef] [PubMed]
34. Cohen, S.; Janicki-Deverts, D.; Miller, G.E. Psychological stress and disease. *JAMA* **2007**, *298*, 1685–1687. [CrossRef]
35. Helfand, S.C.; Modiano, J.F.; Nowell, P.C. Immunophysiological studies of interleukin-2 and canine lymphocytes. *Vet. Immunol. Immunopathol.* **1992**, *33*, 1–16. [CrossRef]
36. Hammarqvist, F.; Wernerman, J.; von der Decken, A.; Vinnars, E. Alanyl-glutamine counteracts the depletion of free glutamine and the postoperative decline in protein synthesis in skeletal muscle. *Ann. Sur.* **1990**, *212*, 637. [CrossRef]
37. Karner, J.; Roth, E. Alanyl-glutamine infusions to patients with acute pancreatitis. *Clin. Nurt.* **1990**, *9*, 43–44. [CrossRef]
38. Weiss, Y.G.; Maloyan, A.; Tazelaar, J.; Raj, N.; Deutschman, C.S. Adenoviral transfer of hsp-70 into pulmonary epithelium ameliorates experimental acute respiratory distress syndrome. *J. Clin. Invest.* **2002**, *110*, 801–806. [CrossRef]
39. Inness, V.; McCartney, A.; Khoo, C.; Gross, K.; Gibson, G. Molecular characterisation of the gut microflora of healthy and inflammatory bowel disease cats using fluorescence in situ hybridisation with special reference to *desulfovibrio* spp. *J. Anim. Physiol. Anim. Nutr.* **2007**, *91*, 48–53. [CrossRef]
40. Janeczko, S.; Atwater, D.; Bogel, E.; Greiter-Wilke, A.; Gerold, A.; Baumgart, M.; Bender, H.; McDonough, P.; McDonough, S.; Goldstein, R. The relationship of mucosal bacteria to duodenal histopathology, cytokine mrna, and clinical disease activity in cats with inflammatory bowel disease. *Vet. Microbiol.* **2008**, *128*, 178–193. [CrossRef]
41. Clavel, T.; Desmarchelier, C.; Haller, D.; Gerard, P.; Rohn, S.; Lepage, P.; Daniel, H. Intestinal microbiota in metabolic diseases from bacterial community structure and functions to species of pathophysiological relevance. *Gut Microbes* **2014**, *5*, 544–551. [CrossRef] [PubMed]
42. Kim, J.; An, J.U.; Kim, W.; Lee, S.; Cho, S. Differences in the gut microbiota of dogs (canis lupus familiaris) fed a natural diet or a commercial feed revealed by the illumina miseq platform. *Gut Pathog.* **2017**, *9*, 68. [CrossRef] [PubMed]
43. Xenoulis, P.G.; Palculict, B.; Allenspach, K.; Steiner, J.M.; Van House, A.M.; Suchodolski, J.S. Molecular-phylogenetic characterization of microbial communities imbalances in the small intestine of dogs with inflammatory bowel disease. *FEMS Microbiol. Ecol.* **2008**, *66*, 579–589. [CrossRef] [PubMed]
44. Thorkildsen, L.T.; Nwosu, F.C.; Avershina, E.; Ricanek, P.; Perminow, G.; Brackmann, S.; Vatn, M.H.; Rudi, K. Dominant fecal microbiota in newly diagnosed untreated inflammatory bowel disease patients. *Gastroenterol. Res. Pract.* **2013**, *2013*, 636785. [CrossRef] [PubMed]
45. Gomez-Arango, L.F.; Barrett, H.L.; Wilkinson, S.A.; Callaway, L.K.; Mcintyre, H.D.; Morrison, M.; Nitert, M.D. Low dietary fiber intake increases collinsella abundance in the gut microbiota of overweight and obese pregnant women. *Gut Microbes* **2018**, *9*, 189–201. [CrossRef]
46. Zhang, H.; DiBaise, J.K.; Zuccolo, A.; Kudrna, D.; Braidotti, M.; Yu, Y.; Parameswaran, P.; Crowell, M.D.; Wing, R.; Rittmann, B.E. Human gut microbiota in obesity and after gastric bypass. *Proc. Natl. Acad. Sci. USA* **2009**, *106*, 2365–2370. [CrossRef]
47. Cox, L.M.; Sohn, J.; Tyrrell, K.L.; Citron, D.M.; Lawson, P.A.; Patel, N.B.; Iizumi, T.; Perez-Perez, G.I.; Goldstein, E.J.; Blaser, M.J. Description of two novel members of the family erysipelotrichaceae: *Ileibacterium valens* gen. Nov., sp. Nov. And *dubosiella newyorkensis*, gen. Nov., sp. Nov., from the murine intestine, and emendation to the description of faecalibacterium rodentium. *Int. J. Syst. Evol. Microbiol.* **2017**, *67*, 1247. [CrossRef]
48. Sheng, K.; Zhang, G.; Sun, M.; He, S.; Kong, X.; Wang, J.; Zhu, F.; Zha, X.; Wang, Y. Grape seed proanthocyanidin extract ameliorates dextran sulfate sodium-induced colitis through intestinal barrier improvement, oxidative stress reduction, and inflammatory cytokines and gut microbiota modulation. *Food Funct.* **2020**, *11*, 7817–7829. [CrossRef]
49. Chamorro, S.; Romero, C.; Brenes, A.; Sanchez-Patan, F.; Bartolome, B.; Viveros, A.; Arija, I. Impact of a sustained consumption of grape extract on digestion, gut microbial metabolism and intestinal barrier in broiler chickens. *Food Funct.* **2019**, *10*, 1444–1454. [CrossRef]
50. Vitek, L.; Haluzik, M. The role of bile acids in metabolic regulation. *J. Endocrinol.* **2016**, *228*, R85–R96. [CrossRef]
51. Ciccarelli, R.; Di Iorio, P.; Giuliani, P.; D'Alimonte, I.; Ballerini, P.; Caciagli, F.; Rathbone, M.P. Rat cultured astrocytes release guanine-based purines in basal conditions and after hypoxia/hypoglycemia. *Glia* **1999**, *25*, 93–98. [CrossRef]
52. Cornforth, D.M.; Foster, K.R. Competition sensing: The social side of bacterial stress responses. *Nat. Rev. Microbiol.* **2013**, *11*, 285–293. [CrossRef] [PubMed]
53. Rowlett, V.W.; Mallampalli, V.K.P.S.; Karlstaedt, A.; Dowhan, W.; Taegtmeyer, H.; Margolin, W.; Vitrac, H. Impact of membrane phospholipid alterations in escherichia coli on cellular function and bacterial stress adaptation. *J. Bacteriol.* **2017**, *199*, e00849-16. [CrossRef] [PubMed]
54. Bange, G.; Bedrunka, P. Physiology of guanosine-based second messenger signaling in bacillus subtilis. *Biol. Chem.* **2020**, *401*, 1307–1322. [CrossRef]
55. Haber, A.; Friedman, S.; Lobel, L.; Burg-Golani, T.; Sigal, N.; Rose, J.; Livnat-Levanon, N.; Lewinson, O.; Herskovits, A.A. L-glutamine induces expression of listeria monocytogenes virulence genes. *PLoS Pathog.* **2017**, *13*, e1006161. [CrossRef]
56. Adén, U.; O'Connor, W.T.; Berman, R.F. Changes in purine levels and adenosine receptors in kindled seizures in the rat. *Neuroreport* **2004**, *15*, 1585–1589. [CrossRef]

57. Bouma, M.G.; van den Wildenberg, F.A.J.M.; Buurman, W.A. The anti-inflammatory potential of adenosine in ischemia-reperfusion injury: Established and putative beneficial actions of a retaliatory metabolite. *Shock* **1997**, *8*, 313–320. [CrossRef]
58. Giriş, M.; Depboylu, B.; Doğru-Abbasoğlu, S.; Erbil, Y.; Olgaç, V.; Alış, H.; Aykaç-Toker, G.; Uysal, M. Effect of taurine on oxidative stress and apoptosis-related protein expression in trinitrobenzene sulphonic acid-induced colitis. *Clin. Exp. Immunol.* **2008**, *152*, 102–110. [CrossRef]
59. Franconi, F.; Loizzo, A.; Ghirlanda, G.; Seghieri, G. Taurine supplementation and diabetes mellitus. *Curr. Opin. Clin. Nutr.* **2006**, *9*, 32–36. [CrossRef]
60. Oliveira, M.W.S.; Minotto, J.B.; de Oliveira, M.R.; Zanotto, A.; Behr, G.A.; Rocha, R.F.; Moreira, J.C.F.; Klamt, F. Scavenging and antioxidant potential of physiological taurine concentrations against different reactive oxygen/nitrogen species. *Pharmacol. Rep.* **2010**, *62*, 185–193. [CrossRef]
61. Kim, C.; Cha, Y.-N. Production of reactive oxygen and nitrogen species in phagocytes is regulated by taurine chloramine. *Taurine* **2009**, *7*, 463–472.
62. Hu, J.; Xu, X.; Yang, J.; Wu, G.; Sun, C.; Lv, Q. Antihypertensive effect of taurine in rat. *Taurine* **2009**, *7*, 75–84.
63. Sudhahar, V.; Shaw, S.; Imig, J.D. Mechanisms involved in oleamide-induced vasorelaxation in rat mesenteric resistance arteries. *Eur. J. Pharmacol.* **2009**, *607*, 143–150. [CrossRef] [PubMed]
64. Moon, S.-M.; Lee, S.A.; Hong, J.H.; Kim, J.-S.; Kim, D.K.; Kim, C.S. Oleamide suppresses inflammatory responses in lps-induced raw264. 7 murine macrophages and alleviates paw edema in a carrageenan-induced inflammatory rat model. *Int. Immunopharmacol.* **2018**, *56*, 179–185. [CrossRef] [PubMed]
65. Kita, M.; Ano, Y.; Inoue, A.; Aoki, J. Identification of p2y receptors involved in oleamide-suppressing inflammatory responses in murine microglia and human dendritic cells. *Sci. Rep.* **2019**, *9*, 3135. [CrossRef]
66. Kim, M.; Kim, M.; Han, J.Y.; Lee, S.-H.; Jee, S.H.; Lee, J.H. The metabolites in peripheral blood mononuclear cells showed greater differences between patients with impaired fasting glucose or type 2 diabetes and healthy controls than those in plasma. *Diabetes Vasc. Dis. Res.* **2017**, *14*, 130–138. [CrossRef]
67. Lombard, G.L.; Dowell, V., Jr. Comparison of three reagents for detecting indole production by anaerobic bacteria in microtest systems. *J. Clin. Mic.* **1983**, *18*, 609–613. [CrossRef]
68. Cai, H.; Li, H.; Yan, X.; Sun, B.; Zhang, Q.; Yan, M.; Zhang, W.-Y.; Jiang, P.; Zhu, R.-H.; Liu, Y.-P. Metabolomic analysis of biochemical changes in the plasma and urine of first-episode neuroleptic-naive schizophrenia patients after treatment with risperidone. *J. Proteomoe Res.* **2012**, *11*, 4338–4350. [CrossRef]
69. Dileep, K.V.; Remya, C.; Tintu, I.; Haridas, M.; Sadasivan, C. Interactions of selected indole derivatives with phospholipase a2: In silico and in vitro analysis. *J. Mol. Model.* **2013**, *19*, 1811–1817. [CrossRef]
70. McNaught, K.S.P.; Carrupt, P.-A.; Altomare, C.; Cellamare, S.; Carotti, A.; Testa, B.; Jenner, P.; Marsden, C.D. Isoquinoline derivatives as endogenous neurotoxins in the aetiology of parkinson's disease. *Biochem. Pharmacol.* **1998**, *56*, 921–933. [CrossRef]

metabolites

Review

The Nexus of Diet, Gut Microbiota and Inflammatory Bowel Diseases in Dogs

Soufien Rhimi [1], Aicha Kriaa [1], Vincent Mariaule [1], Amel Saidi [1], Amandine Drut [1,2], Amin Jablaoui [1], Nizar Akermi [1], Emmanuelle Maguin [1], Juan Hernandez [1,2] and Moez Rhimi [1,*]

1 Microbiota Interaction with Human and Animal Team (MIHA), Micalis Institute, AgroParisTech, Université Paris-Saclay, Institut National de Recherche pour l'Agriculture, l'Alimentation et l'Environnement, 78350 Jouy-en-Josas, France

2 Oniris, Department of Clinical Sciences, Nantes-Atlantic College of Veterinary Medicine and Food Sciences, 44300 Nantes, France

* Correspondence: moez.rhimi@inrae.fr

Abstract: Canine inflammatory bowel diseases (IBD) are of increasing interest in veterinary medicine. They refer to complex and debilitating conditions of dogs' gastrointestinal tract. Although little evidence for causal inferences is currently available, it is believed that IBD pathophysiology entails intricate interactions between environmental factors, the intestinal immune system, and the microbial communities that colonize the gut. To better understand the mechanisms underlying these disorders, leveraging factors associated with the development of these diseases is imperative. Of these factors, emerging evidence supports the role of dietary patterns as key players influencing the composition and function of gut microbes, with subsequent effects on health and disease. In this review, we particularly focus on addressing IBD in dogs and discuss how specific nutrients may elicit or relieve gut inflammation. Gaining mechanistic insights into such interplay and the underpinning mechanisms is key to inferring dietary recommendations, and setting up new and promising therapeutics.

Keywords: canine inflammatory bowel disease (IBD); diet; gut microbiota; holobiont

Citation: Rhimi, S.; Kriaa, A.; Mariaule, V.; Saidi, A.; Drut, A.; Jablaoui, A.; Akermi, N.; Maguin, E.; Hernandez, J.; Rhimi, M. The Nexus of Diet, Gut Microbiota and Inflammatory Bowel Diseases in Dogs. *Metabolites* **2022**, *12*, 1176. https://doi.org/10.3390/metabo12121176

Academic Editors: Lian Li, Baichuan Deng and Giulia Pignataro

Received: 3 November 2022
Accepted: 23 November 2022
Published: 25 November 2022

1. Introduction

As we continue to place increasing focus on our health and wellbeing, this mindset is reflected in our pets' lives. Inflammatory bowel diseases (IBD) are multifactorial and debilitating diseases featuring a chronic immune response, the disruption of intestinal homeostasis, and the altered composition and function of the gut microbiota, referred to as dysbiosis [1,2]. Evidence has shown that even short-term dietary changes may influence gut microbiota composition. Bacterial shifts are likely to be observed in humans within 1–3 days of extreme dietary changes, such as switching from an all-meat to an all-plant diet, being introduced [2]. A few similar studies have been performed to explore the effects of dietary interventions on canine gut microbiota composition and function [3–7]. Most of these studies in dogs only assess the microbial composition changes after a diet adaptation period of 10 days on average, preventing the detection of earlier changes. As dysbiosis has been linked to chronic intestinal inflammation in dogs [8–11], evidence suggests the role of diet in managing the disease. The usefulness of dietetics has been known for a long time in the treatment of IBD in dogs, particularly for the modulation of digestibility and the control of immune reactions [12]. Epidemiological data in humans and pets, and studies in rodent models have shown that low-fiber diets and food additives are likely to compromise the intestinal barrier function and contribute to a myriad of metabolic or inflammatory disorders, including inflammatory bowel diseases (IBD) [13–16]. Recently, the pet food industry has seen significant shifts and growth [17]. Indeed, the global pet food market value is expected to reach 118.83 billion United States Dollar (USD) in 2025, growing by

5.4% in the period of 2021 to 2025 [17]. Notable trends include the emergence of high-protein diets based on new protein sources, high insoluble fiber diets, raw meat diets, as well as insect-based pet food products that have made their way into the market [18]. Such schemes have been shown to impact gut motility and shape the gut microbiota, thereby influencing the overall health of the host [19–21]. Gut microbiota is factually known to play key roles in maintaining gut physiology [22,23]; it comprises a highly complex community that evolves and adapts to its host over the life course, and shows remarkable plasticity to environmental changes, particularly to diet [24,25].

Diet may act as a risk factor when unbalanced or highly processed, but also as a disease management strategy for gastrointestinal (GI), renal or dermatological diseases [26,27]. More recently, the relationship between diet, microbiota, and gastrointestinal inflammation has emerged as a challenging area of research. Therefore, we aim in this review to discuss dietary interventions in dogs, with a scope focused on a better understanding of the dietary-microbiota interplay in IBD.

2. Gut Microbiota in Canine IBD

Growing evidence suggests that bacteria present in a dog's gut may play an essential role in its health and disease [28]. The gut microbiota of healthy dogs is known to comprise three main phyla: Fusobacterium, Bacteroidetes, and Firmicutes [29]. Within this core bacterial community, several taxa are members of the phylum Firmicutes, including bacilli and clostridia, most of which are short-chain fatty acid (SCFA) producers, such as *Faecalibacterium* spp. [30,31]. Bacteroidetes is another prominent phylum and includes the genera *Bacteroides* and *Prevotella* [32]. Similarly, the phylum Fusobacterium has been commonly associated with health in dogs [32].

Key roles of the gut microbiota include protecting against pathogens, shaping the immune system, and providing beneficial metabolites to host epithelial cells through fermentative reactions [28]. Microbial metabolites may influence host health, gut microbes, and multiple interacting communities, thereby maintaining the holobiont symbiosis [33]. They provide other beneficial effects, notably, immunomodulatory, anti-diarrheal and regulatory effects of GI motility [34]. Gut microbiota is also involved in the metabolism of bile acids (BA) as potential mediators linking gut bacteria to metabolic and inflammatory disorders [28].

Links between gut microbiota composition/function and a myriad of diseases have been widely reported. In fact, it was demonstrated in mice that gut microbiota causes several pathologies, including obesity and dyslipidemia [35,36]. Evidence suggests that microbial ecosystem imbalance or dysbiosis has been correlated with several inflammatory diseases in dogs, such as IBD [37]. Intestinal dysbiosis in dogs with IBD is often characterized by a decrease in bacterial richness and diversity [30]. Metagenomic analyses have highlighted a lower abundance of Firmicutes, while Proteobacteria increases in dogs with IBD compared to dogs with a healthy status [38]. The abundance of *Faecalibacterium* spp. and *Fusobacterium* spp. were also significantly decreased in dogs with IBD relative to healthy controls [39]. In addition, higher abundances of adherent and invasive *Escherichia coli* (AIEC) were noted in colonic biopsies from dogs with granulomatous colitis, thus highlighting a potential link with gut inflammation [40]. Metabolic alterations have also been reported, including impaired short chain fatty acids (SCFAs) and tryptophan metabolites production, which may influence intestinal homeostasis and immunological tolerance [41,42]. SCFAs (i.e., acetate, propionate and butyrate) are the main end products of intestinal bacterial fermentation of non-digestible food components, such as dietary fiber. Lower levels of acetate and propionate were detected in fecal samples from dogs with IBD compared to healthy subjects [43]. These SCFAs are known to hold therapeutic promises in IBD as they improve epithelial barrier integrity and alleviate gut inflammation in vivo. In addition to SCFAs deficiency, an altered BA metabolism has been demonstrated in canine IBD [44]. The conversion of primary BA to secondary BA is largely known to be achieved by gut microbes. BA play key roles in the emulsification and absorption of dietary lipids and serve as potent signaling molecules that act through the farnesoid X receptor (FXR) and Takeda G-protein-coupled receptor 5 (TGR5). By activating FXR and TGR5, BA can influence a variety of processes, including inflammation and lipid, glucose and energy metabolism. Accordingly, changes in gut bacterial populations have been suggested to influence inflammatory parameters and pathways through changes in BA metabolism [28,45]. In dogs with IBD, the decrease in the abundance of *Clostridium hiranonis*, a potent BA converter, is correlated with the alteration of the BA metabolism. Conversely, treatment of intestinal inflammation is accompanied by an increase in the abundance of *C. hiranonis* and a normalization of the BA metabolism [39]. The links between dietary interventions, SCFAs, BA metabolism and canine inflammation are yet to be explored. Similarly, the relevance of BA as potential therapeutic targets in dogs would need to be thoroughly addressed as investigations related to this field are still in their infancy.

3. Diet-Microbiota Interactions in Canine IBD

3.1. Dietary Proteins

Diets with high protein levels were associated with a modification of the gut microbiota composition in healthy beagles, mainly characterized by an increase in the genus *Lactobacillus* abundance. This change was linked to high concentrations of butyrate in dogs that were fed a high-protein diet [46]. Furthermore, a high protein diet has been shown to promote the growth of *Clostridium perfringens* and to reduce the abundance of *Clostridium* cluster XIVa (also known as the *Clostridium coccoides* group) in a similar population of healthy dogs [47]. Although the findings from these studies suggest that high-protein diets exert significant effects on the canine gut community, as they elicit the growth of select *Clostridium* species, a major limitation of such trials is the relatively small size of the studied cohort, limited to only twelve and nine Beagles, respectively [46,47]. Other significant differences were observed in the microbiota composition, with a higher Firmicutes:Bacteroidetes ratio in response to a high protein-low carbohydrate (HPLC) diet when compared with a low protein-high carbohydrate (LPHC) diet. Several taxa became detectable in response to diet, such as *Lactobacillus ruminis*, which was detected in 59% of LPHC-fed dogs [48]. In another study, the fecal microbiota of dogs fed a HPLC diet showed an increased abundance of Bacteroidetes in addition to an enrichment in the phylum Firmicutes [49].

In addition to the protein content of food, the protein type (origin, quality) also deserves to be evaluated. To date, few studies have addressed this question. Analysis of the impact of a hydrolyzed soy protein diet combined with oral prednisone on the gut microbiota of dogs with IBD reveals an increase in lactobacilli, *Bifidobacterium* spp., *Faecalibacterium* spp. and *Streptococcus* spp. abundance. This modification of the microbial communities is associated with an enhancement of the intestinal barrier function by increasing mucosal epithelial apical junction protein (AJP) expression [50]. Further studies are required in order to evaluate the role of dietary proteins (content, type, quality) in modeling the gut microbiota, as well as their effect on dog's health, particularly in the context of IBD.

3.2. Dietary Tryptophan and Indole Derivatives

In humans with IBD, reduced availability of tryptophan or tryptophan metabolites has been suggested to contribute to the disease [51,52]. Tryptophan represents a precursor of several microbial and host metabolites, including serotonin and vitamin B3 [53]. Tryptophan metabolites are known as one of the most important endogenous ligands of the aryl hydrocarbon receptor (AhR), a nuclear protein involved in the regulation of gene expression and in maintaining intestinal homeostasis [54]. Microbial metabolites or dietary factors may influence this pathway.

Dogs with IBD and dogs with protein losing enteropathy have also been shown to exhibit lower plasma tryptophan levels than healthy dogs [55,56]. While these studies highlight a potential role of tryptophan in dogs with IBD and protein losing enteropathy, the small cohort size (10 dogs) in the IBD study and the retrospective study design for the protein losing enteropathy pathology represent major limitations. Further prospective studies with larger cohorts are needed. Further analysis of dog's gut microbiota would be beneficial to such studies as the link between the decrease in the plasma concentration of tryptophan and dysbiosis is not yet established. In addition to the functional abnormality of the intestinal microbiota, an absorption defect linked to intestinal inflammation could also be involved.

3.3. Dietary Fibers and SCFAs

Fibers can be defined as non-digestible carbohydrates that come from plants. They can be classified according to their solubility or fermentability. Soluble or fermentable fibers, such as pectin, gum Arabic, and fructooligosaccharides, support normal GI microflora growth and provide fuel for colonocytes. Several human studies showed that they also delay gastric emptying and inhibit absorption in the small intestine [33]. Insoluble fibers, such as cellulose and oat fiber, were shown to increase the volume and water content of stools, to absorb toxins and to normalize colonic motility [33]. SCFAs, including butyrate, acetate and propionate, are well-studied microbial metabolites primarily produced by the bacterial fermentation of non-digestible dietary fibers. Thus far, most human clinical trials investigating the anti-inflammatory effects of dietary fibers have been linked with a higher luminal production of SCFAs following the intake of high-fiber foods [57,58]. It is well demonstrated that SCFAs not only contribute to the regulation of the mucosal barrier function but also provide immune regulatory functions [33]. In addition, their production provides an acidic luminal environment that inhibits the proliferation of pH-sensitive pathogenic bacteria such as Enterobacteriaceae [59]. Furthermore, in human studies, SCFAs are likely to modulate inflammation by increasing the production of anti-inflammatory cytokines, decreasing pro-inflammatory cytokines, and activating the transcription factor Foxp3 [60]. Studies applied to dogs in this regard are still in their infancy and few reports have explored the role of fiber-enriched diets in canine IBD. Interestingly, the intake of high-fiber diets has recently been shown to alleviate acute large-bowel diarrhea and to exhibit significant clinical benefits in dogs [61] (Figure 1). However, the use of antibiotic and antiparasitic treatments and the absence of microbiota analysis are important limitations in this study. More controlled studies are therefore required to confirm these effects. Further metagenomic analysis of dogs' gut microbiomes would shed light on the functional potential of this community, and provide mechanistic knowledge linking dietary fiber, gut microbiota, and the treatment of canine IBD.

The characterization of fecal fatty acids in dogs with IBD has highlighted alterations in SCFAs profiles as well as in gut microbiota composition [34]. A significant decrease in the fecal concentrations of acetate and propionate is demonstrated in the IBD group compared to the control group. A correlation between the decrease in these SCFAs and the abundance of the Bacteroidetes, *Fusobacterium* spp., *Faecalibacterium* spp., *C. hiranonis*, *Blautia* spp., *Streptococcus* spp., Ruminococcaceae, *Bifidobacterium* spp., *C. perfringens*, and *E. coli* is reported by authors [34]. Most of these bacteria are known for their fermentation capacities and as primary (Bacteroidetes) or secondary (*Blautia* spp., *Faecalibacterium* spp.) fermenters of carbohydrate. Mechanistic studies that link changes in gut bacteria composition, SCFAs production and canine intestinal inflammation have yet to be performed.

3.4. Dietary Fat and Bile Acids

Given their direct relationship with intestinal microbiota, BA are promising therapeutic targets in dogs with IBD. Primary BA are synthesized by the liver from cholesterol and conjugated to the amino-acids glycine or taurine. They are further subjected to deconjugation by gut microbes via bile salt hydroxylase (BSH) enzymes and dehydroxylation to yield secondary BA [28]. Higher fecal levels of primary BA were detected in dogs with IBD and correlated with a lower expression of apical sodium-dependent BA transporter proteins (ASBT) in the ileum [62]. This impaired absorption of primary BA due to ASBT downregulation was further suggested to directly contribute to diarrhea in dogs with IBD [62]. The gut microbiota is the sole metabolic pathway for BA metabolism. Thus, intestinal dysbiosis with a decrease in bacteria bearing BSH activity may imply lower BA deconjugation and dehydroxylation. In dogs, a significant decrease in the fecal abundance of *C. hiranonis* is reported during chronic inflammatory enteropathy. Known for its ability to convert primary BA into secondary BA, *C. hiranonis* may illustrate the link between dysbiosis and intestinal inflammation [63].

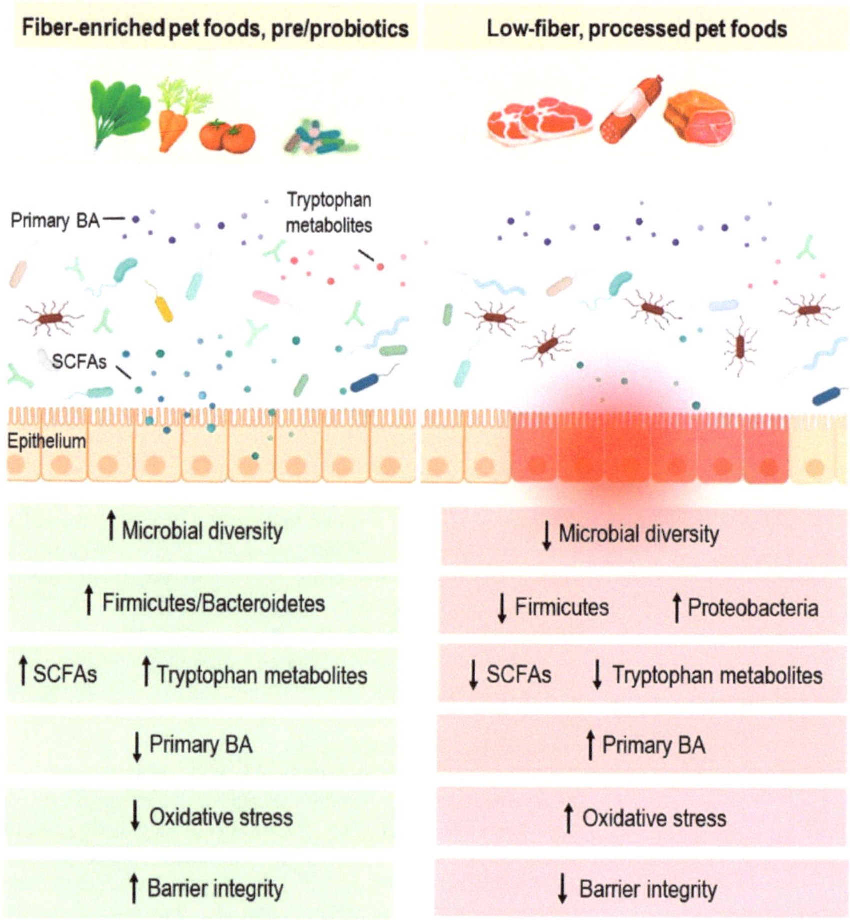

Figure 1. Overview of anti- and pro-inflammatory influences of dietary interventions in dogs. Dietary patterns are linked with changes in the composition and function of the gut microbiota in dogs. Up and down arrows represent increases and decreases, respectively. SCFAs: short-chain fatty acids, BA: bile acids.

Interestingly, fat can be considered a nutrient of concern in some dogs with IBD. Fat malabsorption or reduced fat digestion may lead to an increased passage of fat into the colon, which can be associated with dysbiosis as well as increased intestinal permeability [64].

3.5. Vitamins

Reduced serum concentrations of several vitamins, including folate and cobalamin, have been reported in dogs with IBD. Vitamins are known to play a pivotal role in several cellular processes, such as GI epithelial cell turnover and repair [65]. Folate and cobalamin (also referred to as vitamin B12) are both essential water-soluble vitamins for dogs. Vitamin

B12 is a component of coenzymes and is essential for cell biosynthesis and metabolism in vivo. Vitamin B12 is primarily absorbed in the terminal ileum. Most commercial pet foods are supplemented with cobalamin; nevertheless, dietary cobalamin levels vary among diets [66]. Lower serum levels of cobalamin have been reported in canine IBD as well as in exocrine pancreatic insufficiency and ileal malabsorption [67,68]. Previous reports have linked reduced serum vitamin B12 concentrations with a higher abundance of *E. coli* and enterococci, which gave rise to one of its clinical uses, highlighting it as a marker of dysbiosis [67,69]. In addition, hypocobalaminemia is a negative prognostic factor in dogs with IBD [70]. Folate is primarily absorbed in the duodenum and proximal jejunum and is synthetized by a variety of commensal bacteria. Folate may be increased in the serum of dogs with IBD. It is believed to be the consequence of the proliferation of folate-producing bacteria and to reflect dysbiosis [69]. However, more comprehensive metabolomic studies are needed to elucidate their contribution to the disease.

Interestingly, serum 25-hydroxyvitamin D is a liposoluble vitamin whose concentration is decreased in humans and dogs with IBD for several reasons, including impaired absorption, fat malabsorption, restricted dietary intake, or reduced sunlight exposure [65,71]. Vitamin D supplementation has shown the ability to regulate gut microbiome and to decrease the intensity of intestinal inflammatory lesions in rodent models of IBD [72]. The effects of vitamin D on the gut microbiome appear to be mediated by the expression of the gene encoding cathelicidin antimicrobial peptide (CAMP) by epithelial cells and immune cells [73]. New research aiming at deciphering the effects of dietary vitamin D intake on the function of the microbiota and the possible beneficial effects on the evolution of IBD must be undertaken.

4. The Impact of Nutritional Interventions in Canine IBD

Approximately 50% of dogs with chronic inflammation are responsive to dietary changes and are considered to have Food Responsive Enteropathy (FRE) [74]. FRE is one of the most common forms of chronic inflammatory enteropathy in dogs and includes those with adverse food reactions (i.e., food allergy and food intolerance) and those with intestinal inflammation that benefits from properties of a different diet. Several nutritional interventions are being used to alleviate clinical signs. Highly digestible or hypoallergenic diets and industrial or home-prepared diets are promising candidates [75]. Prevailing large bowel diseases can also be managed with high-fiber diets [76]. Disorders comprising significant lymphangiectasia may be addressed with hyperdigestible low-fat diets [77].

4.1. Hypoallergenic Diets

Commercial hypoallergenic foods may have a role in managing canine IBD. Industrial hypoallergenic foods find their protein in plant sources (most often soybeans), insects, or bird feathers. Animal protein-free diets were shown to increase fecal bacterial richness and diversity in dogs with FRE compared to the control subjects. The fecal microbiota index, a PCR-based assay aiming at assessing the fecal microbiome by quantifying the abundance of predefined bacterial taxa, was significantly higher in dogs with FRE than in healthy controls. No significant differences in the composition of the gut microbiota were detected after the dietary trial [78]. Interestingly, a hydrolyzed protein diet with a probiotic strain *Enterococcus faecium* was associated with a significant increase in bacterial richness and improvement in clinical signs in dogs suffering from FRE [79]. Additionally, the intake of a novel protein diet with cod and rice was associated with significant changes in the abundance of several bacterial taxa, including higher abundances of the genera *Gemella* and *Peptococcus*, in dogs with FRE [80]. The evaluation of the effects of these diets seems to reveal a contribution to the modulation of the dysbiotic canine gut microbiota. Further studies are needed to evaluate and better understand the potential positive effects of such interventions in these dogs.

4.2. Fiber-Enriched Diets/Prebiotics

Prebiotics are defined as non-digestible food ingredients able to promote the growth of beneficial intestinal microorganisms. Fructooligosaccharides (FOS) were demonstrated to affect the gut microbiota in dogs with primary dysbiosis [81]. A diet supplemented with a prebiotic-rich fiber mixture composed of rice bran, banana flakes and deactivated yeasts was shown to be associated with lower abundances of sulfate-reducing bacteria from the order Desulfovibrionales and an increase in *Clostridium* clusters I and II in the fecal samples of dogs with IBD [82]. The addition of resistant starch, β-glucans, mannan oligosaccharides, and chondroitin sulfate to a hydrolyzed diet did not significantly improve the disease activity in dogs with IBD. However, the post-treatment histological score was significantly lower solely in dogs receiving fibers. From a functional perspective, dietary supplements were demonstrated to increase serum paraoxonase-1, total antioxidant capacity and cholesterol, despite the absence of significant differences between the fecal microbiota of dogs with and without fiber enrichment [83]. A metabolomic study underlined additional beneficial effects of prebiotics and glycosaminoglycans on lipid metabolism, and thereby on intestinal membrane integrity in dogs with IBD [84]. Alternatively, supplementation of a hydrolyzed diet with a brown seaweed (*Ascophyllum nodosum*) rich in fermentable fibers drove an increased abundance of the Ruminococcaceae and Rikenellaceae families and higher concentrations of acetate in the feces of dogs with IBD, without improving clinical signs [85]. Overall, these studies suggest that fiber-enriched diets/prebiotics may modulate the gut microbiota and ameliorate oxidative status in canine IBD (Figure 1). To date, no studies are available regarding the effects of highly digestible and low-fat industrial diets and homemade regimens on the gut microbiota of dogs with IBD.

4.3. Industrial or Home-Prepared Diets

Veterinarians and dog owners need to weigh up the available evidence when deciding whether to feed dogs with IBD a commercial or personalized home-prepared diet. According to equivalent macro-nutritional analysis, industrial kibble food differs from home-made diets by the incompressible starch content necessary for the extrusion process. While the authors agree that dog domestication was accompanied by a selection of genes encoding for proteins involved in starch digestion (pancreatic α-amylase 2B-AMY2B), there is still a debate on the individual variation of their expression and on the capabilities to digest starch [86,87]. Resistant starch escapes digestion and substantially impacts the composition of the gut microbiota, depending on the structure of the starches reaching the colon [88]. Many experimental rodent models of IBD document a reduction in inflammatory lesions during moderate starch supplementation compared to the animals fed a starch-free diet [89,90]. While the beneficial effects in humans are generally linked to enhanced butyrate production by bacterial fermentation, significant individual variations are observed. Therefore, it advocates for a personalized approach to starch intake origin, the host's digestive abilities, and microbiome profile [91,92].

5. Conclusions

The pathogenesis of IBD in dogs is poorly characterized; however, recent evidence points to the interplay between diet and gut microbiota. Dietary interventions are likely to play a key role in the management of these diseases. Significant shifts in macronutrient composition, such as high-protein or high-fiber diets, have been associated with changes in the composition and function of the gut microbiota. Dietary fiber, starch, and protein content are known to contribute to such changes in microbiota and metabolome composition. A better understanding of the different dietary strategies available for dogs with IBD would help ensure the selection of the most appropriate diet. Therefore, considerable efforts should be applied to improving our knowledge of diet-microbiota–host molecular interactions. Similarly, further functional studies are required in order to gain mechanistic insights into this intricate loop in health and disease. This will help implement targeted and effective dietary interventions as a means to restoring health and mitigating microbiota-associated disorders, such as IBD. These evidence-based recommendations are increasingly imperative as the burden of inflammatory disorders increases in dogs. The prospect of designing such dietary interventions targeted specifically at increasing key bacterial metabolites to improve inflammatory status would be of interest, considering that formal dietary guidelines are lacking for subjects with IBD.

Author Contributions: A.S., S.R., A.K., V.M., A.D., A.J., N.A., E.M., J.H. and M.R. conceived the scientific ideas. A.S., S.R., J.H., A.K., V.M., A.D., E.M. and M.R. performed and discussed the work and edited the manuscript. All the authors reviewed the manuscript and provided critical feedback. All authors have read and agreed to the published version of the manuscript.

Funding: This work received funding from the PS-Com-Pro Project, CMCU-PHC Utique (No. 19G0819)-Campus France (41786NC) and Twinning European project MICAfrica 952583.

Acknowledgments: The authors would like to express their gratitude to H. Mkaouar for proofreading this manuscript.

Conflicts of Interest: The authors declare no conflict of interest.

References

1. Vindigni, S.M.; Zisman, T.L.; Suskind, D.L.; Damman, C.J. The intestinal microbiome, barrier function, and immune system in inflammatory bowel disease: A tripartite pathophysiological circuit with implications for new therapeutic directions. *Therap. Adv. Gastroenterol.* **2016**, *9*, 606–625. [CrossRef] [PubMed]
2. Nishida, A.; Inoue, R.; Inatomi, O.; Bamba, S.; Naito, Y.; Andoh, A. Gut microbiota in the pathogenesis of inflammatory bowel disease. *Clin. J. Gastroenterol.* **2018**, *11*, 1–10. [CrossRef] [PubMed]
3. Simpson, J.M.; Martineau, B.; Jones, W.E.; Ballam, J.M.; Mackie, R.I. Characterization of fecal bacterial populations in canines: Effects of age, breed and dietary fiber. *Microb. Ecol.* **2002**, *44*, 186–197. [CrossRef]
4. Vanhoutte, T.; Huys, G.; De Brandt, E.; Fahey, G.C., Jr.; Swings, J. Molecular monitoring and characterization of the faecal microbiota of healthy dogs during fructan supplementation. *FEMS Microbiol. Lett.* **2005**, *249*, 65–71. [CrossRef] [PubMed]
5. Biagi, G.; Cipollini, I.; Grandi, M.; Zaghini, G. Influence of some potential prebiotics and fibre-rich foodstuffs on composition and activity of canine intestinal microbiota. *Anim. Feed Sci. Technol.* **2010**, *159*, 50–58. [CrossRef]
6. Beloshapka, A.N.; Dowd, S.E.; Suchodolski, J.S.; Steiner, J.M.; Duclos, L.; Swanson, K.S. Fecal microbial communities of healthy adult dogs fed raw meat-based diets with or without inulin or yeast cell wall extracts as assessed by 454 pyrosequencing. *FEMS Microbiol. Ecol.* **2013**, *84*, 532–541. [CrossRef] [PubMed]
7. Panasevich, M.R.; Kerr, K.R.; Dilger, R.N.; Fahey, G.C., Jr.; Guérin-Deremaux, L.; Lynch, G.L.; Wils, D.; Suchodolski, J.S.; Steer, J.M.; Dowd, S.E.; et al. Modulation of the faecal microbiome of healthy adult dogs by inclusion of potato fibre in the diet. *Br. J. Nutr.* **2015**, *113*, 125–133. [CrossRef] [PubMed]
8. Suchodolski, J.S.; Markel, M.E.; Garcia-Mazcorro, J.F.; Unterer, S.; Heilmann, R.M.; Dowd, S.E.; Kachroo, P.; Ivanov, I.; Minamoto, Y.; Dillman, E.M.; et al. The fecal microbiome in dogs with acute diarrhea and idiopathic inflammatory bowel disease. *PLoS ONE* **2012**, *7*, e51907. [CrossRef]
9. Honneffer, J.B.; Minamoto, Y.; Suchodolski, J.S. Microbiota alterations in acute and chronic gastrointestinal inflammation of cats and dogs. *World J. Gastroenterol.* **2014**, *20*, 16489–16497. [CrossRef]
10. Minamoto, Y.; Otoni, C.C.; Steelman, S.M.; Büyükleblebici, O.; Steiner, J.M.; Jergens, A.E.; Suchodolski, J.S. Alteration of the fecal microbiota and serum metabolite profiles in dogs with idiopathic inflammatory bowel disease. *Gut Microbes* **2015**, *6*, 33–47. [CrossRef]
11. AlShawaqfeh, M.K.; Wajid, B.; Minamoto, Y.; Markel, M.; Lidbury, J.A.; Steiner, J.M.; Serpedin, E.; Suchodolski, J.S. A dysbiosis index to assess microbial changes in fecal samples of dogs with chronic inflammatory enteropathy. *FEMS Microbiol. Ecol.* **2017**, *93*, fix136. [CrossRef] [PubMed]
12. Marchesi, M.C.; Timpano, C.C.; Busechian, S.; Pieramati, C.; Rueca, F. The role of diet in managing inflamatory bowel disease affected dogs: A retrospective cohort study on 76 cases. *Vet. Ital.* **2017**, *53*, 297–302. [CrossRef]
13. Hou, J.K.; Abraham, B.; El-Serag, H. Dietary intake and risk of developing inflammatory bowel disease: A systematic review of the literature. *Am. J. Gastroenterol.* **2011**, *106*, 563–573. [CrossRef] [PubMed]
14. Chassaing, B.; Koren, O.; Goodrich, J.K.; Poole, A.C.; Srinivasan, S.; Ley, R.E.; Gewirtz, A.T. Dietary emulsifiers impact the mouse gut microbiota promoting colitis and metabolic syndrome. *Nature* **2015**, *519*, 92–96. [CrossRef]
15. Desai, M.S.; Seekatz, A.M.; Koropatkin, N.M.; Kamada, N.; Hickey, C.A.; Wolter, M.; Pudlo, N.A.; Kitamoto, S.; Terrapon, N.; Muller, A.; et al. A Dietary Fiber-Deprived Gut Microbiota Degrades the Colonic Mucus Barrier and Enhances Pathogen Susceptibility. *Cell* **2016**, *167*, 1339–1353.e21. [CrossRef] [PubMed]
16. Köhler, I.; Ballhausen, B.D.; Stockhaus, C.; Hartmann, K.; Wehner, A. Prevalence of and risk factors for feline hyperthyroidism among a clinic population in Southern Germany. *Tierärztliche Prax. Kleintiere* **2016**, *44*, 149–157. [CrossRef]

17. Alexander, P.; Berri, A.; Moran, D.; Reay, D.; Rounsevell, M.D.A. The global environmental paw print of pet food. *Glob. Environ. Chang.* **2020**, *65*, 102153. [CrossRef]
18. Valdés, F.; Villanueva, V.; Durán, E.; Campos, F.; Avendaño, C.; Sánchez, M.; Domingoz-Araujo, C.; Valenzuela, C. Insects as Feed for Companion and Exotic Pets: A Current Trend. *Animals* **2022**, *12*, 1450. [CrossRef]
19. Reilly, L.M.; He, F.; Rodriguez-Zas, S.L.; Southey, B.R.; Hoke, J.M.; Davenport, G.M.; de Godoy, M.R.C. Use of Legumes and Yeast as Novel Dietary Protein Sources in Extruded Canine Diets. *Front. Vet. Sci.* **2021**, *8*, 667642. [CrossRef]
20. Fritsch, D.A.; Jackson, M.I.; Wernimont, S.M.; Feld, G.K.; MacLeay, J.M.; Brejda, J.J.; Cochrane, C.Y.; Gross, K.L. Microbiome function underpins the efficacy of a fiber-supplemented dietary intervention in dogs with chronic large bowel diarrhea. *BMC Vet. Res.* **2022**, *18*, 245. [CrossRef]
21. Lee, A.H.; Lin, C.Y.; Do, S.; Oba, P.M.; Belchik, S.E.; Steelman, A.J.; Schauwecker, A.; Swanson, K.S. Dietary supplementation with fiber, "biotics," and spray-dried plasma affects apparent total tract macronutrient digestibility and the fecal characteristics, fecal microbiota, and immune function of adult dogs. *J. Anim. Sci.* **2022**, *100*, skac048. [CrossRef] [PubMed]
22. Kc, D.; Sumner, R.; Lippmann, S. Gut microbiota and health. *Postgrad. Med.* **2020**, *132*, 274. [CrossRef] [PubMed]
23. Fan, Y.; Pedersen, O. Gut microbiota in human metabolic health and disease. *Nat. Rev. Microbiol.* **2021**, *19*, 55–71. [CrossRef]
24. Muegge, B.D.; Kuczynski, J.; Knights, D.; Clemente, J.C.; González, A.; Fontana, L.; Henrissat, B.; Knight, R.; Gordon, J.I. Diet drives convergence in gut microbiome functions across mammalian phylogeny and within humans. *Science* **2011**, *332*, 970–974. [CrossRef]
25. Zmora, N.; Suez, J.; Elinav, E. You are what you eat: Diet, health and the gut microbiota. *Nat. Rev. Gastroenterol. Hepatol.* **2019**, *16*, 35–56. [CrossRef]
26. Lee, D.; Albenberg, L.; Compher, C.; Baldassano, R.; Piccoli, D.; Lewis, J.D.; Wu, G.D. Diet in the pathogenesis and treatment of inflammatory bowel diseases. *Gastroenterology* **2015**, *148*, 1087–1106. [CrossRef]
27. Levine, A.; Boneh, R.S.; Wine, E. Evolving role of diet in the pathogenesis and treatment of inflammatory bowel diseases. *Gut* **2018**, *67*, 1726–1738. [CrossRef] [PubMed]
28. Blake, A.B.; Guard, B.C.; Honneffer, J.B.; Lidbury, J.A.; Steiner, J.M.; Suchodolski, J.S. Altered microbiota, fecal lactate, and fecal bile acids in dogs with gastrointestinal disease. *PLoS ONE* **2019**, *14*, e0224454. [CrossRef] [PubMed]
29. Hand, D.; Wallis, C.; Colyer, A.; Penn, C.W. Pyrosequencing the canine faecal microbiota: Breadth and depth of biodiversity. *PLoS ONE* **2013**, *8*, e53115. [CrossRef]
30. Garcia-Mazcorro, J.F.; Lanerie, D.J.; Dowd, S.E.; Paddock, C.G.; Grützner, N.; Steiner, J.M.; Ivanek, R.; Suchodolski, J.S. Effect of a multi-species synbiotic formulation on fecal bacterial microbiota of healthy cats and dogs as evaluated by pyrosequencing. *FEMS Microbiol. Ecol.* **2011**, *78*, 542–554. [CrossRef]
31. Garcia-Mazcorro, J.F.; Dowd, S.E.; Poulsen, J.; Steiner, J.M.; Suchodolski, J.S. Abundance and short-term temporal variability of fecal microbiota in healthy dogs. *Microbiologyopen* **2012**, *1*, 340–347. [CrossRef] [PubMed]
32. Vázquez-Baeza, Y.; Hyde, E.R.; Suchodolski, J.S.; Knight, R. Dog and human inflammatory bowel disease rely on overlapping yet distinct dysbiosis networks. *Nat. Microbiol.* **2016**, *1*, 16177. [CrossRef]
33. Postler, T.S.; Ghosh, S. Understanding the Holobiont: How Microbial Metabolites Affect Human Health and Shape the Immune System. *Cell Metab.* **2017**, *26*, 110–130. [CrossRef] [PubMed]
34. Minamoto, Y.; Minamoto, T.; Isaiah, A.; Sattasathuchana, P.; Buono, A.; Rangachari, V.R.; McNeely, I.H.; Lidbury, J.; Steiner, J.M.; Suchodolski, J.S. Fecal short-chain fatty acid concentrations and dysbiosis in dogs with chronic enteropathy. *J. Vet. Intern. Med.* **2019**, *33*, 1608–1618. [CrossRef] [PubMed]
35. Ridaura, V.K.; Faith, J.J.; Rey, F.E.; Cheng, J.; Duncan, A.E.; Kau, A.L.; Griffin, N.W.; Lombard, V.; Henrissat, B.; Bain, J.R.; et al. Gut microbiota from twins discordant for obesity modulate metabolism in mice. *Science* **2013**, *341*, 1241214. [CrossRef] [PubMed]
36. Le Roy, T.; Lécuyer, E.; Chassaing, B.; Rhimi, M.; Lhomme, M.; Boudebbouze, S.; Ichou, F.; Haro Barceló, J.; Huby, T.; Guerin, M.; et al. The intestinal microbiota regulates host cholesterol homeostasis. *BMC Biol.* **2019**, *17*, 94. [CrossRef]
37. Schmitz, S.; Glanemann, B.; Garden, O.A.; Brooks, H.; Chang, Y.M.; Werling, D.; Allenspach, K. A prospective, randomized, blinded, placebo-controlled pilot study on the effect of Enterococcus faecium on clinical activity and intestinal gene expression in canine food-responsive chronic enteropathy. *J. Vet. Intern. Med.* **2015**, *29*, 533–543. [CrossRef]
38. Suchodolski, J.S.; Ruaux, C.G.; Steiner, J.M.; Fetz, K.; Williams, D.A. Assessment of the qualitative variation in bacterial microflora among compartments of the intestinal tract of dogs by use of a molecular fingerprinting technique. *Am. J. Vet. Res.* **2005**, *66*, 1556–1562. [CrossRef]
39. Guard, B.C.; Honneffer, J.B.; Jergens, A.E.; Jonika, M.M.; Toresson, L.; Lawrence, Y.A.; Webb, C.B.; Hill, S.; Lidbury, J.A.; Steiner, J.M.; et al. Longitudinal assessment of microbial dysbiosis, fecal unconjugated bile acid concentrations, and disease activity in dogs with steroid-responsive chronic inflammatory enteropathy. *J. Vet. Intern. Med.* **2019**, *33*, 1295–1305. [CrossRef]
40. Simpson, K.W.; Dogan, B.; Rishniw, M.; Goldstein, R.E.; Klaessig, S.; McDonough, P.L.; German, A.J.; Yates, R.M.; Russell, D.G.; Johnson, S.E.; et al. Adherent and invasive *Escherichia coli* is associated with granulomatous colitis in boxer dogs. *Infect. Immun.* **2006**, *74*, 4778–4792. [CrossRef]
41. Suchodolski, J.S.; Camacho, J.; Steiner, J.M. Analysis of bacterial diversity in the canine duodenum, jejunum, ileum, and colon by comparative 16S rRNA gene analysis. *FEMS Microbiol. Ecol.* **2008**, *66*, 567–578. [CrossRef] [PubMed]

42. Vich Vila, A.; Imhann, F.; Collij, V.; Jankipersadsing, S.A.; Gurry, T.; Mujagic, Z.; Kurilshikov, A.; Bonder, M.J.; Jiang, X.; Tigchelaar, E.F.; et al. Gut microbiota composition and functional changes in inflammatory bowel disease and irritable bowel syndrome. *Sci. Transl. Med.* **2018**, *10*, eaap8914. [CrossRef] [PubMed]
43. Derrien, M.; Alvarez, A.S.; de Vos, W.M. The Gut Microbiota in the First Decade of Life. *Trends Microbiol.* **2019**, *27*, 997–1010. [CrossRef] [PubMed]
44. Xenoulis, P.G.; Palculict, B.; Allenspach, K.; Steiner, J.M.; Van House, A.M.; Suchodolski, J.S. Molecular-phylogenetic characterization of microbial communities imbalances in the small intestine of dogs with inflammatory bowel disease. *FEMS Microbiol. Ecol.* **2008**, *66*, 579–589. [CrossRef]
45. Suchodolski, J.S. Diagnosis and interpretation of intestinal dysbiosis in dogs and cats. *Vet. J.* **2016**, *215*, 30–37. [CrossRef]
46. Xu, J.; Verbrugghe, A.; Lourenço, M.; Cools, A.; Liu, D.J.X.; Van de Wiele, T.; Marzorati, M.; Eeckhaut, V.; Van Immerseel, F.; Vanhaecke, L.; et al. The response of canine faecal microbiota to increased dietary protein is influenced by body condition. *BMC Vet. Res.* **2017**, *13*, 374. [CrossRef]
47. Zentek, J.; Marquart, B.; Pietrzak, T.; Ballèvre, O.; Rochat, F. Dietary effects on bifidobacteria and *Clostridium perfringens* in the canine intestinal tract. *J. Anim. Physiol. Anim. Nutr.* **2003**, *87*, 397–407. [CrossRef]
48. Coelho, L.P.; Kultima, J.R.; Costea, P.I.; Fournier, C.; Pan, Y.; Czarnecki-Maulden, G.; Hayward, M.R.; Forslund, S.K.; Schmidt, T.S.B.; Descombes, P.; et al. Similarity of the dog and human gut microbiomes in gene content and response to diet. *Microbiome* **2018**, *6*, 72. [CrossRef]
49. Martínez-López, L.M.; Pepper, A.; Pilla, R.; Woodward, A.P.; Suchodolski, J.S.; Mansfield, C. Effect of sequentially fed high protein, hydrolyzed protein, and high fiber diets on the fecal microbiota of healthy dogs: A cross-over study. *Anim. Microbiome* **2021**, *3*, 42. [CrossRef]
50. Atherly, T.; Rossi, G.; White, R.; Seo, Y.J.; Wang, C.; Ackermann, M.; Breuer, M.; Allenspach, K.; Mochel, J.P.; Jergens, A.E. Glucocorticoid and dietary effects on mucosal microbiota in canine inflammatory bowel disease. *PLoS ONE* **2019**, *14*, e0226780. [CrossRef]
51. Sugihara, K.; Morhardt, T.L.; Kamada, N. The Role of Dietary Nutrients in Inflammatory Bowel Disease. *Front. Immunol.* **2019**, *9*, 3183. [CrossRef]
52. Heinken, A.; Hertel, J.; Thiele, I. Metabolic modelling reveals broad changes in gut microbial metabolism in inflammatory bowel disease patients with dysbiosis. *NPJ Syst. Biol. Appl.* **2021**, *7*, 19. [CrossRef] [PubMed]
53. Alkhalaf, L.M.; Ryan, K.S. Biosynthetic manipulation of tryptophan in bacteria: Pathways and mechanisms. *Chem. Biol.* **2015**, *22*, 317–328. [CrossRef] [PubMed]
54. Ghiboub, M.; Verburgt, C.M.; Sovran, B.; Benninga, M.A.; de Jonge, W.J.; Van Limbergen, J.E. Nutritional Therapy to Modulate Tryptophan Metabolism and Aryl Hydrocarbon-Receptor Signaling Activation in Human Diseases. *Nutrients* **2020**, *12*, 2846. [CrossRef]
55. Kathrani, A.; Allenspach, K.; Fascetti, A.J.; Larsen, J.A.; Hall, E.J. Alterations in serum amino acid concentrations in dogs with protein-losing enteropathy. *J. Vet. Intern. Med.* **2018**, *32*, 1026–1032. [CrossRef]
56. Tamura, Y.; Ohta, H.; Kagawa, Y.; Osuga, T.; Morishita, K.; Sasaki, N.; Takiguchi, M. Plasma amino acid profiles in dogs with inflammatory bowel disease. *J. Vet. Intern. Med.* **2019**, *33*, 1602–1607. [CrossRef] [PubMed]
57. Joossens, M.; De Preter, V.; Ballet, V.; Verbeke, K.; Rutgeerts, P.; Vermeire, S. Effect of oligofructose-enriched inulin (OF-IN) on bacterial composition and disease activity of patients with Crohn's disease: Results from a double-blinded randomised controlled trial. *Gut* **2012**, *61*, 958. [CrossRef] [PubMed]
58. Welters, C.F.; Heineman, E.; Thunnissen, F.B.; van den Bogaard, A.E.; Soeters, P.B.; Baeten, C.G. Effect of dietary inulin supplementation on inflammation of pouch mucosa in patients with an ileal pouch-anal anastomosis. *Dis. Colon Rectum* **2002**, *45*, 621–627. [CrossRef]
59. Sun, Y.; O'Riordan, M.X. Regulation of bacterial pathogenesis by intestinal short-chain Fatty acids. *Adv. Appl. Microbiol.* **2013**, *85*, 93–118. [CrossRef]
60. Smith, P.M.; Howitt, M.R.; Panikov, N.; Michaud, M.; Gallini, C.A.; Bohlooly-Y, M.; Glickman, J.N.; Garrett, W.S. The microbial metabolites, short-chain fatty acids, regulate colonic Treg cell homeostasis. *Science* **2013**, *341*, 569–573. [CrossRef]
61. Lappin, M.R.; Zug, A.; Hovenga, C.; Gagne, J.; Cross, E. Efficacy of feeding a diet containing a high concentration of mixed fiber sources for management of acute large bowel diarrhea in dogs in shelters. *J. Vet. Intern. Med.* **2022**, *36*, 488–492. [CrossRef] [PubMed]
62. Giaretta, P.R.; Rech, R.R.; Guard, B.C.; Blake, A.B.; Blick, A.K.; Steiner, J.M.; Lidbury, J.A.; Cook, A.K.; Hanifeh, M.; Spillmann, T.; et al. Comparison of intestinal expression of the apical sodium-dependent bile acid transporter between dogs with and without chronic inflammatory enteropathy. *J. Vet. Intern. Med.* **2018**, *32*, 1918–1926. [CrossRef] [PubMed]
63. Kitahara, M.; Takamine, F.; Imamura, T.; Benno, Y. *Clostridium hiranonis* sp. nov., a human intestinal bacterium with bile acid 7alpha-dehydroxylating activity. *Int. J. Syst. Evol. Microbiol.* **2001**, *51*, 39–44. [CrossRef] [PubMed]
64. Moinard, A.; Payen, C.; Ouguerram, K.; André, A.; Hernandez, J.; Drut, A.; Biourge, V.C.; Suchodolski, J.S.; Flanagan, J.; Nguyen, P.; et al. Effects of High-Fat Diet at Two Energetic Levels on Fecal Microbiota, Colonic Barrier, and Metabolic Parameters in Dogs. *Front. Vet. Sci.* **2020**, *7*, 566282. [CrossRef]
65. Titmarsh, H.; Gow, A.G.; Kilpatrick, S.; Sinclair, J.; Hill, T.; Milne, E.; Philbey, A.; Berry, J.; Handel, I.; Mellanby, R.J. Association of Vitamin D Status and Clinical Outcome in Dogs with a Chronic Enteropathy. *J. Vet. Intern. Med.* **2015**, *29*, 1473–1478. [CrossRef]

66. Davenport, D.J.; Ching, R.J.; Hunt, J.H.; Bruyette, D.S.; Gross, K.L. The effect of dietary levels of folate and cobalamin on the serum concentration of folate and cobalamin in the dog. *J. Nutr.* **1994**, *124*, 2559S–2562S. [CrossRef] [PubMed]
67. Batt, R.M.; Needham, J.R.; Carter, M.W. Bacterial overgrowth associated with a naturally occurring enteropathy in the German shepherd dog. *Res. Vet. Sci.* **1983**, *35*, 42–46. [CrossRef]
68. Glanemann, B.; Seo, Y.J.; Priestnall, S.L.; Garden, O.A.; Kilburn, L.; Rossoni-Serao, M.; Segarra, S.; Mochel, J.P.; Allenspach, K. Clinical efficacy of prebiotics and glycosaminoglycans versus placebo in dogs with food responsive enteropathy receiving a hydrolyzed diet: A pilot study. *PLoS ONE* **2021**, *16*, e0250681. [CrossRef]
69. Hernandez, J.; Dandrieux, J.R.S. From Bench Top to Clinics: How New Tests can be Helpful in Diagnosis and Management of Dogs with Chronic Enteropathies. *Vet. Clin. N. Am. Small Anim. Pract.* **2021**, *51*, 137–153. [CrossRef]
70. Allenspach, K.; Wieland, B.; Gröne, A.; Gaschen, F. Chronic enteropathies in dogs: Evaluation of risk factors for negative outcome. *J. Vet. Intern. Med.* **2007**, *21*, 700–708. [CrossRef]
71. Dumitrescu, G.; Mihai, C.; Dranga, M.; Prelipcean, C.C. Serum 25-hydroxyvitamin D concentration and inflammatory bowel disease characteristics in Romania. *World J. Gastroenterol.* **2014**, *20*, 2392–2396. [CrossRef] [PubMed]
72. Ooi, J.H.; Li, Y.; Rogers, C.J.; Cantorna, M.T. Vitamin D regulates the gut microbiome and protects mice from dextran sodium sulfate-induced colitis. *J. Nutr.* **2013**, *143*, 1679–1686. [CrossRef] [PubMed]
73. Guo, C.; Sinnott, B.; Niu, B.; Lowry, M.B.; Fantacone, M.L.; Gombart, A.F. Synergistic induction of human cathelicidin antimicrobial peptide gene expression by vitamin D and stilbenoids. *Mol. Nutr. Food Res.* **2014**, *58*, 528–536. [CrossRef] [PubMed]
74. Volkmann, M.; Steiner, J.M.; Fosgate, G.T.; Zentek, J.; Hartmann, S.; Kohn, B. Chronic Diarrhea in Dogs—Retrospective Study in 136 Cases. *J. Vet. Intern. Med.* **2017**, *31*, 1043–1055. [CrossRef] [PubMed]
75. Jugan, M.C. Dietary therapy as a treatment option for dogs with chronic enteropathies. *Vet. Rec.* **2020**, *186*, 23–25. [CrossRef]
76. Lecoindre, P.; Gaschen, F.P. Chronic idiopathic large bowel diarrhea in the dog. *Vet. Clin. N. Am. Small Anim. Pract.* **2011**, *41*, 447–456. [CrossRef]
77. Okanishi, H.; Yoshioka, R.; Kagawa, Y.; Watari, T. The clinical efficacy of dietary fat restriction in treatment of dogs with intestinal lymphangiectasia. *J. Vet. Intern. Med.* **2014**, *28*, 809–817. [CrossRef]
78. Bresciani, F.; Minamoto, Y.; Suchodolski, J.S.; Galiazzo, G.; Vecchiato, C.G.; Pinna, C.; Biagi, G.; Pietra, M. Effect of an extruded animal protein-free diet on fecal microbiota of dogs with food-responsive enteropathy. *J. Vet. Intern. Med.* **2018**, *32*, 1903–1910. [CrossRef]
79. Schmitz, S.; Henrich, M.; Neiger, R.; Werling, D.; Allenspach, K. Stimulation of duodenal biopsies and whole blood from dogs with food-responsive chronic enteropathy and healthy dogs with Toll-like receptor ligands and probiotic Enterococcus faecium. *Scand. J. Immunol.* **2014**, *80*, 85–94. [CrossRef]
80. Kalenyak, K.; Isaiah, A.; Heilmann, R.M.; Suchodolski, J.S.; Burgener, I.A. Comparison of the intestinal mucosal microbiota in dogs diagnosed with idiopathic inflammatory bowel disease and dogs with food-responsive diarrhea before and after treatment. *FEMS Microbiol. Ecol.* **2018**, *94*, fix173. [CrossRef] [PubMed]
81. Willard, M.D.; Simpson, R.B.; Delles, E.K.; Cohen, N.D.; Fossum, T.W.; Kolp, D.; Reinhart, G. Effects of dietary supplementation of fructo-oligosaccharides on small intestinal bacterial overgrowth in dogs. *Am. J. Vet. Res.* **1994**, *55*, 654–659. [PubMed]
82. Jia, J.; Frantz, N.; Khoo, C.; Gibson, G.R.; Rastall, R.A.; McCartney, A.L. Investigation of the faecal microbiota associated with canine chronic diarrhoea. *FEMS Microbiol. Ecol.* **2010**, *71*, 304–312. [CrossRef]
83. Segarra, S.; Martínez-Subiela, S.; Cerdà-Cuéllar, M.; Martínez-Puig, D.; Muñoz-Prieto, A.; Rodríguez-Franco, F.; Rodríguez-Bertos, A.; Allenspach, K.; Velasco, A.; Cerón, J. Oral chondroitin sulfate and prebiotics for the treatment of canine Inflammatory Bowel Disease: A randomized, controlled clinical trial. *BMC Vet. Res.* **2016**, *12*, 49. [CrossRef] [PubMed]
84. Ambrosini, Y.M.; Neuber, S.; Borcherding, D.; Seo, Y.J.; Segarra, S.; Glanemann, B.; Garden, O.A.; Müller, U.; Adam, M.G.; Dang, V.; et al. Treatment with Hydrolyzed Diet Supplemented with Prebiotics and Glycosaminoglycans Alters Lipid Metabolism in Canine Inflammatory Bowel Disease. *Front. Vet. Sci.* **2020**, *7*, 451. [CrossRef] [PubMed]
85. Isidori, M.; Rueca, F.; Massacci, F.R.; Diaferia, M.; Giontella, A.; Caldin, M.; Furlanello, T.; Corbee, R.J.; Mannucci, G.; Pezzotti, G.; et al. The Use of Ascophyllum nodosum and Bacillus subtilis C-3102 in the Management of Canine Chronic Inflammatory Enteropathy: A Pilot Study. *Animals* **2021**, *11*, 3417. [CrossRef] [PubMed]
86. Axelsson, E.; Ratnakumar, A.; Arendt, M.L.; Maqbool, K.; Webster, M.T.; Perloski, M.; Liberg, O.; Arnemo, J.M.; Hedhammar, A.; Lindblad-Toh, K. The genomic signature of dog domestication reveals adaptation to a starch-rich diet. *Nature* **2013**, *495*, 360–364. [CrossRef]
87. Reiter, T.; Jagoda, E.; Capellini, T.D. Dietary Variation and Evolution of Gene Copy Number among Dog Breeds. *PLoS ONE* **2016**, *11*, e0148899. [CrossRef]
88. Warren, F.J.; Fukuma, N.M.; Mikkelsen, D.; Flanagan, B.M.; Williams, B.A.; Lisle, A.T.; Cuív, P.Ó.; Morrison, M.; Gidley, M.J. Food Starch Structure Impacts Gut Microbiome Composition. *mSphere* **2018**, *3*, e00086-18. [CrossRef]
89. Bassaganya-Riera, J.; DiGuardo, M.; Viladomiu, M.; de Horna, A.; Sanchez, S.; Einerhand, A.W.; Sanders, L.; Hontecillas, R. Soluble fibers and resistant starch ameliorate disease activity in interleukin-10-deficient mice with inflammatory bowel disease. *J. Nutr.* **2011**, *141*, 1318–1325. [CrossRef]
90. Moreau, N.M.; Champ, M.M.; Goupry, S.M.; Le Bizec, B.J.; Krempf, M.; Nguyen, P.G.; Dumon, H.J.; Martin, L.J. Resistant starch modulates in vivo colonic butyrate uptake and its oxidation in rats with dextran sulfate sodium-induced colitis. *J. Nutr.* **2004**, *134*, 493–500. [CrossRef]

91. DeMartino, P.; Cockburn, D.W. Resistant starch: Impact on the gut microbiome and health. *Curr. Opin. Biotechnol.* **2020**, *61*, 66–71. [CrossRef] [PubMed]
92. Dey, M. Toward a Personalized Approach in Prebiotics Research. *Nutrients* **2017**, *9*, 92. [CrossRef] [PubMed]

metabolites

MDPI

Article

Effect of a Multistrain Probiotic on Feline Gut Health through the Fecal Microbiota and Its Metabolite SCFAs

Yifei Li [1], Ilyas Ali [2], Zhiqi Lei [1], Yanan Li [1], Min Yang [1], Caixia Yang [1] and Lian Li [1,*]

1 College of Animal Science and Technology, Nanjing Agricultural University, Nanjing 210095, China
2 Department of Medical Cell Biology and Genetics, Health Science Center, Shenzhen University, Shenzhen 518060, China
* Correspondence: lilian@njau.edu.cn

Abstract: With the increasing awareness of raising pets following scientific methods, people are becoming increasingly more interested in the nutrition and health of pets, especially their intestinal health, which has become a research hotspot. Both *Saccharomyces boulardii* and *Pediococcus acidilactici* are probiotics with strong probiotic properties that can maintain the balance of intestinal flora. However, the role of *Saccharomyces boulardii* and *Pediococcus acidilactici* in felines has not been comprehensively studied to date. The aim of this study is to investigate the effect of multistrain probiotics consisting of *Saccharomyces boulardii* and *Pediococcus acidilactici* on the gut health of felines by modulating gut microbes and the production of metabolite SCFAs. The results show that the multistrain probiotic did not alter the intestinal microbial diversity and structure of short-haired domestic cats, promoted the colonization of beneficial bacteria, increased the levels of microbiota-derived SCFAs and fecal antioxidants, and reduced the levels of fecal inflammatory markers. In conclusion, the use of a multistrain probiotic in healthy, short-haired domestic cats can promote gut health by modulating gut microbes, improving microbiota-derived SCFA production, reducing inflammatory conditions, and improving antioxidant status. These results provide new insights for further exploration of the role of probiotics in the gut microbiome of cats.

Keywords: *Saccharomyces boulardii*; *Pediococcus acidilactici*; gut microbiota; SCFAs; cat; gut health

Citation: Li, Y.; Ali, I.; Lei, Z.; Li, Y.; Yang, M.; Yang, C.; Li, L. Effect of a Multistrain Probiotic on Feline Gut Health through the Fecal Microbiota and Its Metabolite SCFAs. *Metabolites* **2023**, *13*, 228. https://doi.org/10.3390/metabo13020228

Academic Editor: Karolina Skonieczna-Żydecka

Received: 12 January 2023
Revised: 30 January 2023
Accepted: 2 February 2023
Published: 3 February 2023

1. Introduction

Pet nutrition and health are attracting increased attention in the literature, as pets play a greater role in the daily lives of human beings. The gastrointestinal microbiome of pets is closely related to pet health. Gut microbes, which are typically categorized as commensal, pathogenic, or opportunistic bacteria [1], are observed in the animal's digestive tract, and each of them has its own distinct flora balance. Intestinal microorganisms participate in the digestion, absorption, and metabolism of nutrients in the host body and play an important regulatory role in the metabolism of nutrients, such as lipid metabolism, carbohydrate metabolism, protein metabolism, and vitamin synthesis. The composition and status of the gut microbiome are closely associated with the intestinal health of the host, and the imbalance of the host intestinal microbiome may lead to a series of disorders in the body, such as gastrointestinal diseases, metabolic diseases, obesity, and immune system diseases.

Probiotics are defined as "live microorganisms that are beneficial to the host when ingested in sufficient amounts" [2], and the most commonly studied in the literature are *Lactobacillus*, *Bifidobacterium*, and yeasts [3]. Probiotics are capable of lowering the pH value, creating an acidic environment, reducing the reproduction of harmful microorganisms, increasing the number of beneficial bacteria, interacting directly with immune cells [4], and participating in regulating the immune balance of digestive tract flora, thereby contributing to the prevention and treatment of gastrointestinal-related diseases. Studies have proven that probiotics are effective in the treatment of acute infectious diarrhea, antibiotic-associated diarrhea, *Clostridium difficile*-associated diarrhea, hepatic encephalopathy, ulcerative colitis, irritable bowel syndrome, functional gastrointestinal disorders, and necrotizing enterocolitis [5–9].

The function of probiotics is closely related to the species of microorganisms colonized in the intestines. The interaction occurring among probiotics, host cells, and intestinal flora is a key factor affecting host health. Probiotics can regulate intestinal mucosal immunity, interact with symbiotic bacteria or potentially harmful pathogens, and produce metabolites, such as short-chain fatty acids (SCFAs), which help to inhibit and eliminate potential pathogens, thereby improving the intestinal microbial environment, strengthening the intestinal barrier, and reducing inflammation, with the purpose of enhancing antigen-specific immune responses [10]. An aqueous probiotic suspension was studied in in vitro human intestinal models, and it was observed that probiotics could increase the production of SCFAs and anti-inflammatory cytokines [11]. Studies using *Lactobacillus paracasei* SSP in d-galactose-induced aging mice determined that probiotics improved memory and learning ability in mice by increasing SCFA production and inhibiting cell apoptosis and brain injury [12]. The growth of bacteria in microorganisms affects the type and amount of fermentation products. After sugar is phosphorylated, bacteria can be converted to pyruvate or pyruvate and additional acetyl phosphates by glycolysis (the Embden–Meyerhof–Parnassian pathway), the Enter–Doudoroff pathway, or the *Bifidobacterium* pathway [13].

Saccharomyces boulardii is the only yeast among intestinal probiotics. Unlike other Saccharomyces cerevisiae strains, *Saccharomyces boulardii* is physiologically and metabolically tolerant to low pH levels and resistant to the gastric environment [14]. *Saccharomyces boulardii* exhibits distinct probiotic properties, including improved intestinal barrier function, competitive exclusion of pathogens, production of antimicrobial peptides, and immunomodulatory and nutritional effects [15]. *Saccharomyces boulardii* significantly ameliorates colon damage and regulates inflammatory responses in mice with colitis induced by dextran sulfate sodium (DSS) through changes in the microbiome composition and SCFA metabolism [16]. *Pediococcus acidilactici* is a facultative anaerobic bacterium of Firmicutes with a high salt tolerance, and it plays an important role in enhancing immunity, producing bacteriocins, maintaining the balance of intestinal flora, and improving the intestinal barrier [17]. The intervention of *Pediococcus acidilactici* FZU106 was observed to obviously increase SCFA levels in high-fat diet (HFD)-induced hyperlipidemic rats, which was strongly associated with changes in gut microbial composition and metabolism [18]. *Saccharomyces boulardii* has been observed to be safe for use in dogs with chronic enteritis and to control clinical symptoms better than standard treatment alone [19]. The use of *Enterococcus faecium* SF68 in cats infected with latent feline herpesvirus 1 (FHV-1) reduced morbidity and maintained fecal microbiota diversity [20]. Studies on *Saccharomyces boulardii* and *Pediococcus acidilactici* in pet cats have rarely been reported, and this trial investigated the effect of a multistrain probiotic preparation added to the diet on intestinal fecal microbial diversity, microbiota-derived SCFA production, antioxidative status, and fecal inflammatory markers in healthy short-haired domestic cats. This study provides an innovative platform to further investigate the significance of *Saccharomyces boulardii* and *Pediococcus acidilactici* with respect to the intestinal health of companion animals.

2. Materials and Methods

2.1. Study Probiotic

The multistrain probiotic preparation contained approximately 2.0×10^{10} CFU/g of *Saccharomyces boulardii* and approximately 2.5×10^{10} CFU/g of *Pediococcus acidilactici* mixed in a 1:1 ratio. *Saccharomyces boulardii* was supplied by the Lallemand Animal Nutrition Group and is the strain used in Levucell SB®. *Pediococcus acidilactici* (Mito 5051) was provided by Imagilin Technology LLC (Frederick, MD, USA). In the test, both the multistrain probiotic preparation and a basic diet were provided by Nanjing Xintengrui Biotechnology Co., Ltd. (Nanjing, China).

2.2. Animals

A total of 12 short-haired domestic cats between the ages of 2 and 4 years with an average body condition score of 3.5 points (on a 5-point scale) were used in this study. On day 10 of the trial, 1 of the 12 cats in the control group (male) suffered trauma and withdrew from the study; additionally, a cat (male) in the probiotic group dropped out of the study due to a skin disease that developed during the trial. Therefore, 10 cats (4 males and 6 females) were fully involved in the entire trial. They were randomly divided into control and probiotic groups, each containing two male and three female cats (n = 5). Animal experiments were approved by the Laboratory Animal Welfare Ethics Review Committee of Nanjing Agricultural University.

2.3. Experimental Design

A total of 12 short-haired domestic cats were fed with a cat basal diet and acclimatized for 1 week with free access to water. After 1 week of acclimatized feeding, the cats were randomly divided into 2 groups of 6 replicates each, with 1 cat per replicate. The control group was fed a basic diet, and the probiotic group was fed a multistrain probiotic preparation on the basis of the control group. The dosage of probiotics was added according to the cat's weight (0.5 g/kg).

2.4. Sample Collection and Fecal Microbiota Analysis

Fresh fecal specimens obtained from all subjects were collected on day 28. According to previously described methods [21], DNA extraction and sequencing were completed by Shanghai Biozeron Biotechnology Co., Ltd. (Shanghai, China). The V3–V4 region of the bacterial 16S rRNA gene was amplified by PCR using primers 341F and 806R. The purification of amplification products was performed using an AxyPrepDNA gel extraction kit (Axygen Biosciences, Union City, CA, USA). Based on the preliminary results of electrophoresis, PCR products were detected and quantified with a QuantiFluor-ST™ Blue Fluor Quantitative System (Promega Corporation, Madison, WI, USA). Isomolar pooled and paired-end sequences of purified amplification products on the Illumina platform (2×250) were obtained according to the procedure described in [22].

2.5. Fermentation Metabolite Analysis

Concentrations of short-chain fatty acids (SCFAs) were determined by gas chromatography according to the method described in a previous study [23].

2.6. Enzyme-Linked Immunosorbent Assay (ELISA) for Fecal Inflammatory Markers

After mixing feces with normal saline in equal amounts, fecal supernatant samples were collected for further analysis. The concentration of inflammatory markers, such as fecal calprotectin (FC), lactoferrin (LF), matrix metalloproteinase-9 (MMP-9), and myeloperoxidase (MPO), in the supernatant were determined by an enzyme-linked immunosorbent assay (ELISA); then, the assay procedure was performed using a cat CALP ELISA kit (ANG-E71001C), cat LTF ELSA kit (ANG-E71005C), cat MMP-9 ELISA kit (ANG-E71013C), and cat MPO ELISA kit (ANG-E71014C) according to manufacturer's instructions (Nanjing Jiancheng Bioengineering Institute, Nanjing, China).

2.7. Fecal Supernatant Antioxidant Capacity Analysis

The contents of glutathione (GSH), superoxide dismutase (SOD), and malondialdehyde (MDA) in fecal supernatants on day 28 were determined using GSH (A006), SOD (A001), and MDA (A003) commercial assay kits, respectively (Nanjing Jiancheng Bioengineering Institute, Nanjing, China).

2.8. Statistical Analysis

The data were analyzed using SPSS 23.0 statistical software with an independent t-test and expressed as mean $\pm$ standard deviation (SD). The difference between the two means was considered statistically significant when $p < 0.05$. Data visualization was performed using GraphPad Prism version 8.0.2.

3. Results

3.1. Analysis of Intestinal Flora Structure of Healthy Cats

As shown in Figure 1, the number of OTUs in the control group was 435, and the number of OTUs in the probiotic group was 450, with a total of 369 OTUs in common between the two groups. The number of OTUs in the probiotic group was 81, compared to 66 in the control group alone.

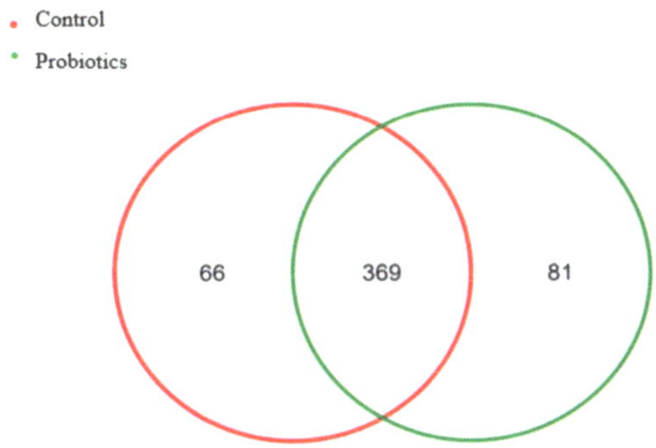

Figure 1. OTU Venn diagram.

3.2. Phylum-Level Structural Analysis

A few species (genus level) with the highest sampling frequency (the 30 most dominant species) were selected, and the phylum taxonomic information corresponding to the species was displayed; species with the same color indicating their origin from the same phylum were also displayed. The results show that there were five dominant phyla (Figure 2A), namely Actinobacteriota, Bacteroidota, Desulfobacterota, Firmicutes, and Fusobacteriota.

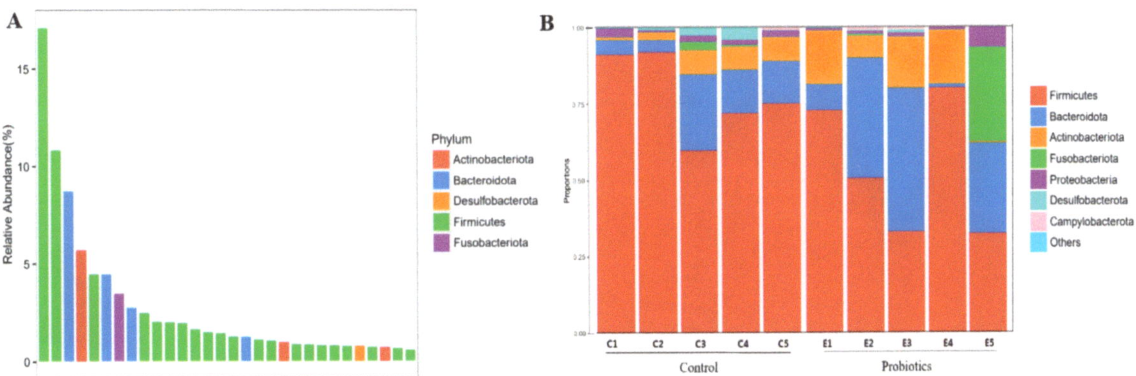

Figure 2. Phylum-level structural analysis. (**A**) The 30 most dominant species in all samples. (**B**) Relative abundance of the gut microbiota at the phylum level.

The results show (Figure 2B) that the control and probiotic groups were mainly composed of Actinobacteriota, Bacteroidota, and Firmicutes, whereas the control group was dominated by Firmicutes. The phylum Firmicutes had the highest relative phylum-level frequency in the probiotic group, followed by the phylum Bacteroidota. Table 1 shows that there was no significant difference in the relative abundance of microorganisms at the phylum level between the control and probiotic groups ($p > 0.05$).

Table 1. Relative abundance of major flora at the phylum level (%).

Phylum	Control	Probiotics
Bacteroidota	12.30 ± 8.43	25.08 ± 19.68
Firmicutes	78.07 ± 13.66	53.93 ± 22.14

Data are presented as mean ± SD. There was no significant difference between the control and probiotic groups ($p > 0.05$).

3.3. Genus-Level Structural Analysis

At the genus level (Figure 3), the microbes with a high abundance in the control group were Peptoclostridium and Blautia, whereas the content of Bacteroides was higher in the probiotic group. There was no significant difference in the relative abundance of the top-10 microbial species between the control and probiotic groups ($p > 0.05$).

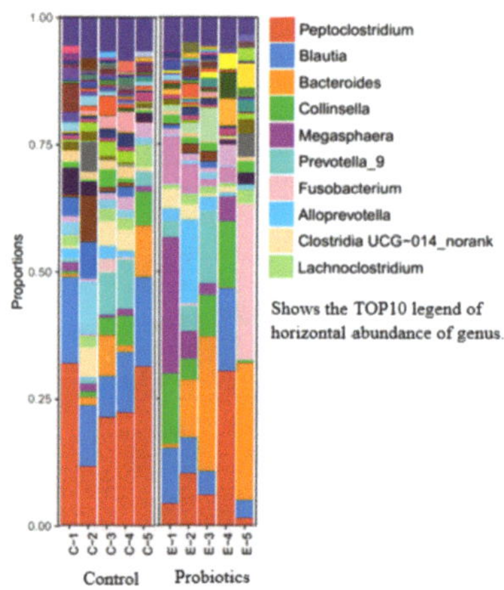

Figure 3. Relative abundance of the gut microbiota at the genus level.

3.4. Alpha Diversity Analysis

The Chao index indicates species abundance. The ACE, Shannon, and Simpson indices are used to calculate species richness, and the evenness index indicates the homogeneity of the microbial community. As shown in Table 2, there were no significant differences ($p > 0.05$).

Table 2. Statistical analysis of microbial alpha diversity.

Group	Chao	ACE	Shannon	Simpson	Evenness
Control	316.68 ± 27.52	315.58 ± 29.33	3.95 ± 0.25	0.05 ± 0.02	0.70 ± 0.03
Probiotics	284.35 ± 71.53	283.44 ± 70.47	3.55 ± 0.41	0.07 ± 0.03	0.64 ± 0.04

Data are presented as mean ± SD. There was no significant difference between control and probiotic groups ($p > 0.05$).

3.5. Dimension Reduction Analysis

Principal coordinate analysis (PCoA) is based on sample information reflected in the form of points on a two-dimensional plane. Small sample intervals indicate species with similar compositional structures; therefore, samples with a high community structure similarity tend to cluster together, whereas samples with considerable community differences are placed far apart. In Figure 4A,B, the PCoA analysis shows that there was no significant separation between fecal samples of the control and probiotic groups (ANOSIM similarity analysis, weighted UniFrac distance $p = 0.052$, and unweighted UniFrac distance $p = 0.265$).

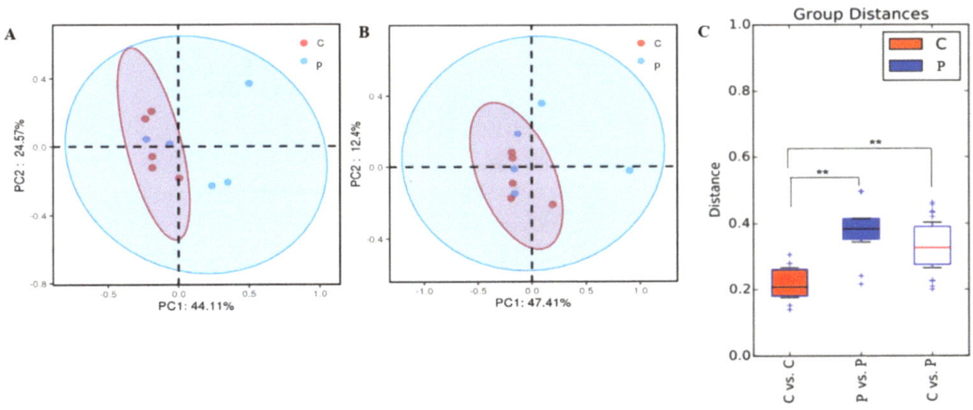

Figure 4. Dimension reduction analysis. (**A**) Principal coordinate analysis (PCoA) of weighted UniFrac distances of 16S rRNA genes. (**B**) Principal coordinate analysis (PCoA) of unweighted UniFrac distances of 16S rRNA genes. (**C**) UniFrac distance matrix plot. The pink circles represent the control group, while the blue circles represent the probiotic group. **: the difference between the control and probiotic groups was extremely significant ($p < 0.01$).

The weighted UniFrac distance reflects the similarity of gut microbial communities in terms of phylogenetic tree overlap. The greater the distance, the greater the species and abundance of microbial communities between the two groups. As shown in Figure 4C, the weighted UniFrac distances between cats in the probiotic group were greater than those between the control groups, and the difference was highly significant ($p < 0.01$).

3.6. LEFSe Difference Analysis

Figure 5A presents species with significant differences in abundance between groups. The differences are represented by LDA (linear discriminant analysis) scores, and the histogram length represents the contribution of different species (LDA score). Figure 5A shows that 16 species of Firmicutes and Proteobacteria play an important role in the control group. These 16 groups of bacteria are mainly *Bacillus* and *Clostridium* of the phylum Firmicutes and Proteobacteria. The three groups of bacteria that play an important role in the probiotic group are *Pediococcus*, Bacillaceae, and *Bacillus* of Firmicutes.

Furthermore, Figure 5B shows the clade of different species, with circles radiating from the inside to the outside representing the taxonomic levels ranging from phylum to genus (or species). The red nodes indicate microbial taxa that play an important role in the control group, and green nodes indicate microbial taxa that play an important role in the probiotic group. Microbiota that played an important role in the control group were Erysipelatoclostridiaceae, Erysipelotrichales, Lachnospiraceae, Lachnospirales, Peptostreptococcaceae, Peptostreptococcales_Tissierellales, and Clostridia. An important player in the probiotic group is Bacillaceae.

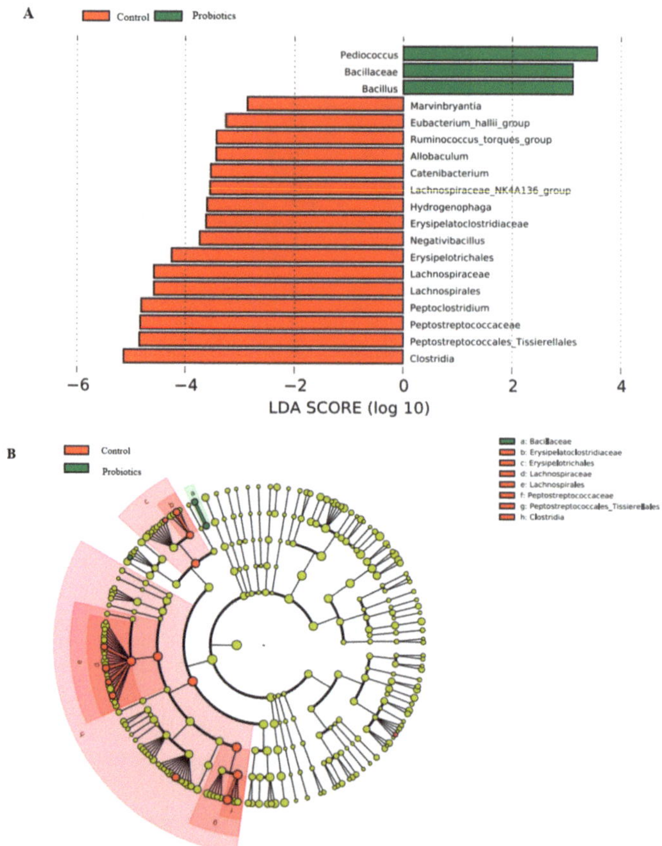

Figure 5. LEFSe difference analysis. (**A**) LDA scores for microbial taxa with significant roles in both groups. (**B**) 16S rRNA sequencing analysis of differences in functional gene abundance in the guts of cats in the two groups.

3.7. Microbiota-Derived SCFA Production

The effect of a multistrain probiotic on fecal fermentation metabolites is presented in Figure 6. Both fecal butyric acid and total SCFA concentrations were elevated in the probiotic group compared to the control group ($p < 0.05$). There were no differences in acetic and propionic acid quantities as compared to the control group.

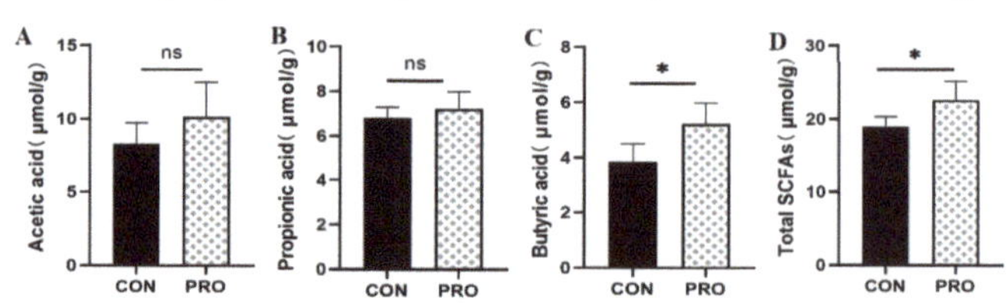

Figure 6. Effect of a multistrain probiotic preparation on fecal fermentation metabolites in healthy short-haired domestic cats. (**A–D**) Concentrations of acetic acid (**A**), propionic acid (**B**), butyric acid (**C**), and total SCFAs (**D**) detected by gas chromatography. Data are presented as mean ± SD. * $p < 0.05$; ns, $p > 0.05$ compared with the control group.

3.8. Fecal-Related Inflammatory Indicators

Concentrations of inflammatory indicators in cat feces at each stage were determined using ELISA assay kits. As shown in Figure 7, the probiotic group presented less myeloperoxidase (MPO) and fecal calprotectin (FC) activities on day 28 than the control group ($p < 0.05$).

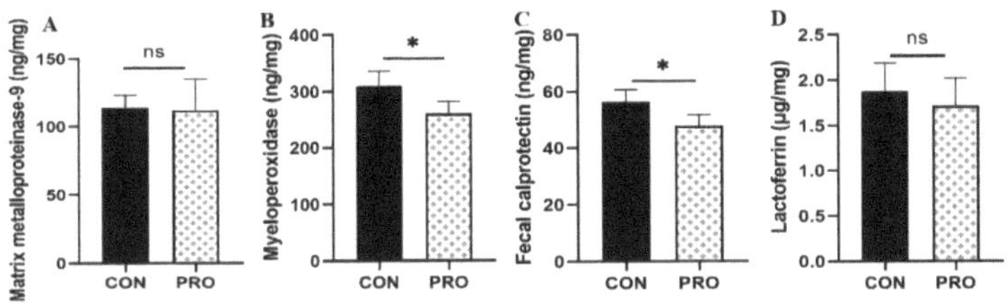

Figure 7. Effect of a multistrain probiotic preparation on inflammatory markers in feces in healthy short-haired domestic cats. (**A–D**) Concentrations of matrix metalloproteinase-9 (**A**), myeloperoxidase (**B**), calprotectin (**C**), and lactoferrin (**D**) in feces. Data are presented as mean ± SD. * $p < 0.05$; ns, $p > 0.05$ compared with the control group.

3.9. Fecal Antioxidant Capacity

As can be observed in Figure 8, compared with the control group, the probiotic group presented higher ($p < 0.05$) superoxide dismutase (SOD) and ($p < 0.01$) glutathione (GSH) activity. There was no significant difference in malondialdehyde (MDA) activity between the control and probiotic groups.

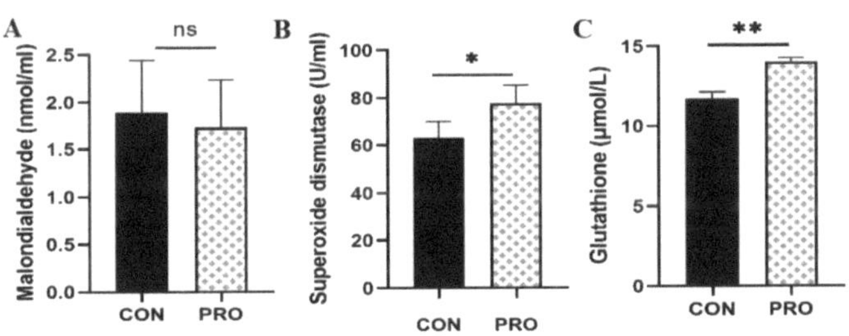

Figure 8. Effect of a multistrain probiotic preparation on fecal antioxidant capacity in healthy short-haired domestic cats. (**A–C**) Concentrations of malondialdehyde (**A**), superoxide dismutase (**B**), and glutathione (**C**) detected by commercial assay kits. Data are presented as mean ± SD. * $p < 0.05$; ** $p < 0.01$; ns, $p > 0.05$ compared with the control group.

4. Discussion

Despite the fact that probiotics can improve host health by modulating the gut microbiota, typically through metabolic interactions mediated by stimulation of resident bacteria to alter the abundance of pathogenic bacteria, they indirectly alter the resident microbiota through metabolic interactions or by interacting with host epithelial cells and the epithelial immune system [24]. However, probiotics have not been well studied in the literature in relation to cats. In this study, we investigated the effect of a multistrain probiotic containing *Saccharomyces boulardii* and *Pediococcus acidilactici* in the diet on the intestinal flora of healthy short-haired domestic cats. The presented results indicate that the colonization of beneficial bacteria promotes gut health by modulating the gut microbiota, reducing inflammatory conditions, improving microbiota-derived SCFA production, and improving antioxidant status.

Saccharomyces boulardii belongs to *Saccharomyces cerevisiae*, with strong probiotic properties and antipathogen abilities. Studies have determined that *Saccharomyces boulardii* plays a significant role in the prevention of antibiotic-associated diarrhea [25]. Mixtures of probiotics such as *Saccharomyces boulardii* and *Bifidobacterium* reduce the severity of *E. coli* diarrhea in rats [26]. *Saccharomyces boulardii* mediates proinflammatory responses through yeast-secreted factors and signaling molecules that control inflammation, thereby regulating gastrointestinal inflammation [27]. *Pediococcus acidilactici* has the remarkable characteristics of lactic acid bacteria and can colonize the intestinal tract. Its metabolite organic acids can participate in physiological responses, such as by improving the balance of intestinal flora, promoting digestion and absorption, and improving immune performance [17]. *Pediococcus acidilactici* isolated from the feces of unweaned puppies was fed to weaned puppies, and it was observed that *Pediococcus acidilactici* can promote digestion and improve serum antioxidant capacity, thereby restoring the puppies' appetite and reducing transportation stress [28]. Adding *Pediococcus acidilactici* to the diet of weaned piglets can increase the number of intestinal lactic acid bacteria, reduce the severity of diarrhea [29], and improve intestinal health. Studies have determined that a single- or multistrain probiotics can regulate intestinal function and treat gastrointestinal diseases [30]. How to correctly select the type and combination of probiotic strains is the key to achieving such outcomes.

OTUs are designed to facilitate analytical phylogenetic or population genetics studies, classifying and distinguishing all sequences according to similarity levels. In this study, compared with the control group, the probiotic group had 81 unique OTUs. Alpha refers to the diversity of fecal microbial communities and represents the richness and uniformity of species within the sample. In this study, there was no significant difference in the alpha diversity between the control and probiotic groups, which is consistent with Lin's study of dogs using a fermentation product of *Saccharomyces cerevisiae*, which showed no change in the fecal alpha diversity [31]. Minamoto observed that the bacterial diversity and microbial community of dogs with idiopathic inflammatory bowel disease (IBD) were significantly reduced compared to those of the healthy control group [32]. In this study, a multistrain probiotic did not change the diversity of the gut microbiome of short-haired domestic cats, suggesting that a multistrain probiotic can be used safely in short-haired domestic cats.

The results of principal coordinate analysis (PCoA) in this study indicated no change in the microbial community structure between the control and probiotic groups. The weighted UniFrac distance map shows that the weighted UniFrac distance between the cats in the probiotic group is significantly greater than that in the control group, indicating that the differences in species and abundance of microbial communities between the two groups are great, which may be the result of individual differences between short-haired domestic cats. In a previous study, when multiple synbiotic preparations were administered to healthy cats and dogs, there was no significant difference in fecal microflora; however, it led to an increase in the abundance of probiotics in the feces of healthy cats and dogs [33]. Rossi studied the effect of multistrain probiotic SLAB51TM on cats suffering from chronic constipation and Hirschsprung's disease and observed that it could significantly improve their clinical condition, with a significant increase in *Lactobacillus* spp. and Bacteroidetes; however, there was no significant difference in the microbiota between healthy controls and constipated cats [34]. LEFSe differential analysis showed that the 16 types of bacteria of the phylum Firmicutes and Proteobacteria played an important role in the control group. The study determined that the abundance of Proteobacteria is related to the increase in *Escherichia coli*, which belongs to Proteobacteria, has adhesion ability, invades the body, produces toxins, stimulates inflammatory cytokines, and is closely related to gastrointestinal diseases [35,36]. *Pediococcus* and Bacillaceae played an important role in the probiotic group. Bacillaceae have strong probiotic properties, such as the activity of secreted enzymes, which can help the host digest nutritional compounds and facilitate intestinal colonization [37]. The significant increase in the probiotic group may be associated with the use of a multistrain probiotic containing *Pediococcus acidilactici*, which can produce bacteriocin from ribosomes, colonize the intestine, regulate the composition of the intestinal microbiota, improve the host immune response, and enhance intestinal barrier function [38,39]. Studies have determined that bacteriocin-producing *Pediococcus acidilactici* can relieve constipation in mice and regulate intestinal flora [40]. The microencapsulation of probiotic strains by lyophilization can maintain microbial viability in cats and improve

Lactobacillus count, exerting a positive effect on cat gut microbiota [41]. The supplementation of *Lactobacillus acidophilus* in healthy cat diets resulted in an increase the number of *Lactobacillus* in their feces and a decrease in the number of *E. coli*, which can effectively improve the fecal quality parameters and intestinal health of healthy adult cats [42]. In this study, the addition of *Saccharomyces boulardii* and *Pediococcus acidilactici* to the diet of healthy cats did not change the fecal microbial community structure, which varied due to individual differences; however, it promoted the colonization and abundance of beneficial bacterial strains and may have a positive effect on the gut microbiota of healthy cats.

The presented results show that there were no significant differences at the phylum and genus levels of microorganisms between the control and probiotics groups, and the dominant intestinal microbiota of the two groups are Firmicutes, Bacteroidetes, Actinobacteria, etc. Previous studies have also shown that the dominant microbiota of dogs' and cats' intestinal microorganisms are Firmicutes and Bacteroidetes [43], and their abundance values differ according to individual differences. The treatment of dextran sulfate sodium (DSS)-induced rats with *Saccharomyces boulardii* resulted in a decrease in Firmicutes/Bacteroidetes [44], whereas colitis was associated with increased numbers of Firmicutes and decreased numbers of Bacteroidetes. Bacteroidetes are abundant in healthy microbiomes. A decreased microbial abundance of Bacteroidetes was observed in cats with chronic enteropathy (CE) [45]. Firmicutes are associated with the production of short-chain fatty acids and other metabolites [46,47]. The host gut microbiota produces metabolites such as short-chain fatty acids (SCFAs) through fermentation, which play an important role in host/pathogen interactions [48]. The most common SCFAs are acetic, propionic, and butyric acids, which are involved in immunomodulatory and anti-inflammatory responses [49]. In this study, the concentrations of butyric acid and total SCFAs in the probiotic group were significantly higher than those in the control group. Coincidentally, previous studies have shown that DSS-induced rats have significantly increased acetic, propionic, and butyric acid levels, as well as total SCFA production, following the addition of multistrain probiotics, with a tendency to restore normal rats' short-chain fatty acid levels (unpublished). Butyric acid has the strongest anti-inflammatory effect among all SCFAs [50]. Studies have shown that SCFAs play a crucial role in alleviating intestinal dysbiosis and maintaining intestinal homeostasis [51]. The gut microbiota is closely linked to host gut health, and an imbalance in the microbiota may cause host dybiosis, as well as intestinal and systemic diseases [52]. A characteristic feature of gut microbial dysbiosis is a decrease in obligate anaerobic flora, such as Firmicutes and Bacteroidetes, whereas there is an increase in facultative anaerobic bacteria, such as Enterobacteriaceae and *Escherichia coli* [53]. This study shows that a multistrain probiotic can improve the intestinal immune regulation and anti-inflammatory ability of healthy cats by adjusting the proportion of microbial compositions and increasing the production of SCFAs. Paradoxically, the multistrain probiotic did not alter the feline gut microbial diversity and structure presented in this study, although SCFA production was increased. It has been suggested that this may be due to probiotics playing a more significant role in the gut functional level compared to gut microbial composition [33]. Not coincidentally, a previous study reported that *Lactobacillus helveticus* Bar13 and *B. longum* Bar33 intake in humans promoted a significant increase in acetate and valeric acid levels but did not alter the structure of the fecal microbiota [54]. Meanwhile, inulin or short-chain fructooligosaccharides were reported to alter the short-chain fatty acid content in dog feces and modestly alter fecal flora abundance [55]. Therefore, further studies on the effect of probiotics on gut microbial function are necessary.

The degree of intestinal inflammation can be manifested by neutrophil derivatives, such as fecal calprotectin (FC), myeloperoxidase (MPO), and matrix metalloproteinases (MMPs) in feces [56]. Fecal calprotectin has an antibacterial effect and is proportional to the severity of inflammation [57]. Myeloperoxidase is known to catalyze the production of hypochlorous acid and promote the production of reactive oxygen species (ROS) and reactive nitrogen species (RNS), mediating oxidative stress and inflammation [58]. Matrix metalloproteinase-9 also activates underlying inflammatory molecules to modulate the inflammatory response [59]. Compared with the control group, the concentration of calprotectin and myeloperoxidase was lower in the probiotic group. Interestingly, Peptostreptococcaceae and Peptostreptococcales_Tissierellales, which are microbiota associated with inflammation, were relatively more abundant in the control group [60]. The content of MPO and FC of inflammatory markers in the probiotic group was significantly reduced, indicating that the addition of probiotics can regulate the intestinal inflammation-related flora and induce a decrease in the level of inflammatory markers, thereby reducing the inflammatory response. In our previous study, we observed that in DSS-induced rats, Peptostreptococcales and Peptostreptococcales_Tissierellales were relatively abundant, with the highest levels of inflammatory markers MPO, MMP-9, and FC, suggesting a relationship between inflammatory markers and inflammation-related gut flora (unpublished). Moreover, the antioxidant activity of superoxide dismutase (SOD) and glutathione (GSH) in the probiotic group was significantly increased. Studies have shown that the probiotic mixture affects the production of SCFAs in aging mice, and SCFAs may regulate antioxidant enzymes by inducing the expression of Nrf2 or HO-1, thereby directing antioxidant effects [12]. Studies have determined that probiotic-treated fermented sausages can not only reduce human inflammatory markers but also improve antioxidant plasma markers and increase butyric acid production [61]. The administration of *Saccharomyces boulardii*, which are associated with inflammation, has been observed to reduce FC levels in healthy adult dogs and maintain the stability of the intestinal environment [62]. *Lactobacillus* fermentum reduced the production of MDA and MPO and inhibited the release of proinflammatory cytokine tumor necrosis factor-α (TNF-α), IFN-γ, IL-1β, IL-6, and IL-12, effectively alleviating the symptoms of dextran sulfate sodium (DSS)-induced rats [63]. In a previous study, we observed that treatment administered with a multistrain probiotic in DSS-induced rats significantly reduced the levels of fecal inflammatory markers MPO, MMP-9, and FC; inhibited the activation of proinflammatory cytokines IL-6, IL-1β, interferon IFN-γ, and tumor necrosis factor TNF-α; increased the levels of antioxidant stress parameters SOD and GSH; and reduced the inflammatory state of rats (unpublished), consistent with the results obtained in this study. Following multistrain probiotic treatment, in this study, we observed that the metabolites of SCFA production may play a crucial role in the anti-inflammatory and antioxidant effects of cats.

5. Conclusions

In the present study, feeding a multistrain probiotic to healthy short-haired domestic cats did not alter their gut microbial structure and diversity. In addition, it promoted the colonization of beneficial bacteria (*Pediococcus* and Bacillaceae) and decreased the proportion of Firmicutes/Bacteroidetes. It was observed that the production of butyric acid and total SCFAs in feces increased, which was associated with Firmicutes and Bacteroidetes. Furthermore, SCFAs may have induced anti-inflammatory and antioxidant effects, not only reducing the concentration of fecal inflammatory markers MPO and FC but also increasing the activity of antioxidant enzymes SOD and GSH. Therefore, the presented results indicate that the colonization of beneficial bacteria promotes gut health by modulating gut microbes, improving microbiota-derived SCFA production, reducing the inflammatory state, and improving antioxidant status. Further studies on the effect of probiotics on the intestinal functional level (i.e., production of SCFAs) are needed. However, due to the influence of individual differences among cats and owner compliance, few studies have been conducted on cat gastrointestinal health in the literature. Importantly, it is necessary to explore the role of ideal probiotic strains in cat intestinal diseases to provide a theoretical basis to improve cat intestinal health.

Author Contributions: Conceptualization, L.L.; methodology, L.L.; software, Y.L. (Yifei Li); validation, Z.L., Y.L. (Yanan Li), C.Y. and M.Y.; formal analysis, Z.L.; investigation, Z.L. and Y.L. (Yifei Li); resources, L.L.; data curation, Y.L. (Yifei Li); writing—original draft preparation, Y.L. (Yifei Li); writing—review and editing, L.L., I.A. and Y.L. (Yifei Li); supervision, L.L.; project administration, L.L., Z.L. and Y.L. (Yifei Li); funding acquisition, L.L. All authors have read and agreed to the published version of the manuscript.

Funding: This research received no external funding.

Institutional Review Board Statement: The animal study protocol was approved by the Laboratory Animal Welfare Ethics Review Committee of Nanjing Agricultural University (SYXK (SU) 2017-0027).

Informed Consent Statement: Not applicable.

Data Availability Statement: The data underlying this article will be shared upon reasonable request to the corresponding author. The data are not publicly available due to the probiotics product is under development and patent application.

Acknowledgments: The authors would like to express their gratitude to Nanjing Xintengrui Biotechnology Co., Ltd., for this research.

Conflicts of Interest: The authors declare no conflict of interest.

References

1. Zhang, R.J. Absorption and metabolism of microorganisms and host nutrients in the digestive tract. *China Feed* **2003**, *2*, 11–14. [CrossRef]
2. Food and Agriculture Organization of the United Nations; World Health Organization. *Probiotics in Food: Health and Nutritional Properties and Guidelines for Evaluation*; Food and Agriculture Organization of the United Nations; World Health Organization: Rome, Italy, 2006; p. viii. 50p.
3. Kligler, B.; Cohrssen, A. Probiotics. *Am. Fam. Physician* **2008**, *78*, 1073–1078.
4. Ricci, A.; Tagliacarne, S.C.; Valsecchi, C.; Boggini, T.; Cattaneo, F.; Licari, A.; Caimmi, S.; Castellazzi, A.M. Probiotics and Inflammatory Bowel Diseases. *J. Biol. Regul. Homeost. Agents* **2015**, *29*, 96–113.
5. Collinson, S.; Deans, A.; Padua-Zamora, A.; Gregorio, G.V.; Li, C.; Dans, L.F.; Allen, S.J. Probiotics for treating acute infectious diarrhoea. *Cochrane Database Syst. Rev.* **2020**, *12*, CD003048. [CrossRef] [PubMed]
6. Shan, L.S.; Hou, P.; Wang, Z.J.; Liu, F.R.; Chen, N.; Shu, L.H.; Zhang, H.; Han, X.H.; Han, X.X.; Cai, X.X.; et al. Prevention and treatment of diarrhoea with Saccharomyces boulardii in children with acute lower respiratory tract infections. *Benef. Microbes* **2013**, *4*, 329–334. [CrossRef] [PubMed]
7. Shen, J.; Zuo, Z.X.; Mao, A.P. Effect of probiotics on inducing remission and maintaining therapy in ulcerative colitis, Crohn's disease, and pouchitis: Meta-analysis of randomized controlled trials. *Inflamm. Bowel. Dis.* **2014**, *20*, 21–35. [CrossRef]
8. Korterink, J.J.; Ockeloen, L.; Benninga, M.A.; Tabbers, M.M.; Hilbink, M.; Deckers-Kocken, J.M. Probiotics for childhood functional gastrointestinal disorders: A systematic review and meta-analysis. *Acta Paediatr.* **2014**, *103*, 365–372. [CrossRef] [PubMed]
9. AlFaleh, K.; Anabrees, J. Probiotics for prevention of necrotizing enterocolitis in preterm infants. *Cochrane Database Syst. Rev.* **2014**, *9*, CD005496. [CrossRef]
10. Markowiak, P.; Slizewska, K. Effects of Probiotics, Prebiotics, and Synbiotics on Human Health. *Nutrients* **2017**, *9*, 1021. [CrossRef]
11. Moens, F.; Van den Abbeele, P.; Basit, A.W.; Dodoo, C.; Chatterjee, R.; Smith, B.; Gaisford, S. A four-strain probiotic exerts positive immunomodulatory effects by enhancing colonic butyrate production in vitro. *Int. J. Pharm.* **2019**, *555*, 1–10. [CrossRef]
12. Ho, S.T.; Hsieh, Y.T.; Wang, S.Y.; Chen, M.J. Improving effect of a probiotic mixture on memory and learning abilities in d-galactose-treated aging mice. *J. Dairy Sci.* **2019**, *102*, 1901–1909. [CrossRef] [PubMed]
13. Markowiak-Kopec, P.; Slizewska, K. The Effect of Probiotics on the Production of Short-Chain Fatty Acids by Human Intestinal Microbiome. *Nutrients* **2020**, *12*, 1107. [CrossRef] [PubMed]
14. McFarland, L.V. Chapter 18—Common Organisms and Probiotics: Saccharomyces boulardii. In *The Microbiota in Gastrointestinal Pathophysiology*; Floch, M.H., Ringel, Y., Allan Walker, W., Eds.; Academic Press: Boston, MA, USA, 2017; pp. 145–164.
15. Pais, P.; Almeida, V.; Yilmaz, M.; Teixeira, M.C. Saccharomyces boulardii: What Makes It Tick as Successful Probiotic? *J. Fungi* **2020**, *6*, 78. [CrossRef]
16. Li, B.; Zhang, H.; Shi, L.; Li, R.; Luo, Y.; Deng, Y.; Li, S.; Li, R.; Liu, Z. Saccharomyces boulardii alleviates DSS-induced intestinal barrier dysfunction and inflammation in humanized mice. *Food Funct.* **2022**, *13*, 102–112. [CrossRef] [PubMed]
17. Wang, Y.N.; Hu, S.Y.; Yu, L.Z. Research progress of Pediococcus acidilactici in animal production. *Chin. J. Vet. Med.* **2021**, *57*, 54–57.
18. Zhang, Q.; Guo, W.L.; Chen, G.M.; Qian, M.; Han, J.Z.; Lv, X.C.; Chen, L.J.; Rao, P.F.; Ai, L.Z.; Ni, L. Pediococcus acidilactici FZU106 alleviates high-fat diet-induced lipid metabolism disorder in association with the modulation of intestinal microbiota in hyperlipidemic rats. *Curr. Res. Food Sci.* **2022**, *5*, 775–788. [CrossRef]
19. D'Angelo, S.; Fracassi, F.; Bresciani, F.; Galuppi, R.; Diana, A.; Linta, N.; Bettini, G.; Morini, M.; Pietra, M. Effect of Saccharomyces boulardii in dogs with chronic enteropathies: Double-blinded, placebo-controlled study. *Vet. Rec.* **2018**, *182*, 258. [CrossRef]
20. Lappin, M.R.; Veir, J.K.; Satyaraj, E.; Czarnecki-Maulden, G. Pilot study to evaluate the effect of oral supplementation of Enterococcus faecium SF68 on cats with latent feline herpesvirus. *J. Feline Med. Surg.* **2009**, *11*, 650–654. [CrossRef]
21. Yan, X.; Zhang, H.; Lin, A.; Su, Y. Antagonization of Ghrelin Suppresses Muscle Protein Deposition by Altering Gut Microbiota and Serum Amino Acid Composition in a Pig Model. *Biology* **2022**, *11*, 840. [CrossRef] [PubMed]

22. Lin, C.; Wan, J.; Su, Y.; Zhu, W. Effects of Early Intervention with Maternal Fecal Microbiota and Antibiotics on the Gut Microbiota and Metabolite Profiles of Piglets. *Metabolites* **2018**, *8*, 89. [CrossRef]
23. Yu, M.; Zhang, C.; Yang, Y.; Mu, C.; Su, Y.; Yu, K.; Zhu, W. Long-term effects of early antibiotic intervention on blood parameters, apparent nutrient digestibility, and fecal microbial fermentation profile in pigs with different dietary protein levels. *J. Anim. Sci. Biotechnol.* **2017**, *8*, 60. [CrossRef]
24. Derrien, M.; van Hylckama Vlieg, J.E. Fate, activity, and impact of ingested bacteria within the human gut microbiota. *Trends Microbiol.* **2015**, *23*, 354–366. [CrossRef] [PubMed]
25. McFarland, L.V. Systematic review and meta-analysis of Saccharomyces boulardii in adult patients. *World J. Gastroenterol.* **2010**, *16*, 2202–2222. [CrossRef]
26. Bisson, J.F.; Hidalgo, S.; Rozan, P.; Messaoudi, M. Preventive effects of different probiotic formulations on travelers' diarrhea model in wistar rats: Preventive effects of probiotics on TD. *Dig. Dis. Sci.* **2010**, *55*, 911–919. [CrossRef]
27. Pothoulakis, C. Review article: Anti-inflammatory mechanisms of action of Saccharomyces boulardii. *Aliment. Pharmacol. Ther.* **2009**, *30*, 826–833. [CrossRef]
28. Wang, C.F.; Wang, J.Z.; Yang, G.L.; Shi, C.W.; Li, N.; Shan, B.L.; Gu, W.; Wang, H. A Strain of Canine Pediococcus Acidilactici and Its Application. China Patent CN109182184B, 20 July 2021.
29. Liu, G.Y.; Sun, H.T.; Liu, C.; Wang, Y.P.; Zhao, X.Y.; Wang, C. Effects of Pediococcus acidilactici on Growth Performance and Intestinal Microflora of Weaned Piglets. *Shandong Agric. Sci.* **2020**, *52*, 131–135. [CrossRef]
30. Thomas, C.M.; Versalovic, J. Probiotics-host communication: Modulation of signaling pathways in the intestine. *Gut Microbes* **2010**, *1*, 148–163. [CrossRef]
31. Lin, C.Y.; Alexander, C.; Steelman, A.J.; Warzecha, C.M.; de Godoy, M.R.C.; Swanson, K.S. Effects of a Saccharomyces cerevisiae fermentation product on fecal characteristics, nutrient digestibility, fecal fermentative end-products, fecal microbial populations, immune function, and diet palatability in adult dogs1. *J. Anim. Sci.* **2019**, *97*, 1586–1599. [CrossRef]
32. Minamoto, Y.; Otoni, C.C.; Steelman, S.M.; Buyukleblebici, O.; Steiner, J.M.; Jergens, A.E.; Suchodolski, J.S. Alteration of the fecal microbiota and serum metabolite profiles in dogs with idiopathic inflammatory bowel disease. *Gut Microbes* **2015**, *6*, 33–47. [CrossRef] [PubMed]
33. Garcia-Mazcorro, J.F.; Lanerie, D.J.; Dowd, S.E.; Paddock, C.G.; Grützner, N.; Steiner, J.M.; Ivanek, R.; Suchodolski, J.S. Effect of a multi-species synbiotic formulation on fecal bacterial microbiota of healthy cats and dogs as evaluated by pyrosequencing. *FEMS Microbiol. Ecol.* **2011**, *78*, 542–554. [CrossRef]
34. Rossi, G.; Jergens, A.; Cerquetella, M.; Berardi, S.; Di Cicco, E.; Bassotti, G.; Pengo, G.; Suchodolski, J.S. Effects of a probiotic (SLAB51™) on clinical and histologic variables and microbiota of cats with chronic constipation/megacolon: A pilot study. *Benef. Microbes* **2018**, *9*, 101–110. [CrossRef]
35. De la Fuente, M.; Franchi, L.; Araya, D.; Díaz-Jiménez, D.; Olivares, M.; Álvarez-Lobos, M.; Golenbock, D.; González, M.J.; López-Kostner, F.; Quera, R.; et al. Escherichia coli isolates from inflammatory bowel diseases patients survive in macrophages and activate NLRP3 inflammasome. *Int. J. Med. Microbiol. IJMM* **2014**, *304*, 384–392. [CrossRef]
36. Kotlowski, R.; Bernstein, C.N.; Sepehri, S.; Krause, D.O. High prevalence of Escherichia coli belonging to the B2+D phylogenetic group in inflammatory bowel disease. *Gut* **2007**, *56*, 669–675. [CrossRef] [PubMed]
37. Jeżewska-Frąckowiak, J.; Seroczyńska, K.; Banaszczyk, J.; Jedrzejczak, G.; Żylicz-Stachula, A.; Skowron, P.M. The promises and risks of probiotic Bacillus species. *Acta Biochim. Pol.* **2018**, *65*, 509–519. [CrossRef] [PubMed]
38. Heeney, D.D.; Zhai, Z.; Bendiks, Z.; Barouei, J.; Martinic, A.; Slupsky, C.; Marco, M.L. Lactobacillus plantarum bacteriocin is associated with intestinal and systemic improvements in diet-induced obese mice and maintains epithelial barrier integrity in vitro. *Gut Microbes* **2019**, *10*, 382–397. [CrossRef] [PubMed]
39. Kommineni, S.; Bretl, D.J.; Lam, V.; Chakraborty, R.; Hayward, M.; Simpson, P.; Cao, Y.; Bousounis, P.; Kristich, C.J.; Salzman, N.H. Bacteriocin production augments niche competition by enterococci in the mammalian gastrointestinal tract. *Nature* **2015**, *526*, 719–722. [CrossRef]
40. Qiao, Y.; Qiu, Z.; Tian, F.; Yu, L.; Zhao, J.; Zhang, H.; Zhai, Q.; Chen, W. Pediococcus acidilactici Strains Improve Constipation Symptoms and Regulate Intestinal Flora in Mice. *Front. Cell. Infect. Microbiol.* **2021**, *11*, 655258. [CrossRef]
41. Rodrigues, B.M.; Olivo, P.M.; Osmari, M.P.; Vasconcellos, R.S.; Ribeiro, L.B.; Bankuti, F.I.; Pozza, M.S.S. Microencapsulation of Probiotic Strains by Lyophilization Is Efficient in Maintaining the Viability of Microorganisms and Modulation of Fecal Microbiota in Cats. *Int. J. Microbiol.* **2020**, *2020*, 1293481. [CrossRef]
42. Fusi, E.; Rizzi, R.; Polli, M.; Cannas, S.; Giardini, A.; Bruni, N.; Marelli, S.P. Effects of Lactobacillus acidophilus D2/CSL (CECT 4529) supplementation on healthy cat performance. *Vet. Rec. Open* **2019**, *6*, e000368. [CrossRef] [PubMed]
43. He, S.S.; Jian, S.J.; Wen, C.Y.; Wen, J.W.; Kuang, T.; Tong, A.; Deng, B.C. Research advances on gut microbiota of cats. *China Feed* **2022**, *13*, 92–98. [CrossRef]
44. Rodriguez-Nogales, A.; Algieri, F.; Garrido-Mesa, J.; Vezza, T.; Utrilla, M.P.; Chueca, N.; Garcia, F.; Rodriguez-Cabezas, M.E.; Galvez, J. Intestinal anti-inflammatory effect of the probiotic Saccharomyces boulardii in DSS-induced colitis in mice: Impact on microRNAs expression and gut microbiota composition. *J. Nutr. Biochem.* **2018**, *61*, 129–139. [CrossRef] [PubMed]
45. Marsilio, S.; Pilla, R.; Sarawichitr, B.; Chow, B.; Hill, S.L.; Ackermann, M.R.; Estep, J.S.; Lidbury, J.A.; Steiner, J.M.; Suchodolski, J.S. Characterization of the fecal microbiome in cats with inflammatory bowel disease or alimentary small cell lymphoma. *Sci. Rep.* **2019**, *9*, 19208. [CrossRef]

46. Rossi, G.; Pengo, G.; Caldin, M.; Palumbo Piccionello, A.; Steiner, J.M.; Cohen, N.D.; Jergens, A.E.; Suchodolski, J.S. Comparison of microbiological, histological, and immunomodulatory parameters in response to treatment with either combination therapy with prednisone and metronidazole or probiotic VSL#3 strains in dogs with idiopathic inflammatory bowel disease. *PLoS ONE* **2014**, *9*, e94699. [CrossRef]
47. Suchodolski, J.S.; Camacho, J.; Steiner, J.M. Analysis of bacterial diversity in the canine duodenum, jejunum, ileum, and colon by comparative 16S rRNA gene analysis. *FEMS Microbiol. Ecol.* **2008**, *66*, 567–578. [CrossRef]
48. Wei, Z.; Xiao, C.; Guo, C.; Zhang, X.; Wang, Y.; Wang, J.; Yang, Z.; Fu, Y. Sodium acetate inhibits Staphylococcus aureus internalization into bovine mammary epithelial cells by inhibiting NF-κB activation. *Microb. Pathog.* **2017**, *107*, 116–121. [CrossRef]
49. Ratajczak, W.; Rył, A.; Mizerski, A.; Walczakiewicz, K.; Sipak, O.; Laszczyńska, M. Immunomodulatory potential of gut microbiome-derived short-chain fatty acids (SCFAs). *Acta Biochim. Pol.* **2019**, *66*, 1–12. [CrossRef] [PubMed]
50. Kuczyńska, B.; Wasilewska, A.; Biczysko, M.; Banasiewicz, T.; Drews, M. Short chain fatty acids—Mechanisms of action, potential clinical indications and nutritional indications. *Now. Lek.* **2011**, *80*, 299–304.
51. Parada Venegas, D.; De la Fuente, M.K.; Landskron, G.; González, M.J.; Quera, R.; Dijkstra, G.; Harmsen, H.J.M.; Faber, K.N.; Hermoso, M.A. Short Chain Fatty Acids (SCFAs)-Mediated Gut Epithelial and Immune Regulation and Its Relevance for Inflammatory Bowel Diseases. *Front. Immunol.* **2019**, *10*, 277. [CrossRef]
52. DuPont, A.W.; DuPont, H.L. The intestinal microbiota and chronic disorders of the gut. *Nat. Rev. Gastroenterol. Hepatol.* **2011**, *8*, 523–531. [CrossRef] [PubMed]
53. Wilson, M. *Microbial Inhabitants of Humans: Their Ecology and Role in Health and Disease*; Cambridge University Press: Cambridge, UK, 2004. [CrossRef]
54. Vitali, B.; Ndagijimana, M.; Cruciani, F.; Carnevali, P.; Candela, M.; Guerzoni, M.E.; Brigidi, P. Impact of a synbiotic food on the gut microbial ecology and metabolic profiles. *BMC Microbiol.* **2010**, *10*, 4. [CrossRef]
55. Barry, K.A.; Hernot, D.C.; Middelbos, I.S.; Francis, C.; Dunsford, B.; Swanson, K.S.; Fahey, G.C., Jr. Low-level fructan supplementation of dogs enhances nutrient digestion and modifies stool metabolite concentrations, but does not alter fecal microbiota populations. *J. Anim. Sci.* **2009**, *87*, 3244–3252. [CrossRef] [PubMed]
56. Jakubczyk, D.; Leszczyńska, K.; Górska, S. The Effectiveness of Probiotics in the Treatment of Inflammatory Bowel Disease (IBD)-A Critical Review. *Nutrients* **2020**, *12*, 1973. [CrossRef] [PubMed]
57. Manceau, H.; Chicha-Cattoir, V.; Puy, H.; Peoc'h, K. Fecal calprotectin in inflammatory bowel diseases: Update and perspectives. *Clin. Chem. Lab. Med.* **2017**, *55*, 474–483. [CrossRef] [PubMed]
58. Chen, S.; Chen, H.; Du, Q.; Shen, J. Targeting Myeloperoxidase (MPO) Mediated Oxidative Stress and Inflammation for Reducing Brain Ischemia Injury: Potential Application of Natural Compounds. *Front. Physiol.* **2020**, *11*, 433. [CrossRef]
59. Deleon-Pennell, K.Y.; Altara, R.; Yabluchanskiy, A.; Modesti, A.; Lindsey, M.L. The circular relationship between matrix metalloproteinase-9 and inflammation following myocardial infarction. *IUBMB Life* **2015**, *67*, 611–618. [CrossRef]
60. Jin, S.; Zhao, D.; Cai, C.; Song, D.; Shen, J.; Xu, A.; Qiao, Y.; Ran, Z.; Zheng, Q. Low-dose penicillin exposure in early life decreases Th17 and the susceptibility to DSS colitis in mice through gut microbiota modification. *Sci. Rep.* **2017**, *7*, 43662. [CrossRef]
61. Pérez-Burillo, S.; Pastoriza, S.; Gironés, A.; Avellaneda, A.; Pilar Francino, M.; Rufián-Henares, J.A. Potential probiotic salami with dietary fiber modulates metabolism and gut microbiota in a human intervention study. *J. Funct. Foods* **2020**, *66*, 103790. [CrossRef]
62. Meineri, G.; Martello, E.; Atuahene, D.; Miretti, S.; Stefanon, B.; Sandri, M.; Biasato, I.; Corvaglia, M.R.; Ferrocino, I.; Cocolin, L.S. Effects of Saccharomyces boulardii Supplementation on Nutritional Status, Fecal Parameters, Microbiota, and Mycobiota in Breeding Adult Dogs. *Vet. Sci.* **2022**, *9*, 389. [CrossRef]
63. Zhou, X.; Liu, H.; Zhang, J.; Mu, J.; Zalan, Z.; Hegyi, F.; Takács, K.; Zhao, X.; Du, M. Protective effect of Lactobacillus fermentum CQPC04 on dextran sulfate sodium-induced colitis in mice is associated with modulation of the nuclear factor-κB signaling pathway. *J. Dairy Sci.* **2019**, *102*, 9570–9585. [CrossRef]

metabolites

Article

Antimicrobial Peptides Relieve Transportation Stress in Ragdoll Cats by Regulating the Gut Microbiota

Shansong He [1], Kang Yang [1], Jiawei Wen [1], Tao Kuang [1], Zhihao Cao [1], Lingna Zhang [1], Sufang Han [1], Shiyan Jian [1], Xin Chen [2,*], Limeng Zhang [3], Jinping Deng [1] and Baichuan Deng [1,*]

1 Guangdong Provincial Key Laboratory of Animal Nutrition Control, National Engineering Research Center for Breeding Swine Industry, College of Animal Science, South China Agricultural University, Guangzhou 510642, China
2 School of Medicine, Foshan University, Foshan 528000, China
3 Research Center of Pet Nutrition, Guangzhou Qingke Biotechnology Co., Ltd., Guangzhou 510642, China
* Correspondence: chenxin1210@163.com (X.C.); dengbaichuan@scau.edu.cn (B.D.)

Abstract: Transportation is common in cats and often causes stress and intestinal disorders. Antimicrobial peptides (AMPs) exhibit a broad spectrum of antibacterial activity, and they may have the capacity for antioxidant and immune regulation. The objective of this study was to investigate the effects of dietary supplementation with AMPs on stress response, gut microbiota and metabolites of cats that have undergone transport stress. A total of 14 Ragdoll cats were randomly allocated into 2 treatments: basal diet (CON) and a basal diet supplemented with 0.3% AMPs. After a 6-week feeding period, all cats were transported for 3 h and, then, fed for another week. The results show that the diarrhea rate of cats was markedly reduced by supplementation with AMPs throughout the trial period ($p < 0.05$). In addition, AMPs significantly reduced serum cortisol and serum amyloid A ($p < 0.05$) and increased apolipoprotein 1 after transportation ($p < 0.05$). Moreover, AMPs reduced the level of inflammatory factors in the serum caused by transportation stress, including tumor necrosis factor-alpha (TNF-α) and interleukin-1 beta (IL-1β) ($p < 0.05$). The AMPs enhanced the activities of glutathione peroxidase ($p < 0.01$) and superoxide dismutase ($p < 0.05$). Furthermore, cats fed AMPs had higher levels of branched chain fatty acids (BCFAs) and a relative abundance of *Blautia* and a lower relative abundance of *Negativibacillus* after transportation ($p < 0.05$). The serum metabolome analysis further revealed that AMPs markedly regulated lipid metabolism by upregulating cholic acid expression. In conclusion, AMP supplementation alleviated oxidative stress and inflammatory response in transportation by regulating the gut microbiota and metabolites, thereby relieving stress-induced diarrhea and supporting gut and host health in cats.

Keywords: transportation stress; antimicrobial peptides; cats; pet food; microbiota; metabolome

Citation: He, S.; Yang, K.; Wen, J.; Kuang, T.; Cao, Z.; Zhang, L.; Han, S.; Jian, S.; Chen, X.; Zhang, L.; et al. Antimicrobial Peptides Relieve Transportation Stress in Ragdoll Cats by Regulating the Gut Microbiota. *Metabolites* **2023**, *13*, 326. https://doi.org/10.3390/metabo13030326

Academic Editor: Karolina Skonieczna-Żydecka

Received: 14 January 2023
Revised: 16 February 2023
Accepted: 17 February 2023
Published: 22 February 2023

1. Introduction

Cats as companion animals have become increasingly popular worldwide in recent years. According to the reports, the number of domestic cats is 58.06 million in China and 113.59 million in Europe in 2021 [1,2]. Furthermore, transportation is common for cats during vet visits or outings. Previous studies revealed that transportation can be particularly stressful due to the novel environment, engine noise and the irregular movement of the vehicle [3,4]. Therefore, reducing the stress caused by transportation has become a big concern for pet owners.

Stress is broadly defined as the physiological response when homeostasis is threatened [5]. Stress induced by transportation may result in diarrhea, weight loss, and the increased production of cell-free radicals, therefore causing oxidative stress [3,6]. Previous studies have shown that transportation stress activates the hypothalamic–pituitary–adrenal (HPA) axis and promotes the release of glucocorticoids, mainly cortisol (COR) and corticosterone [7]. At the same time, macrophages and lymphocytes release multiple cytokines (interleukins, interferons, tumor necrosis factors, etc.), among which proinflammatory cytokines stimulate the liver to synthesize acute phase proteins (APPs), such as serum amyloid A (SAA) and apolipoprotein 1 (Apo-A1) [8–10]. In addition, a growing number of studies have found a close link between stress and gut microbiota [11]. Transportation stress can change the intestinal flora of animals, reduce intestinal microbial richness, increase intestinal permeability [12,13], and cause gastrointestinal diseases, including diarrhea and acute intestinal infection [14].

Feline intestinal flora is mainly composed of *Firmicutes*, *Bacteroidetes*, *Actinobacteria*, and *Proteobacteria* [15]. The intestinal microbiota can regulate the host's health by producing metabolites. Commensal bacteria in the colon use dietary fiber to synthesize short-chain fatty acids (SCFAs) to regulate immune function and intestinal mucosal barrier function [16]. *Atoposipes* and *Pseudomonas* are dominant branched-chain fatty acids (BCFAs) that produce bacteria, which can fight against inflammation and increase the expression of anti-inflammatory cytokines [17].

Antimicrobial peptides (AMPs), as a kind of polypeptide, are widely distributed in insects, amphibians and mammals. They have strong cationic properties, high stability, and broad-spectrum antibacterial activity against a variety of microorganisms (bacteria, fungi, viruses and parasites) [18,19]. Furthermore, AMPs have been shown to effectively prevent diarrhea in animals by eliminating harmful bacteria, including *Escherichia coli* (*E. coli*) [20,21]. A previous study proposed that AMPs could alleviate inflammatory bowel disease by modulating the gut microbiota in mice [22]. Furthermore, AMPs have been reported to have beneficial effects on intestinal morphology, enhancing the intestinal barrier in broilers [23]. However, few studies have investigated the effects of AMPs on domestic cats. Thus, it is of great significance to study the protection of cat health by adding AMPs to diets.

Ragdoll cats are common in pet-owning families. The objectives of this study were to determine the potential of AMPs as food additives to alleviate transportation stress. To be specific, we investigated the effects of an AMP supplement diet on the diarrhea rate, serum hormone, inflammatory factors, antioxidant indicators, fecal microbiota and metabolites of cats before and after transportation. The aim of this study is to provide a reference for the further future study studies about the significance of AMPs with respect to the intestinal health of cats.

2. Materials and Methods

All experimental procedures were approved by the Experiment Animal Ethics Committee of South China Agricultural University.

2.1. Animals and Diets

Fourteen ragdolls (four males and ten females, age 1–2 years) were allotted to two treatments (n = 7/group) on the basis of body weight (BW) and sex, including a basal diet (CON) group and a basal diet supplemented with the 0.3% AMP group. The BW of the CON group is 4.18 $\pm$ 0.84 Kg, the BW of the AMPs group is 4.08 $\pm$ 1.16 Kg, and there are no differences between the two groups ($p > 0.05$). The AMP products were purchased from Guangdong Genuizymes Animal Health Co., Ltd. (Guangzhou, China), which were isolated from chicken intestines and cultured by *Bacillus subtilis*. The dose of dietary AMPs was based on the findings of our previously unpublished data, showing that 0.3% of AMPs was effective in preventing diarrhea in cats. The composition and nutrition levels of a basal diet are shown in Table 1, which met the requirements of the National Research Council (NCR, 2006).

Table 1. Experimental diet and nutrition composition.

Ingredients	Basis %	Analytical Composition	DM Basis %
Fish	23.00	DM	93.38
Chicken meat meal	24.00	CP	40.17
Duck meat meal	12.00	CF	22.26
Fish meal	8.00	Ash	8.13
Chicken oil	6.00	OM	91.87
Tapioca	5.00		
Tapioca flour	4.00		
Alfalfa grass grain	4.00		
Chicken liver	2.70		
Chicken heart	2.00		
Flavoring paste	1.60		
Salmon oil	1.50		
Beer yeast power	1.20		
Egg power	1.00		
Kelp power	1.00		
Yucca	1.00		
Semen plantaginis	1.00		
Madder	1.00		
Total	100.00		

DM, dry matter; CP, crude protein; CF, crude fat; OM, organic matter.

2.2. Experimental Design

This study included two periods: the feeding period and transportation period. In the feeding period, each ragdoll was raised individually in cat cages (1.1 × 0.7 × 0.55 m) in a temperature-controlled room for 42 days, and was fed at 8:30 am each day. Ragdolls had free access to fresh water. All cats were dewormed quarterly and no drugs (such as antibiotics) were given 1 month before the study. Feed intake and fecal score (FS) were recorded once a day. FS is graded on a scale from 1 to 5, where grade 1 represented dry crumbly feces, grade 2 represented dry feces with little water, grade 3 represented normal moist feces, grade 4 represented wet feces, and grade 5 represented diarrhea [24]. BW was recorded once a week.

On day 43, all cats were transported back and forth from Guangzhou city to Zhaoqing city (Guangdong province, China), and were transported for 3 h with an average speed of 70 Km/h. The truck size is 5.2 × 2.2 × 2.1 m, and the temperature was 23 °C with 60% relative humidity. Transportation started at 9:20 am, and the roads include city roads and highways. After transportation, FS recordings were continued for 7 days. The study's design is depicted in Figure 1.

2.3. Blood Sample Collection and Analysis

Blood samples were obtained at the time right before transportation (BT0h), right after transportation (AT0h) and then 4 h (AT4h), 1 d (AT1d) and 7 d post-transportation (AT7d). Approximately 3 mL of blood was taken into a 10 mL anticoagulant-free vacutainer tube. After holding a 45° angle for 30 min, the serum was collected by centrifugation at 3500× g rpm for 15 min at room temperature, and stored at −80 °C for serum biochemical parameter measurements. COR, SAA, Apo-A1, interleukin-1beta (IL-1β), tumor necrosis factor-alpha (TNF-α) and interferon-gamma (IFN-γ) were analyzed by using commercial ELISA kits (MEIMIAN, Jiangsu Meimian Industrial Co., Ltd., Yancheng, China). Serum catalase (CAT), total antioxidant capacity (T-AOC), glutathione peroxidase (GSH-Px), superoxide dismutase (SOD) and malondialdehydes (MDA) were measured using commercial kits (Nanjing Jiancheng Bioengineering Institute, Nanjing, China) according to the manufacturer's protocol.

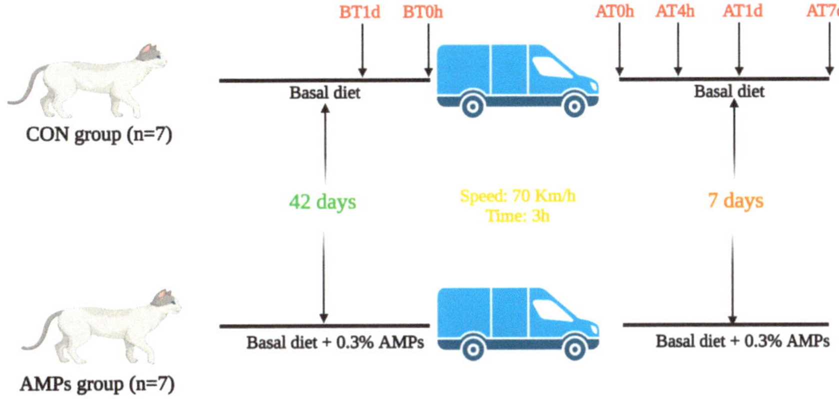

Figure 1. The sketch map of the study's design. The CON group was fed a basal diet (n = 7), and the AMPs group was fed a basal diet with 0.3% AMPs (n = 7). BT1d, 1 day before transportation; BT0h, right before transportation; AT0h, right after transportation; AT4h, 4 h after transportation; AT1d, 1 day after transportation; AT7d, 7 days after transportation.

2.4. Fresh Fecal Sample Collection and Analysis

On BT1d and AT1d, cats' fresh feces were collected within 15 min of defecation, and stored at −80 °C for further analyses. An aliquot was collected for fecal short-chain fatty acids (SCFAs) and branched-chain fatty acids (BCFAs) measurements. An aliquot was collected for microbiota measurements. Lastly, an aliquot was collected for metabolomics analysis.

2.4.1. Fecal SCFAs and BCFAs Analysis

The new fecal samples gathered on BT1d and AT1d were preprocessed, and SCFAs and BCFAs were extracted according to the method described in our previous work [6]. The quantitative analysis of SCFAs and BCFAs was carried out using the GCMS-QP2020 system (Shimadzu, Kyoto, Japan). Instrumental parameters were obtained with the method established in our laboratory [6].

2.4.2. 16S rRNA High-Throughput Sequencing

Fresh fecal samples were collected on BT1d and AT1d, and transported to Hangzhou Lianchuan Biotechnology Co., Ltd. (Hangzhou, China) for gut microbial analysis and sequenced on the Illumina NovaSeq platform. In brief, genome DNA was extracted using the cetyltrimethylammonium bromide method, and then the highly variable region V3-V4 of the 16S rDNA gene was amplified; the quality of the obtained PCR amplification products was monitored on 1% agarose gels. Finally, the samples were subjected to the high-throughput sequencing of 16S rRNA genes on the Illumina MiSeq platform of Lianchuan Biologicals. After the completion of sequencing, raw data were obtained. Later, the splicing of double-ended data with overlap was carried out, while high-quality clean data were obtained via the quality control and filtering of chimera. Afterward, the feature table and sequences for further analyses of diversity, species taxonomic annotation and difference analysis were obtained. Finally, the Amplicon Sequence Variant's (ASV) feature table and sequence were obtained.

The QIIME2 process was used to analyze alpha diversity (α-diversity) and beta diversity (β-diversity), and the pictures were displayed with R software (Version 3. 5.2). Linear discriminant analysis (LDA) and effect size (LEfSe) were processed by using the LDA score of >2.5 using LEfSe software.

2.5. Serum Untargeted Metabolomics Analysis

The serum metabolome was assessed using an untargeted approach via UPLC-MS/MS (Thermo Fisher Scientific, Waltham, MA, USA). The sample pretreatment methods and UPLC-MS/MS analysis method were carried out as described previously [6,25], with minor modification. In brief, we removed the protein from the serum sample and then dried the supernatant for testing. Raw data were preprocessed by compound discoverer 2.1 software (Thermo Fisher Scientific). Pathway enrichment analyses were processed in MetaboAnalyst 5.0 (https://www.metaboanalyst.ca (accessed on 10 September 2022)). The visualization results of the models were obtained with MetaboAnalyst 5.0.

2.6. Statistical Analysis

Original data were initially processed by Excel 2019. All data were expressed as means$\pm$standard deviation and analyzed using an unpaired Student's *t*-test in SPSS 26.0. $p < 0.05$ was defined as statistical significance, and $0.05 < p < 0.10$ was defined as a trend of difference. Graphs were drawn using GraphPad Prism 8.0 software. Microbe were compared using LEfSe, and significant differences were denoted by an LDA score > 2.5, Kruskal–Wallis rank sum test < 0.05 and Wilcoxon rank sum test < 0.05. The differential microbial were calculated by the Wilcoxon rank sum test. The volcano plot was constructed using OmicStudio tools available at https://www.omicstudio.cn/tool (accessed on 10 September 2022). For metabolome data, an unpaired Student's *t*-test was performed between the CON and AMPs groups, with $p < 0.05$ and VIP > 1 indicating significantly different metabolites. A clustering correlation heatmap with signs was drawn using the OmicStudio tools and Spearman's correlation analysis (R version 3.6.1).

3. Results

3.1. Effect of AMPs on Diarrhea Rate and Serum Hormone in Ragdolls

Compared with the CON group, the AMPs group had a significantly lower diarrhea rate during the feeding period and transportation period ($p < 0.05$, Figure 2A). Compared with the CON group, AMPs reduced the diarrhea rate by as much as 12.3% and 16.3% in the feeding period and transportation period, respectively. On AT1d, ragdolls fed with AMPs had a lower COR concentration ($p < 0.05$, Figure 2B). The AMPs group had a trend toward lower SAA levels on AT1d ($p = 0.062$, Figure 2C) and significantly decreased SAA levels on AT7d ($p < 0.05$, Figure 2C). Moreover, the AMPs group had a higher Apo-A1 concentration on AT1d ($p < 0.05$, Figure 2D).

3.2. Effect of AMPs on Serum Inflammatory Factors and Antioxidant Capacity in Ragdolls

At AT0h and AT4h, the AMPs group had a decreasing TNF-α level over the CON group ($p < 0.05$, Figure 3A). Similarly, the AMPs group tended to reduce the INF-γ level at AT0h ($p = 0.075$, Figure 3B) and AT4h ($p = 0.059$, Figure 3B). Ragdolls fed with AMPs had a lower IL-1β level than the CON group at BT0h ($p < 0.05$, Figure 3C), and they exhibited an increasing trend at AT4h ($p = 0.099$, Figure 3C).

After feeding AMPs for 42 days, ragdolls had increased GSH-Px levels over the CON group before transportation ($p < 0.01$, Figure 3D). After transportation, the AMPs group had a higher GSH-Px level than the CON group at BT0h ($p < 0.01$, Figure 3D). Additionally, the AMPs group had a higher SOD level than the CON group at AT4h ($p < 0.05$, Figure 3G). Serum MDA, CAT and T-AOC levels had no obvious difference between two groups ($p > 0.05$, Figure 3E,F,H).

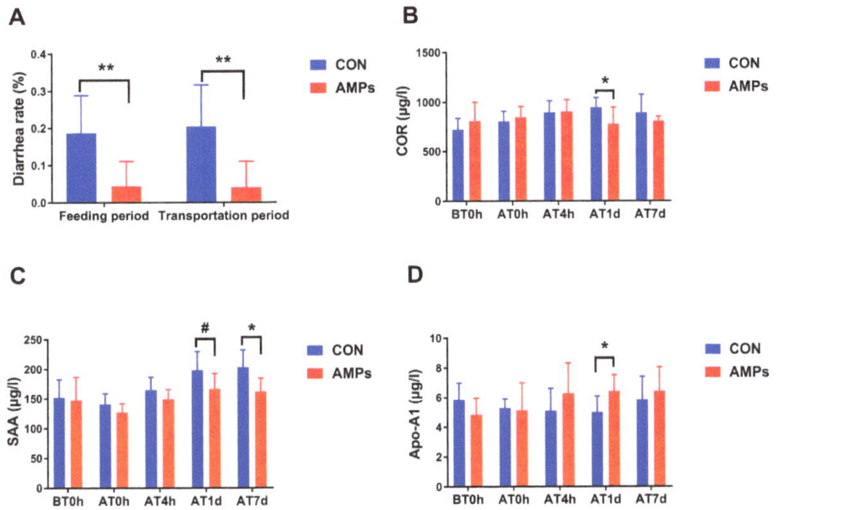

Figure 2. Effect of AMPs on diarrhea rate (**A**) and serum hormone (**B–D**) in ragdolls. Symbol (*) indicates statistically significant differences between two groups (* $p < 0.05$ and ** $p < 0.01$), and symbol (#) indicates difference tendencies ($0.05 <$ # $p < 0.10$). BT0h, right before transportation; AT0h, right after transportation; AT4h, 4 h after transportation; AT1d, 1 day after transportation; AT7d, 7 days after transportation.

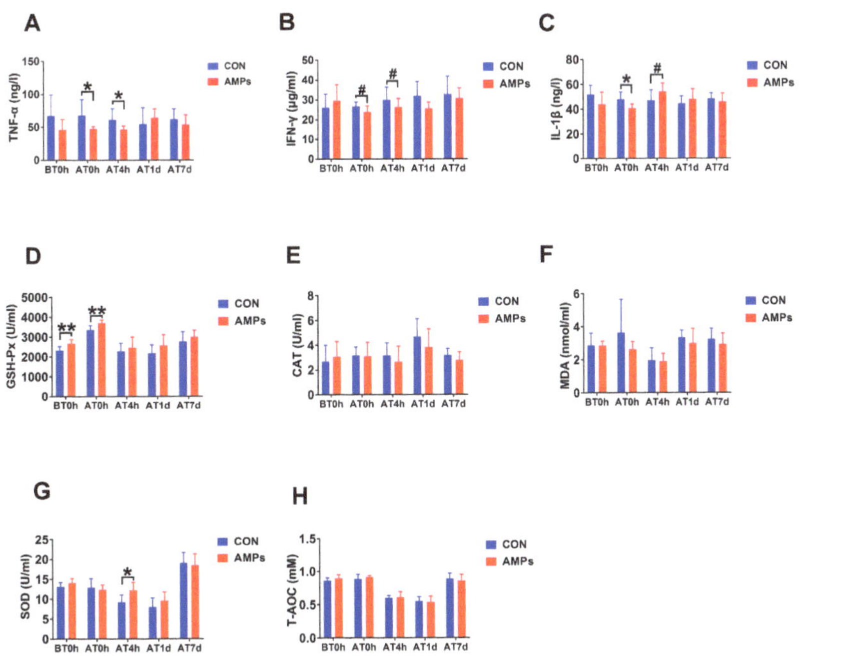

Figure 3. Effect of AMPs on inflammatory factors (**A–C**) and antioxidant capacity (**D–H**) in ragdolls. Symbol (*) indicates statistically significant differences between two groups (* $p < 0.05$ and ** $p < 0.01$), and symbol (#) indicates difference tendencies ($0.05 <$ # $p < 0.10$) between two groups. BT0h, right before transportation; AT0h, right after transportation; AT4h, 4 h after transportation; AT1d, 1 day after transportation; AT7d, 7 days after transportation.

3.3. Effect of AMPs on Fecal Microbiota in Ragdolls

After feeding AMPs for 42 days, ragdolls had a markedly decreased chao1 index and observed_species compared with the CON group on BT1d ($p < 0.01$, Figure 4A), but there was no significant difference between two groups after transportation ($p > 0.1$, Figure 4A). Both Shannon and Simpson indices show no difference on BT1d and AT1d ($p > 0.1$, Figure 4A). From the beta diversity index based on weighted UniFrac distances, the CON group shows a trend toward significant separation relative to the AMPs group on BT1d ($p = 0.05$, Figure 4A), whereas no separation was observed between two groups on AT1d ($p = 0.408$, Figure 4A).

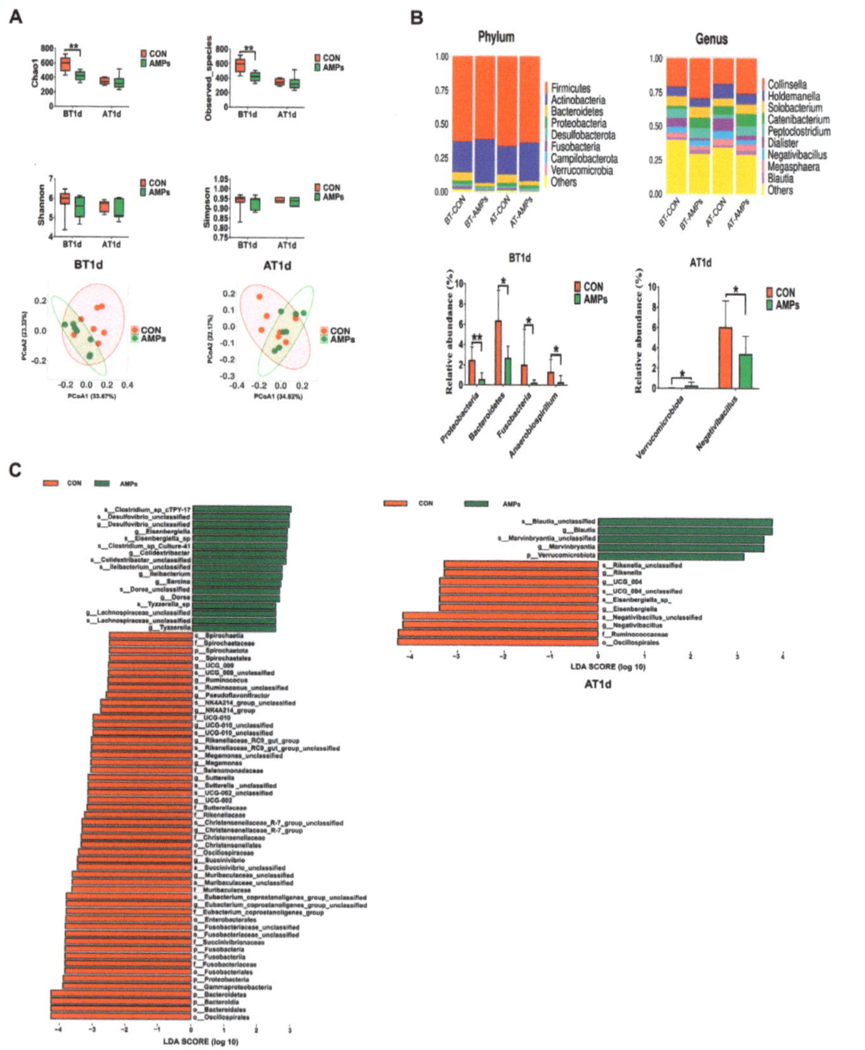

Figure 4. Effect of AMPs on fecal microbial. α-diversity and principal coordinate analysis (PCoA) based on weighted UniFrac distances (**A**); predominant fecal microbial communities (**B**); the linear discriminant analysis effect size (LEfSe) analysis identified gut bacterial biomarkers (**C**). Symbol (*) indicates statistically significant differences between two groups (* $p < 0.05$ and ** $p < 0.01$). BT1d, 1 day before transportation; AT1d, 1 day after transportation.

As shown in Figure 4B, the predominant fecal phyla included *Firmicutes*, *Actinobacteria*, *Bacteroidetes*, *Proteobacteria* and *Desulfobacterota* in ragdolls at different time points. Furthermore, the most abundant genera were *Collinsella*, *Holdemanella*, *Solobacterium*, *Catenibacterium*, *Peptoclostridium*, *Dialister*, *Negativibacillus*, *Megasphaera* and *Blautia*. Wilcoxon test results revealed that cats fed withAMPs had lower Proteobacteria ($p < 0.01$), *Bacteroidetes* ($p < 0.05$), *Fusobacteria* ($p < 0.05$) and *Anaerostipes* ($p < 0.05$) compared with the CON group on BT1d. Furthermore, the AMPs group had a higher relative abundance of *Verrucomicrobiota* ($p < 0.05$) and a lower relative abundance of *Negativibacillus* than the CON group on AT1d ($p < 0.05$).

We used the LEfSe analysis to further distinguish differential taxon abundances. On BT1d, the LEfSe analysis indicated that *Bacteroidetes* and *Proteobacteria* dominated in the CON group, while *Eisenbergiella* was the most abundant in the AMPs group. Furthermore, *Negativibacillus* was the predominant bacterial strain in the CON group on AT1d, whereas *Blautia* numbered the highest in the AMPs group.

3.4. Effect of AMPs on Fecal SCFAs and BCFAs in Ragdolls

As shown in Figure 5, after feeding with AMPs for 42 days, ragdolls had a lower acetic acid level compared with the CON group on BT1d ($p < 0.05$, Figure 5A). After transportation, the AMPs group had an increasing trend toward total BFCAs compared with the CON group ($p = 0.055$, Figure 5B).

3.5. Effect of AMPs on Serum Metabolic Profiles in Ragdolls

In our study, a total of 366 serum metabolites were monitored at all stages. Due to the fact that very few differential serum metabolites were identified on BT1d, thus, we didn't obtain the data onserum metabolites on BT1d. Differential serum metabolites between the CON and AMPs groups were screened out using the standard of VIP > 1 and $p < 0.05$ on AT1d. The volcano plot (fold change >2/<0.5) shows that there were 16 metabolites upregulated and 9 metabolites downregulated in the AMPs group compared with the CON group. The primary significant metabolites were cholic acid, rifamycin S, bis (glutathionyl) spermine, hydroquinone, lofexidine and cuscohygrine (Figure 6A). To determine the functions of these differential metabolites, we performed the KEGG enrichment analysis; the results indicated that bile acid biosynthesis, glycerolipid metabolism, riboflavin metabolism and glycerol phosphate shuttle were significantly enriched (Figure 6B).

To further clarify the association among gut microbiota, serum metabolites and differential parameters, we performed Spearman's correlation analysis among differential serum metabolites, the differential serum/fecal parameters and the differential fecal microbiota. The results are visualized as heatmaps in Figure 6. As shown in Figure 6C, SOD was positively associated with hexobarbital, cholic acid and L-carnitine but negatively associated with hordenine, cinnamyl alcohol and 3-dechloroethylifosfamide. Apo-A1 was positively associated with cerulenin but negatively associated with dimethyl trisulfide. GSH-Px was positively associated with ophthalmic acid, but negatively associated with 4-hydroxybenzaldehyde and cuscohygrine. SAS was negatively associated with indoleacrylic acid, indolelactic acid and forskolin. Cortisol was positively associated with dimethyl trisulfide and cuscohygrine. IFN-γ was positively associated with mestranol, cyclohexylamine, cuscohygrine, kynurenic acid, L-glutamine, (1R)-hydroxy-(2R)-n-acetyl-l-cysteinyl-1,2-dihydronaphthalene, yangonin, cinnamyl alcohol and 3-dechloroethylifosfamide, but it was negatively associated with indolelactic acid, fenthion, sesartemin, bis(glutathionyl)spermine and ophthalmic acid. As shown in Figure 6D, *Blautia* was positively associated with pregnanediol-3-glucuronide but negatively associated with phenylethylamine. *Negativibacillus* was positively associated with cyclohexylamine, homo-L-arginine and trimethylamine n-oxide, but it was negatively associated with sesartemin, cerulenin, indolelactic acid and indoleacrylic acid. As shown in Figure 6E, *Blautia* was positively associated with SOD and the diarrhea rate, and *Negativibacillus* was positively associated with SAA.

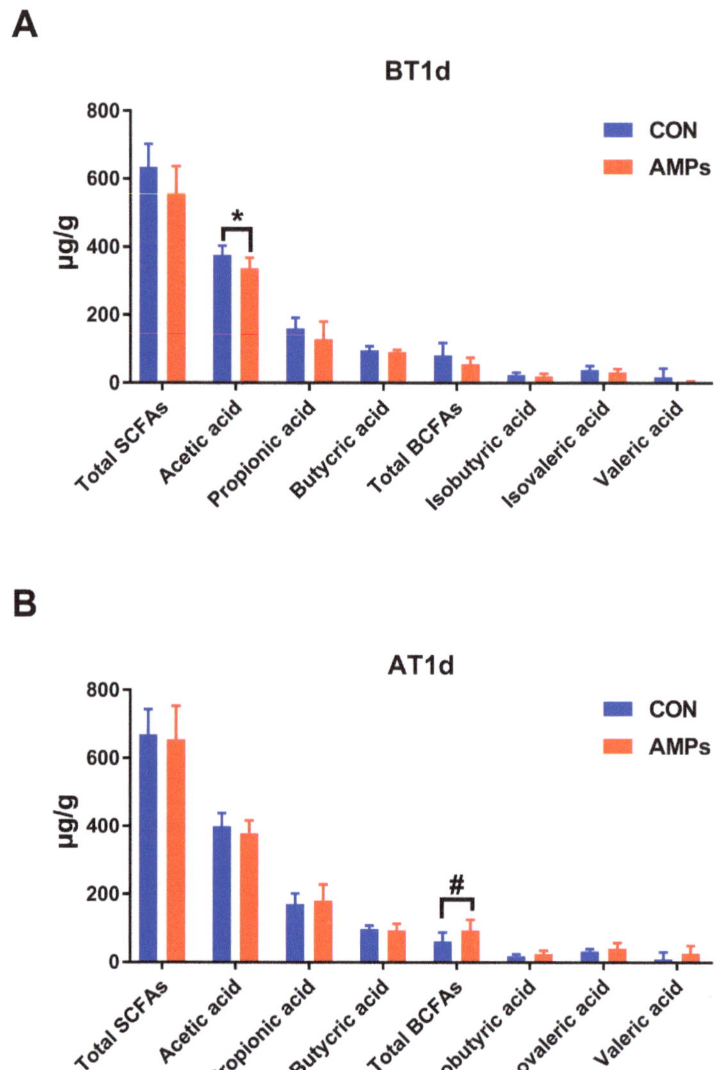

Figure 5. Effect of AMPs on the production of SCFAs (**A**) and BCFAs (**B**) in ragdolls. Symbol (*) indicates statistically significant differences between two groups (* $p < 0.05$), and symbol (#) indicates difference tendencies ($0.05 < {}^{\#} p < 0.10$) between two groups. BT1d, 1 day before transportation; AT1d, 1 day after transportation.

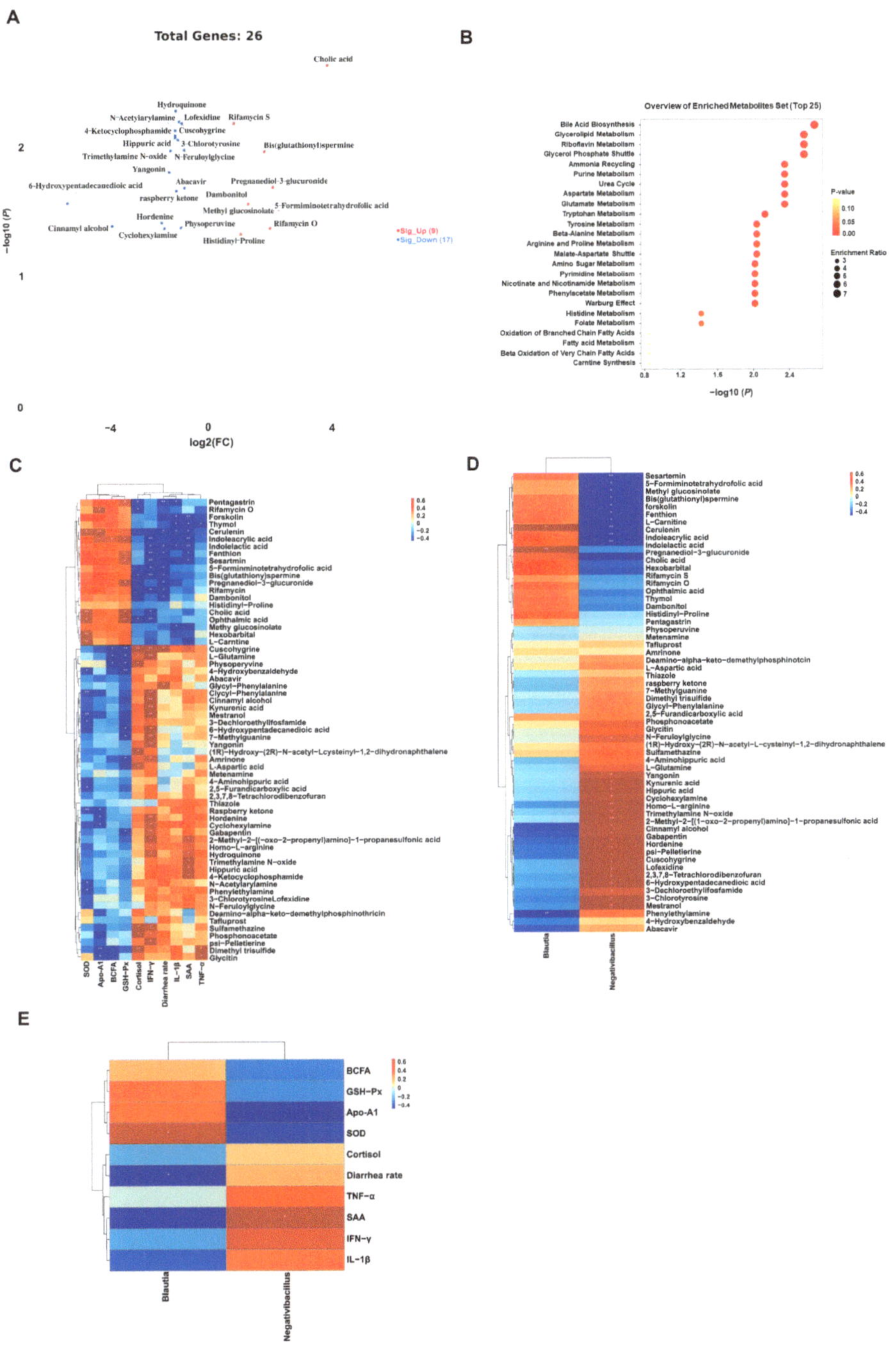

Figure 6. Serum metabolic profiles of cats, and Spearman's correlation analysis. Volcano plot (**A**), KEGG metabolic pathways enrichment analysis (**B**), heatmaps of Spearman's correlation analysis between (**C**) differential serum metabolites and serum/fecal parameters, (**D**) differential serum metabolites and fecal microbiota, (**E**) differential fecal microbiota and serum/fecal parameters. Red and blue boxes represent positive and negative correlations.

4. Discussion

4.1. AMPs Relieve Transportation Stress in Ragdoll Cats

In this study, we have demonstrated that AMPs could alleviate diarrhea symptoms, suppress inflammation and enhance antioxidant capacities in ragdolls under road transportation. Feng et al. confirmed that the antimicrobial peptide cathelicidin-BF had therapeutic effects for post-weaning diarrhea in piglets by improving immune response and attenuating intestinal inflammation [21]. A previous study conducted by Wu et al. also found that AMPs treatments reduced the incidence of diarrhea in weaned piglets challenged with *E. coli* [26]. Similar results were also obtained in our study. Our results suggested that dietary AMPs supplementation reduced diarrhea in cats before and after transportation.

As many studies indicated, transportation stress activates the HPA axis and promotes the release of glucocorticoids, including COR [27]. Our study shows an increasing serum COR level after transportation, indicating that cats have suffered from stress. We found cats fed with 0.3% AMPs had a lower serum COR level on AT1d. Furthermore, stress can lead to an increase in acute phase protein (APP). According to the increase or decrease in concentration, APP can be divided into positive acute proteins and negative acute proteins. Positive acute proteins mainly include SAA and C-reactive proteins, and negative acute proteins mainly include Apo-A1 and albumin [10,28]. Increasing serum SAA and decreasing Apo-A1 levels were observed in this study. Nevertheless, AMPs reduced SAA levels on AT1d and AT7d and increased Apo-A1 levels at AT4h, indicating that AMPs have great potential toward relieving stress.

Previous studies revealed that AMPs not only have antimicrobial activities but also play an important role in innate immunity [29]. AMPs maintain the homeostasis of the immune microenvironment by regulating cytokines such as interleukins, tumor necrosis factors, interferons and the activities of immune cells (e.g., dendritic cells, monocytes and mast cells) [18]. Sorrells et al. and Khan et al. observed that acute stress could enhance the body's immune function [30,31]. Zhai et al. found that AMPs could downregulate the mRNA expression of TNF-α, IL-6 and IL-8 [32]. Similarly, our results showed that AMPs effectively alleviated inflammatory responses in cats by downregulating the release of pro-inflammatory cytokines (e.g., TNF-α, IFN-γ and IL-1β), indicating that dietary supplementation with 0.3% AMPs could improve anti-inflammatory function in cats. Moreover, stress induces excessive ROS production, which can cause oxidative stress and lipid peroxidation [33]. MDA is the ending product that serves as a marker of lipid peroxidation and oxidative stress, [34] and the levels of antioxidant enzymes, such as SOD, GSH-Px, CAT and T-AOC, can reflect the ability of the organism in eliminating free radicals [35]. In the present study, we found that AMPs increased the antioxidant activities in cats by upregulating GSH-Px and SOD. Consistent with previous studies, we found that compared with the CON group, the concentration of SOD increased in the AMPs group [33,36], which indicates that dietary supplements with AMPs at 0.3% protected cats from oxidative stress induced by transportation.

4.2. AMPs Alter Intestinal Microbiota and Metabolite in Ragdoll Cats

Previous studies revealed dietary supplementation with AMPs could regulate gut microbiota, reduce pathogenic bacteria and alter the diversity of microorganisms [22,37–39]. In our study, AMPs reduced the richness of bacterial flora; decreased the relative abundances of *Bacteroidetes*, *Proteobacteria*, *Fusobacteria* and *Anaerobiospirillum* on BT1d; and inhibited the growth of *Negativibacillus* on AT1d. Previous studies found that *Bacteroides* are positively correlated with the disease activity index, and *Bacteroidetes* are associated with intestinal inflammation [40,41]. *Proteobacteria* include many pathogenic bacteria, such as *E. coli*, *Salmonella* and *Campylobacter*, which have been associated with dysbiosis and inflammatory disorders [42,43]. *Fusobacteria* was reported to be linked with colon cancer [44]. The genus *Anaerobiospirillum* consists of *Anaerobiospirillum thomasii* and *Anaerobiospirillum succiniciproducens*, which can cause infectious human diseases, such as bacteremia and abdominal infection [45,46]. Previous studies show that the genus *Negativibacillus* is associated with intestinal disorders and many other diseases such as obesity-related disorders and pediatric Crohn's disease [47–50]. Our results were further verified with LEfSe analyses. The AMPs group was mainly enriched with beneficial bacteria, including *Eisenbergiella* and *Blautia*. *Eisenbergiella* is the main producer of SCFAs [51]. *Blautia* was reported to alleviate inflammatory diseases and metabolic diseases, and it has potential probiotic functions [52,53]. Collectively, we can infer that AMPs can alleviate the imbalance of microbiota from transportation stress by reducing harmful bacteria and increasing beneficial bacteria.

SCFAs are produced by the colonic bacterial fermentation of non-digestible carbohydrates, and BCFAs are produced by the fermentation of branched-chain amino acids [54]. These microbiota-derived metabolites play important roles in maintaining epithelial function and participate in various metabolic pathways (e.g., lipid and glucose metabolism) [54,55]. In our study, AMPs reduced the fecal acetic acid concentration on BT1d, possibly due to the decreased abundance of *Bacteroides*, which are the main acetic-acid-producing bacteria [56]. However, fecal total BCFAs levels were higher in cats fed with AMPs on AT1d. BCFAs have beneficial effects on preventing intestinal inflammation and increasing the expression of anti-inflammatory cytokines [17]. These results reveal that AMPs might provide protection against transportation-induced inflammation by improving the intestinal's microbial structure.

Gut microbiota contribute to the host's metabolism and the regulation of the immune system, thus significantly affecting the host's physiological functions [57]. We found a high level of serum cholic acid and rifamycin S in the AMPs group. Cholic acid is the primary bile acid, which is synthesized from cholesterol in the liver and reabsorbed by the ileum into the blood's circulation [58]. Rifamycin S belongs to the ansamycin family, which bears antimicrobial activity [59,60]. The enrichment analysis further revealed that 0.3% AMPs mainly affected lipid metabolism (i.e., bile acid biosynthesis, glycerolipid metabolism and glycerol phosphate shuttle).

Spearman's correlation analysis found that strong correlations between fecal microbiota and serum metabolites could influence stress parameters, antioxidant capacity and inflammatory responses. *Blautia* was negatively associated with serum phenylethylamine, which was negatively associated with SOD. Furthermore, *Blautia* was positively associated with serum pregnanediol-3-glucuronide, which was negatively associated with diarrhea rates. *Negativibacillus* was negatively associated with serum indoleacrylic acid and indolelactic acid, which were negatively associated with IFN-γ and SAA. The results indicated that phenylethylamine, pregnanediol-3-glucuronide, indoleacrylic acid and indolelactic acid could serve as potential biomarkers of transportation stress. However, the connections between gut microbiota and serum metabolic changes still need to be further explored and validated.

4.3. Future Prospects and Drawbacks of the Present Study

AMPs have potential of clinic applications with respect to relieving diarrhea and transportation stress in cats. However, the dosage should be determined more precisely according to the different degrees of symptoms.

In this study, we used ragdoll cats as research subjects partly because the ragdoll cat is a representative and widely owned breed of domestic cats. In addition, a previous study convinced us that the variation between cat breeds is relatively small compared to species such as dogs [61]. However, it is still of interest to carry out studies on other cat breeds in the future to provide more data for clinical use.

5. Conclusions

In conclusion, our study indicated that AMPs could beneficially reduce the incidence of diarrhea in cats induced by transportation. In addition, AMPs could decrease the level of inflammatory factors caused by transportation stress, including TNF-α and IL-1β, and enhance the activities of GSH-Px and SOD. Furthermore, AMPs could reduce the abundance of *Negativibacillus* and increase the abundance of *Blautia* and the number of BCFAs. Therefore, the presented data indicated that AMPs supplementation may alleviate diarrhea and oxidative stress in transportation by regulating the gut microbiota and metabolites, thereby supporting gut and host health in cats. In a future study, we will focus on the connection between gut microbiota and metabolites to better understand the precise mechanisms of AMPs on cat health.

Author Contributions: Conceptualization, S.H. (Shansong He) and B.D.; methodology, S.H. (Shansong He); formal analysis, S.H. (Shansong He) and K.Y.; investigation, S.H. (Shansong He); resources, S.H. (Shansong He); data curation, S.H. (Shansong He), J.W., T.K. and Z.C.; writing—original draft preparation, S.H. (Shansong He); writing—review and editing, S.H. (Shansong He), K.Y., B.D., L.Z. (Lingna Zhang), S.H. (Sufang Han) and L.Z. (Limeng Zhang); visualization; S.H. (Shansong He) and S.J.; supervision, B.D., J.D. and X.C.; project administration, S.H. (Shansong He) All authors have read and agreed to the published version of the manuscript.

Funding: This research was funded by National Natural Science Foundation of China (Grant No. 32002186), National Key R&D Program of China (Grant No. 2021YFD1300400), Natural Science Foundation of Guangdong Province (Grant No. 2020A1515010322) and Guangzhou Basic and Applied Basic Research Foundation (Grant No. 202102020850).

Institutional Review Board Statement: All experimental procedures were approved by the Experiment Animal Ethics Committee of South China Agricultural University (2021A030).

Informed Consent Statement: Not appliable.

Data Availability Statement: The data presented in this study are available in the manuscript.

Acknowledgments: We gratefully appreciate Guangzhou Qingke Biotechnology Co., Ltd. (Guangzhou, China).

Conflicts of Interest: Limeng Zhang is an employee of Guangzhou Qingke Biotechnology Co., Ltd. The paper reflects the views of the scientists and not the company. The authors declare there is no conflict of interest.

References

1. Ma, Y. Number of Domestic Cats in China from 2018 to 2021 (in Millions). Available online: https://www.statista.com/statistics/992420/china-number-of-cats/ (accessed on 15 February 2022).
2. Bedford, E. Number of Pet Cats in Europe from 2010 to 2021 (in 1000s). Available online: https://www.statista.com/statistics/516041/cat-population-europe-europe/ (accessed on 27 September 2022).
3. Zhang, L.; Bian, Z.; Liu, Q.; Deng, B. Dealing With Stress in Cats: What Is New About the Olfactory Strategy? *Front. Vet. Sci.* **2022**, *9*, 928943. [CrossRef]
4. Amat, M.; Camps, T.; Manteca, X. Stress in owned cats: Behavioural changes and welfare implications. *J. Feline Med. Surg.* **2016**, *18*, 577–586. [CrossRef]
5. Moberg, G.P.; Mench, J.A. The biology of animal stress: Basic principles and implications for animal welfare. *Vet. J.* **2002**, *164*, 77.
6. Yang, K.; Deng, X.; Jian, S.; Zhang, M.; Wen, C.; Xin, Z.; Zhang, L.; Tong, A.; Ye, S.; Liao, P.; et al. Gallic Acid Alleviates Gut Dysfunction and Boosts Immune and Antioxidant Activities in Puppies Under Environmental Stress Based on Microbiome–Metabolomics Analysis. *Front. Immunol.* **2022**, *12*, 813890. [CrossRef]
7. Bailey, M.T. Influence of stressor-induced nervous system activation on the intestinal microbiota and the importance for immunomodulation. *Adv. Exp. Med. Biol.* **2014**, *817*, 255–276.
8. Fazio, F.; Casella, S.; Giannetto, C.; Giudice, E.; Piccione, G. Characterization of acute phase proteins and oxidative stress response to road transportation in the dog. *Exp. Anim.* **2015**, *64*, 19–24. [CrossRef]
9. Wernicki, A.S.A.S.; Urban-Chmiel, R.S.A.S.; Kankofer, M.S.B.S.; Mikucki, P.S.S.; Puchalski, A.S.A.S.; Tokarzewski, S.S.A.S. Evaluation of plasma cortisol and TBARS levels in calves after short-term transportation. *Rev. Med. Vet.-Toulouse* **2006**, *157*, 30–34.
10. Paltrinieri, S. The feline acute phase reaction. *Vet. J.* **2008**, *177*, 26–35. [CrossRef]
11. Molina-Torres, G.; Rodriguez-Arrastia, M.; Roman, P.; Sanchez-Labraca, N.; Cardona, D. Stress and the gut microbiota-brain axis. *Behav. Pharmacol.* **2019**, *30*, 187–200. [CrossRef]
12. Jacobs, L.; Delezie, E.; Duchateau, L.; Goethals, K.; Ampe, B.; Buyse, J.; Tuyttens, F.A.M. Impact of transportation duration on stress responses in day-old chicks from young and old breeders. *Res. Vet. Sci.* **2017**, *112*, 172–176. [CrossRef]
13. Jiang, G.; Zhang, X.; Gao, W.; Ji, C.; Wang, Y.; Feng, P.; Feng, Y.; Zhang, Z.; Li, L.; Zhao, F. Transport stress affects the fecal microbiota in healthy donkeys. *J. Vet. Intern. Med.* **2021**, *35*, 2449–2457. [CrossRef]

14. Guard, B.C.; Barr, J.W.; Reddivari, L.; Klemashevich, C.; Jayaraman, A.; Steiner, J.M.; Vanamala, J.; Suchodolski, J.S. Characterization of Microbial Dysbiosis and Metabolomic Changes in Dogs with Acute Diarrhea. *PLoS ONE* **2015**, *10*, e0127259. [CrossRef]
15. Lyu, Y.; Su, C.; Verbrugghe, A.; Van de Wiele, T.; Martos Martinez-Caja, A.; Hesta, M. Past, Present, and Future of Gastrointestinal Microbiota Research in Cats. *Front. Microbiol.* **2020**, *11*, 1661. [CrossRef]
16. Schwiertz, A.; Lehmann, U.; Jacobasch, G.; Blaut, M. Influence of resistant starch on the SCFA production and cell counts of butyrate-producing Eubacterium spp. in the human intestine. *J. Appl. Microbiol.* **2002**, *93*, 157–162. [CrossRef]
17. Ran-Ressler, R.R.; Glahn, R.P.; Bae, S.; Brenna, J.T. Branched-chain fatty acids in the neonatal gut and estimated dietary intake in infancy and adulthood. *Importance Immunonutrition* **2013**, *77*, 133–143.
18. Zhang, Q.; Yan, Z.; Meng, Y.; Hong, X.; Shao, G.; Ma, J.; Cheng, X.; Liu, J.; Kang, J.; Fu, C. Antimicrobial peptides: Mechanism of action, activity and clinical potential. *Mil. Med. Res.* **2021**, *8*, P231–P258. [CrossRef]
19. Valdez-Miramontes, C.E.; De Haro-Acosta, J.; Aréchiga-Flores, C.F.; Verdiguel-Fernández, L.; Rivas-Santiago, B. Antimicrobial peptides in domestic animals and their applications in veterinary medicine. *Peptides* **2021**, *142*, 170576. [CrossRef]
20. Sierra, J.M.; Vinas, M. Future prospects for Antimicrobial peptide development: Peptidomimetics and antimicrobial combinations. *Expert Opin. Drug Discov.* **2021**, *16*, 601–604. [CrossRef]
21. Feng, J.; Wang, L.; Xie, Y.; Chen, Y.; Yi, H.; He, D. Effects of antimicrobial peptide cathelicidin-BF on diarrhea controlling, immune responses, intestinal inflammation and intestinal barrier function in piglets with postweaning diarrhea. *Int. Immunopharmacol.* **2020**, *85*, 106658. [CrossRef]
22. Zhai, Z.; Zhang, F.; Cao, R.; Ni, X.; Xin, Z.; Deng, J.; Wu, G.; Ren, W.; Yin, Y.; Deng, B. Cecropin A Alleviates Inflammation Through Modulating the Gut Microbiota of C57BL/6 Mice With DSS-Induced IBD. *Front. Microbiol.* **2019**, *10*, 1595. [CrossRef]
23. Choi, S.C.; Ingale, S.L.; Kim, J.S.; Park, Y.K.; Kwon, I.K.; Chae, B.J. Antimicrobial peptide-A3: Effects on growth performance, nutrient retention, intestinal and faecal microflora and intestinal morphology of broilers. *Brit. Poultry Sci.* **2013**, *54*, 738–746. [CrossRef]
24. Baillon, M.A.; Marshall-Jones, Z.V.; Butterwick, R.F. Effects of probiotic Lactobacillus acidophilus strain DSM13241 in healthy adult dogs. *Am. J. Vet. Res.* **2004**, *65*, 338–343. [CrossRef]
25. Xin, Z.; Ma, S.; Ren, D.; Liu, W.; Han, B.; Zhang, Y.; Xiao, J.; Yi, L.; Deng, B. UPLC-Orbitrap-MS/MS combined with chemometrics establishes variations in chemical components in green tea from Yunnan and Hunan origins. *Food Chem.* **2018**, *266*, 534–544. [CrossRef]
26. Wu, S.; Zhang, F.; Huang, Z.; Liu, H.; Xie, C.; Zhang, J.; Thacker, P.A.; Qiao, S. Effects of the antimicrobial peptide cecropin AD on performance and intestinal health in weaned piglets challenged with Escherichia coli. *Peptides* **2012**, *35*, 225–230. [CrossRef]
27. De, A.K.; Sawhney, S.; Ponraj, P.; Sunder, J.; Banik, S.; Bhattacharya, D. Physiological and immune responses to long road transportation in Andaman local pigs. *Trop. Anim. Health Pro.* **2021**, *53*, 247. [CrossRef]
28. Cray, C.; Zaias, J.; Altman, N.H. Acute Phase Response in Animals: A Review. *Comp. Med.* **2009**, *59*, 517–526.
29. Shooshtarizadeh, P.; Zhang, D.; Chich, J.; Gasnier, C.; Schneider, F.; Haïkel, Y.; Aunis, D.; Metz-Boutigue, M. The antimicrobial peptides derived from chromogranin/secretogranin family, new actors of innate immunity. *Regul. Pept.* **2010**, *165*, 102–110. [CrossRef]
30. Khan, S.R.; van der Burgh, A.C.; Peeters, R.P.; van Hagen, P.M.; Dalm, V.A.S.H.; Chaker, L. Determinants of Serum Immunoglobulin Levels: A Systematic Review and Meta-Analysis. *Front. Immunol.* **2021**, *12*, 664526. [CrossRef]
31. Sorrells, S.F.; Sapolsky, R.M. An inflammatory review of glucocorticoid actions in the CNS. *Brain Behav. Immun.* **2007**, *21*, 259–272. [CrossRef]
32. Zhai, Z.; Ni, X.; Jin, C.; Ren, W.; Li, J.; Deng, J.; Deng, B.; Yin, Y. Cecropin A Modulates Tight Junction-Related Protein Expression and Enhances the Barrier Function of Porcine Intestinal Epithelial Cells by Suppressing the MEK/ERK Pathway. *Int. J. Mol. Sci.* **2018**, *19*, 1941. [CrossRef]
33. Wang, S.; Xie, S.; Zhou, A.; Zhang, C.; Wen, L.; Xu, G.; Zou, J. Effects of mixed antimicrobial peptide on the growth performance, antioxidant and immune responses and disease resistance of Pengze crucian carp (*Carassius auratus* var. Pengze). *Fish Shellfish Immun.* **2021**, *114*, 112–118. [CrossRef]
34. Todorova, I.; Simeonova, G.; Kyuchukova, D.; Dinev, D.; Gadjeva, V. Reference values of oxidative stress parameters (MDA, SOD, CAT) in dogs and cats. *Comp. Clin. Pathol.* **2005**, *13*, 190–194. [CrossRef]
35. Cecerska-Heryc, E.; Surowska, O.; Heryc, R.; Serwin, N.; Napiontek-Balinska, S.; Dolegowska, B. Are antioxidant enzymes essential markers in the diagnosis and monitoring of cancer patients—A review. *Clin. Biochem.* **2021**, *93*, 1–8. [CrossRef]
36. Liu, S.; Wang, S.; Liu, X.; Wen, L.; Zou, J. Effects of dietary antimicrobial peptides on intestinal morphology, antioxidant status, immune responses, microbiota and pathogen disease resistance in grass carp Ctenopharyngodon idellus. *Microb. Pathog.* **2022**, *165*, 105386. [CrossRef]
37. Liu, N.; Ma, X.; Jiang, X. Effects of Immobilized Antimicrobial Peptides on Growth Performance, Serum Biochemical Index, Inflammatory Factors, Intestinal Morphology, and Microbial Community in Weaning Pigs. *Front. Immunol.* **2022**, *13*, 872990. [CrossRef]
38. Ren, Z.; Yao, R.; Liu, Q.; Deng, Y.; Shen, L.; Deng, H.; Zuo, Z.; Wang, Y.; Deng, J.; Cui, H.; et al. Effects of antibacterial peptides on rumen fermentation function and rumen microorganisms in goats. *PLoS ONE* **2019**, *14*, e0221815. [CrossRef]

39. Dai, Z.; Shang, L.; Wang, F.; Zeng, X.; Yu, H.; Liu, L.; Zhou, J.; Qiao, S. Effects of Antimicrobial Peptide Microcin C7 on Growth Performance, Immune and Intestinal Barrier Functions, and Cecal Microbiota of Broilers. *Front. Vet. Sci.* **2022**, *8*, 1542. [CrossRef]
40. Shang, L.; Yu, H.; Liu, H.; Chen, M.; Zeng, X.; Qiao, S. Recombinant antimicrobial peptide microcin J25 alleviates DSS-induced colitis via regulating intestinal barrier function and modifying gut microbiota. *Biomed. Pharmacother.* **2021**, *139*, 111127. [CrossRef]
41. Schwab, C.; Berry, D.; Rauch, I.; Rennisch, I.; Ramesmayer, J.; Hainzl, E.; Heider, S.; Decker, T.; Kenner, L.; Muller, M.; et al. Longitudinal study of murine microbiota activity and interactions with the host during acute inflammation and recovery. *ISME J.* **2014**, *8*, 1101–1114. [CrossRef]
42. Garrigues, Q.; Apper, E.; Chastant, S.; Mila, H. Gut microbiota development in the growing dog: A dynamic process influenced by maternal, environmental and host factors. *Front. Vet. Sci.* **2022**, *9*, 964649. [CrossRef]
43. Minamoto, Y.; Otoni, C.C.; Steelman, S.M.; Buyukleblebici, O.; Steiner, J.M.; Jergens, A.E.; Suchodolski, J.S. Alteration of the fecal microbiota and serum metabolite profiles in dogs with idiopathic inflammatory bowel disease. *Gut Microbes* **2015**, *6*, 33–47. [CrossRef]
44. King, M.; Hurley, H.; Davidson, K.R.; Dempsey, E.C.; Barron, M.A.; Chan, E.D.; Frey, A. The Link between Fusobacteria and Colon Cancer: A Fulminant Example and Review of the Evidence. *Immune Netw.* **2020**, *20*, e30. [CrossRef]
45. Fox, B.; Berger, M.; Roncallo, M.; Pinoche, L.; Ibáñez, M.; Gonzalez-Fraga, S.; Fernández-Canigia, L. MALDI-TOF MS in Anaerobiospirillum succiniciproducens bacteremia: A report of 4 cases in different hosts. *Anaerobe* **2018**, *154*, 267–270. [CrossRef]
46. Schaumburg, F.; Dieckmann, R.; Auml, T.S.; Kling; Becker, K.; Idelevich, E.A. First description of an Anaerobiospirillum succiniciproducens prosthetic joint infection. *New Microbes New Infect.* **2017**, *18*, 1–2. [CrossRef]
47. Tang, S.; Zhong, R.; Yin, C.; Su, D.; Xie, J.; Chen, L.; Liu, L.; Zhang, H. Exposure to High Aerial Ammonia Causes Hindgut Dysbiotic Microbiota and Alterations of Microbiota-Derived Metabolites in Growing Pigs. *Front. Nutr.* **2021**, *8*, 689818. [CrossRef]
48. Wang, Y.; Gao, X.; Zhang, X.; Xiao, F.; Hu, H.; Li, X.; Dong, F.; Sun, M.; Xiao, Y.; Ge, T.; et al. Microbial and metabolic features associated with outcome of infliximab therapy in pediatric Crohn's disease. *Gut Microbes* **2021**, *13*, 1865708. [CrossRef]
49. Dovrolis, N.; Michalopoulos, G.; Theodoropoulos, G.E.; Arvanitidis, K.; Kolios, G.; Sechi, L.A.; Eliopoulos, A.G.; Gazouli, M. The Interplay between Mucosal Microbiota Composition and Host Gene-Expression is Linked with Infliximab Response in Inflammatory Bowel Diseases. *Microorganisms* **2020**, *8*, 438. [CrossRef]
50. Wang, Q.; Wang, X.F.; Xing, T.; Li, J.; Zhang, L.; Gao, F. The combined impact of xylo-oligosaccharides and gamma-irradiated *Astragalus polysaccharides* on the immune response, antioxidant capacity, and intestinal microbiota composition of broilers. *Poultry Sci.* **2022**, *101*, 101996. [CrossRef]
51. Zhao, Q.; Fu, Y.; Zhang, F.; Wang, C.; Yang, X.; Bai, S.; Xue, Y.; Shen, Q. Heat-Treated Adzuki Bean Protein Hydrolysates Reduce Obesity in Mice Fed a High-Fat Diet via Remodeling Gut Microbiota and Improving Metabolic Function. *Mol. Nutr. Food Res.* **2022**, *66*, 2100907. [CrossRef]
52. Jenq, R.R.; Taur, Y.; Devlin, S.M.; Ponce, D.M.; Goldberg, J.D.; Ahr, K.F.; Littmann, E.R.; Ling, L.; Gobourne, A.C.; Miller, L.C.; et al. Intestinal Blautia Is Associated with Reduced Death from Graft-versus-Host Disease. *Biol. Blood Marrow Tr.* **2015**, *21*, 1373–1383. [CrossRef]
53. Liu, X.; Mao, B.; Gu, J.; Wu, J.; Cui, S.; Wang, G.; Zhao, J.; Zhang, H.; Chen, W. Blautia-a new functional genus with potential probiotic properties? *Gut Microbes* **2021**, *13*, 1875796. [CrossRef]
54. Heimann, E.; Nyman, M.; Palbrink, A.K.; Lindkvist-Petersson, K.; Degerman, E. Branched short-chain fatty acids modulate glucose and lipid metabolism in primary adipocytes. *Adipocyte* **2016**, *5*, 359–368. [CrossRef]
55. Ouyang, Z.; Niu, X.; Wang, W.; Zhao, J. The role of short-chain fatty acids in Clostridioides difficile infection: A review. *Anaerobe* **2022**, *75*, 102585. [CrossRef]
56. Wang, Z.; Liu, J.; Li, F.; Luo, Y.; Ge, P.; Zhang, Y.; Wen, H.; Yang, Q.; Ma, S.; Chen, H. The gut-lung axis in severe acute Pancreatitis-associated lung injury: The protection by the gut microbiota through short-chain fatty acids. *Pharmacol. Res.* **2022**, *182*, 106321. [CrossRef]
57. Yang, K.; Jian, S.; Wen, C.; Guo, D.; Liao, P.; Wen, J.; Kuang, T.; Han, S.; Liu, Q.; Deng, B. Gallnut Tannic Acid Exerts Anti-stress Effects on Stress-Induced Inflammatory Response, Dysbiotic Gut Microbiota, and Alterations of Serum Metabolic Profile in Beagle Dogs. *Front. Nutr.* **2022**, *9*, 847966. [CrossRef]
58. Wang, N.; Wang, J.; Zhang, T.; Huang, L.; Yan, W.; Lu, L.; Jia, J.; Tao, Y.; Cai, W.; Wang, Y. Alterations of gut microbiota and serum bile acids are associated with parenteral nutrition-associated liver disease. *J. Pediatr. Surg.* **2021**, *56*, 738–744. [CrossRef]
59. Huang, H.; Wu, X.; Yi, S.; Zhou, Z.; Zhu, J.; Fang, Z.; Yue, J.; Bao, S. Rifamycin S and its geometric isomer produced by a newly found actinomycete, Micromonospora rifamycinica. *Antonie Van Leeuwenhoek* **2009**, *95*, 143–148. [CrossRef]
60. Arora, S.K. Correlation of structure and activity in ansamycins: Structure, conformation, and interactions of antibiotic rifamycin S. *J. Med. Chem.* **1985**, *28*, 1099–1102. [CrossRef]
61. Handl, S.; Dowd, S.E.; Garcia-Mazcorro, J.F.; Steiner, J.M.; Suchodolski, J.S. Massive parallel 16S rRNA gene pyrosequencing reveals highly diverse fecal bacterial and fungal communities in healthy dogs and cats. *FEMS Microbiol. Ecol.* **2011**, *76*, 301–310. [CrossRef]

metabolites

MDPI

Article

Chitosan Enhances Intestinal Health in Cats by Altering the Composition of Gut Microbiota and Metabolites

Ruixia Mo, Mingrui Zhang, Haotian Wang, Tianyi Liu, Pan Liu and Yi Wu *

State Key Laboratory of Animal Nutrition, College of Animal Science and Technology, China Agricultural University, Beijing 100193, China
* Correspondence: wuyi2013@cau.edu.cn; Tel.: +86-136-1130-1596

Abstract: The interaction between gut microbiota and the health of the host has gained increasing attention. Chitosan is a natural alkaline polysaccharide with a wide range of beneficial effects. However, rare studies have been observed on the effects of dietary chitosan supplementation on intestinal health in cats. A total of 30 cats with mild diarrhea were divided into three groups, receiving a basic diet with 0 (CON), 500 (L-CS) or 2000 (H-CS) mg/kg chitosan. Samples of blood and feces were collected and analyzed for serology and gut microbiota composition. The results demonstrated that chitosan alleviated symptoms of diarrhea, with enhanced antioxidant capability and decreased inflammatory biomarker levels in serum. Chitosan reshaped the composition of gut microbiota in cats that the beneficial bacteria *Allobaculum* was significantly increased in the H-CS group. Acetate and butyrate contents in feces were significantly higher in the H-CS group in comparison to the CON group ($p < 0.05$). In conclusion, the addition of dietary chitosan in cats enhanced intestinal health by modulating their intestinal microbes and improved microbiota-derived SCFA production. Our results provided insights into the role of chitosan in the gut microbiota of felines.

Keywords: cat; chitosan; intestinal health; gut microbiota; SCFAs

Citation: Mo, R.; Zhang, M.; Wang, H.; Liu, T.; Liu, P.; Wu, Y. Chitosan Enhances Intestinal Health in Cats by Altering the Composition of Gut Microbiota and Metabolites. *Metabolites* **2023**, *13*, 529. https://doi.org/10.3390/metabo13040529

Academic Editor: Bruno Stefanon

Received: 19 February 2023
Revised: 10 March 2023
Accepted: 4 April 2023
Published: 6 April 2023

1. Introduction

The intestine of mammals is a vital organ for digestion and absorption, while also being the largest immune organ [1], thus illustrating an intricate association between intestinal health and host health status [2]. In the intestine, the most abundant and diverse microbial systems coexist, including interacting bacteria, fungi, and viruses coexisting in the intestine [3]. Following the rapid development of microbial analysis techniques, such as 16S rRNA analysis, the gut microbiota has been a critical frontier in understanding animal intestinal homeostasis and disease progression [4].

Nutritional interventions in the diet play a critical role in altering the composition and function of the host's gut microbiota [4,5]. The structure of the gut microbiota and its metabolites in cats, which are carnivores by nature, is different from that of omnivores and herbivores [6,7]. Despite the fact that cats have a short and less functional colon, fermenting microbes are abundant in their posterior intestine [8]. Moreover, numerous studies have reported that complex bacterial communities experience significant alterations during intestinal diseases in cats [9]. Consequently, the gut microbiota and its metabolites caused by diets could affect the intestinal health of felines [10].

Therefore, it is important to investigate and understand the relationship between gut microbiota and feline diets. Dietary polysaccharides are carbohydrates synthesized by the polymerization of more than ten monosaccharides. Complex dietary polysaccharides could be metabolized into rich short-chain fatty acids (SCFAs) by intestinal microbes capable of passing through an enzyme system that is not available in mammals [11]. Acetate, propionate, and butyrate were the most abundant SCFAs [12]. SCFAs possess many beneficial effects, including providing energy to the intestine, maintaining intestinal pH, stimulating the secretion of gastrointestinal hormones, and protecting the functioning of the intestinal barrier [11].

Chitosan is a dietary polysaccharide derived from the deacetylation or partial deacetylation of chitin in arthropod and mollusk cell membranes [13]. Chitosan has many physiological properties, including antibacterial, anti-inflammatory, and immune-regulatory properties [14,15]. In addition, chitosan has been found to affect the microbial diversity in the intestine of mice, especially by lowering the ratio of Firmicutes/Bacteroidetes and decreasing Proteobacteria [16]. Concomitantly, previous research has suggested that the supplementation of dietary chitosan has decreased the abundance of pathogenic bacteria, such as *Escherichia coli, Shigella* spp., and *Desulfovibrio* spp. [17]. Meanwhile, in a study on rats, chitosan has been found to increase the serum concentrations of propionate and butyrate derived from gut microbiota [18]. However, research on the effects of dietary chitosan on the intestinal health of felines is scarce.

Thus, the aim of this study was to examine the effects of chitosan on immunity, intestinal health, the composition of the gut microbiota, and fecal SCFAs, and illustrate the potential dietary administration to improve the intestinal health of felines.

2. Materials and Methods

2.1. Animal Ethics Statement

The experimental protocol and animal care were approved by the CAU Animal Care and Use Committee (AW60203202-1-1). All feeding practices were performed strictly following the National Research Council's (NRC) Guide.

2.2. Animals and Experimental Treatments

At the beginning of the research, a medical examination was performed on a colony of cats, which included appetite, mental status, parasites, and body condition. Thirty adult British shorthair cats (half male and half female) with an average age of 2.5 years and an average body weight (BW) of 8.33 $\pm$ 0.53 kg that had been spayed or neutered were included in the study. These cats experienced mild diarrhea with fecal scores ranging between 4 and 6, which is depicted in Appendix A Table A1. No chronic systemic or immune-mediated diseases were observed in these cats, as indicated by the absence of parasites in feces, urine, and blood. The cats had not received any antibiotics for three months before the experiment and were on the same basic diet for one month prior to the experiment.

Thirty cats were randomly assigned to three treatments with 10 replicates per group while ensuring that five males and five females were included in each treatment. The cats received a basal diet with 0 (CON), 500 (L-CS) or 2000 (H-CS) mg/kg chitosan for a period of 60 days. The diet was formulated to meet the nutrient recommendation of NRC (2006) (Table 1) [19]. The chitosan (S24914) used in this study was obtained from Shanghai Yuanye Biotechnology Co., Ltd. (Shanghai, China) and the deacetylation of chitosan was 85%. The samples of diet and feces were dried at 65 °C for 48 h, then ground through a 40 mesh (425 μm) sieve. According to the previous method [20], the samples of feed were analyzed in terms of crude protein (CP), dry matter (DM), ether extract (EE), and organic matter (OM). Gross energy (GE) was evaluated using an automatic calorimeter (Parr 6400, Moline, IL, USA). The samples of feces were analyzed in terms of fecal water content (FWC). All the samples were measured in duplicate.

Table 1. Ingredients and nutrient levels of basal diet.

Ingredients (As-Fed Basis)	%
Chicken meal	54.6
Fish meal	3.21
Rice	5.35
Poultry fat	9.73
Fish fat	2.15
Potato starch	18.2
Potatoes	4.85
Taurine	0.24
Chicken liver powder	0.60
Salt	0.49
Vitamin premix [1]	0.36
Mineral premix [2]	0.24
Total	100
Analyzed nutrient levels (on a DM basis)	
Dry matter, %	90.4
Total energy, MJ/kg	23.4
Crude protein, %	41.3
Ether extract, %	22.2
Ash, %	8.67

Note: [1] Vitamin premix provided the following per kilogram of feed: vitamin A (15,000 IU), vitamin B_1 (30 mg), vitamin B_2 (28 mg), vitamin B_3 (110 mg), vitamin B_5 (85 mg), vitamin B_6 (12 mg), vitamin B_{12} (0.19 mg), vitamin D_3 (15 IU), and vitamin E (75,300 IU). [2] Mineral premixes provided the following per kilogram of feed: Ca (CaI_2) 20 mg, Co ($CoSO_4$) 0.10 mg, Cu ($CuSO_4$) 3 mg, Fe ($FeSO_4$) 50 mg, I (CaI_2) 40 mg, Mn ($MnSO_4$) 18 mg, Na (Na_2SeO_3) 0.05 mg, Zn ($ZnSO_4$) 38 mg, and Se (Na_2SeO_3) 260 mg.

2.3. Feeding and Sample Collection

The cats were provided *ad libitum* access to water and food throughout the entire experiment. The cat was housed in a cage (1.5 m × 1.5 m × 2.0 m) individually. Humidity and temperature were maintained at 50–60% and 23–26 °C, respectively. The cats were weighed, and their feces were collected on days 0, 30, and 60 to determine FWC, while feces were scored according to the fecal condition score (Appendix A Table A1). The feed supply and feed refusals were recorded daily to calculate the average daily feed intake (ADFI). The water consumption by the cat was also recorded daily to evaluate the average daily water intake (ADWI).

On days 0 and 60, blood samples were drawn from the forelimb veins into vacutainer tubes following 12 h of fasting. All blood samples were centrifuged at 3000× *g* for 30 min at 4 °C, and serum was collected and stored at − 20 °C for further analysis. On day 60, all the feces of the cats were collected immediately following defecation. Fecal samples were thereafter immediately snap-frozen in liquid nitrogen and stored at −80 °C for further examination of the microbial composition and concentration of SCFAs.

2.4. Serum Parameters Measurement

The concentrations of cytokines, such as interleukin-6 (IL-6; DY2305), interleukin-10 (IL-10; DY736), interleukin-1β (IL-1β; DY1796), tumor necrosis factor-α (TNF-α; DY2586), and superoxide dismutase (SOD; DYC3419–2) in the serum were determined using assay kits from R&D Systems (Minneapolis, USA). The levels of immunoglobulin A (IgA; OKIA00194) and malondialdehyde (MDA; OKEH02548) were assessed using the enzyme-linked immunosorbent assay (ELISA) kit from Aviva Systems Biology (San Diego, CA, USA). The concentrations of diamine oxidase (DAO; E-EL-H1241c), lipopolysaccharide (LPS; E-EL-H6108), and D-lactic acid (D-LA; E-BC-K002-M) in the serum were determined using Elabscience (Wuhan, China) ELISA kits. The levels of immunoglobulin G (IgG; FGG91-K01) were assessed using the ELISA kit from Eagle Biosciences (Amherst, NH, USA). All indicators were in accordance with the instructions of the kit manufacturer.

2.5. Hematological Parameters

On days 0 and 60 of the trial, the blood samples were drawn from cats, and their hematological parameters were analyzed using an automated animal blood cell analyzer (Mindray BC2800, Shenzhen, China).

2.6. Determination of Fecal SCFAs

The contents of fecal SCFAs were determined using gas chromatography according to our previous study [21]. These SCFAs include acetate, propionate, butyrate, iso-butyrate, valerate, iso-valerate, and caproate.

2.7. Fecal Microbiota Analysis

The total microbial genomic DNA of the fecal samples was extracted using the Fecal Genomic DNA Extraction Kit (Tiangen Biochemical Technology Co, Beijing, China). The V3-V4 region of the 16S rRNA gene was amplified gene was amplified with primer pairs 338F (5′-ACTCCTACGGGAGGCAGCAG-3′) and 806R(5′-GGACTACHVGGGTWTCTAAT-3′), subsequently pooled into equimolar amounts, and sequenced on the Illumina MiSeq platform to generate paired-end reads of 300 base pairs (bp). The high-quality sequences were clustered into ASVs using DADA2 in QIIME 2. Representative sequences of the 16S rRNA gene were annotated taxonomically against the SILVA 138 database with a confidence threshold of 70%. The raw data was uploaded to the database of the National Centre for Biotechnology Information (NCBI) with Sequence Read Archive (SRA) accession number PRJNA933768.

2.8. Statistical Analysis

Data analysis was performed using IBM SPSS Statistics 25 (Chicago, IL, USA). One-way analysis of variance (ANOVA) and Tukey's test were utilized to determine the difference among groups. $p < 0.05$ was considered statistically significant, and $0.05 \leq p < 0.10$ was considered a tendency. Data were expressed as the mean and pooled standard error of means (SEM). Bar plots were generated using GraphPad Prism 8 (San Diego, CA, USA).

Microbial community analysis was performed using Quantitative Insights into Microbial Ecology 2 software. Only ASVs with a minimum abundance of two reads observed in more than two samples were retained. The differences between the microbiota in the principal coordinate analysis (PCoA) plots were calculated using the Abund-Jaccard distance. Kruskal–Wallis rank sum tests were used to analyze the α and β diversity and relative abundance (RA), as well as to statistically compare differences across various datasets. PCoA based on Abund-Jaccard distance was used to calculate the community similarity across subgroups, and analysis of similarities (ANOSIM) was used to test the differences between groups.

3. Results

3.1. Physical Characteristics

The results of dietary chitosan supplementation on the physical characteristics of cats are depicted in Table 2. On day 60, the FWC and fecal scores of cats in the CON group were higher in comparison to L-CS and H-CS ($p < 0.05$). The ADFI was significantly higher in cats on the H-CS diet in comparison to the CON group ($p < 0.05$). Between the three treatments, no differences were observed in ADWI and BW during the entire experimental period ($p \geq 0.05$).

Table 2. Effect of chitosan on physical characteristics of cats.

Items	CON	L-CS	H-CS	SEM	*p*-Value
BW, kg					
d 0	8.340	8.310	8.326	0.097	0.993
d 30	8.451	8.763	8.656	0.118	0.563
d 60	8.557	8.862	8.811	0.096	0.391
FCW, %					
d 0	78.12	78.51	77.51	0.750	0.878
d 30	76.87	73.33	74.37	0.642	0.064
d 60	75.81 ^{a}	70.04 ^{b}	71.42 ^{b}	0.599	<0.01
Fecal scores					
d 0	5.300	5.300	5.300	0.098	1.000
d 30	5.100	4.300	4.600	0.146	0.074
d 60	4.700 ^{a}	3.100 ^{b}	3.700 ^{b}	0.198	0.001
ADFI, g/d	90.62 ^{b}	94.24 ^{ab}	99.04 ^{a}	1.427	0.048
ADWI, mL/d	137.2	148.7	152.0	2.924	0.092

Note: CON represents basic diet; L-CS, basic diet containing 500 mg/kg chitosan; H-CS, basic diet containing 2000 mg/kg chitosan; SEM, standard error of means; BW, body weight gain; d, day; FWC, fecal water content; ADFI, average daily feed intake; and ADWI, average daily water intake. ^{a,b} Within a row, means without a common superscript differ at $p < 0.05$. Values are presented as means and pooled SEMs, $n = 10$.

3.2. Hematological Parameters

The results pertaining to the hematological parameters of cats are presented in Table 3. On day 0, there was no difference in any of the items among the three groups ($p \geq 0.05$), except that the MCH in the L-CS group was higher in comparison to the CON group ($p < 0.05$). Meanwhile, on day 60, no significant difference was observed in hematological parameters between the three groups ($p \geq 0.05$).

Table 3. Hematological parameters of cats on d 0 and 60 of the experiment.

Items	CON	L-CS	H-CS	SEM	*p*-Value
d 0					
WBC, 10^9/L	9.810	9.210	10.25	0.359	0.510
Lymphocyte, 10^9/L	3.600	3.900	3.780	0.277	0.912
Monocyte, 10^9/L	0.500	0.480	0.410	0.034	0.550
Granulocyte, 10^9/L	6.070	6.470	6.820	0.446	0.801
Lymphocyte, %	30.88	28.18	27.07	1.696	0.657
Monocyte, %	5.080	4.590	4.870	0.280	0.786
Granulocyte, %	51.97	49.63	49.23	2.282	0.877
RBC, 10^{12}/L	7.691	7.275	6.493	0.298	0.258
Hemoglobin, g/L	121.1	118.3	120.6	3.085	0.930
Hematocrit, %	39.41	37.97	37.99	0.968	0.796
MCV, fL	45.40	46.05	45.05	0.578	0.786
MCH, pg	14.31 ^{b}	15.50 ^{a}	15.38 ^{ab}	0.212	0.035
MCHC, g/L	318.7	328.5	331.3	3.466	0.307
RDW, %	16.02	15.51	15.71	0.182	0.532
PLT, 10^9/L	250.6	246.0	238.8	19.80	0.972
MPV, fL	9.600	9.020	8.830	0.285	0.532
PDW	15.92	15.33	14.92	0.181	0.072
Plateletcrit, %	0.282	0.253	0.237	0.024	0.759
Eosinophil, %	2.860	3.710	3.970	0.358	0.431
d 60					
WBC, 10^9/L	9.560	10.37	10.79	0.480	0.584
Lymphocyte, 10^9/L	3.700	3.790	3.100	0.298	0.606
Monocyte, 10^9/L	0.600	0.560	0.570	0.046	0.937
Granulocyte, 10^9/L	6.320	6.720	6.580	0.440	0.936
Lymphocyte, %	31.70	32.48	27.57	1.498	0.368
Monocyte, %	5.330	5.090	5.220	0.250	0.931
Granulocyte, %	49.48	50.85	49.08	1.817	0.922
RBC, 10^{12}/L	7.334	7.369	7.134	0.244	0.919
Hemoglobin, g/L	121.7	120.2	119.6	3.107	0.963
Hematocrit, %	38.46	37.41	38.82	1.053	0.860

Note: CON represents basic diet; L-CS, basic diet containing 500 mg/kg chitosan; H-CS, basic diet containing 2000 mg/kg chitosan; SEM, standard error of means; d, day; WBC, white blood cell; RBC, red blood cell; MCV, mean corpuscular volume; MCH, mean corpuscular hemoglobin; MCHC, mean corpuscular hemoglobin concentration; RDW, red blood cell volume distribution width; PLT, platelet count; MPV, mean platelet volume; and PDW, platelet distribution width. ^{a,b} Within a row, means without a common superscript differ at $p < 0.05$. Values are presented as means and pooled SEMs, $n = 10$.

3.3. Fecal SCFAs

The concentrations of SCFAs in the fecal samples of cats among the three dietary treatments are depicted in Table 4. The concentrations of acetate in fecal samples decreased in the CON group in comparison to the L-CS group and the H-CS group ($p < 0.05$). The levels of butyrate and total SCFAs in the feces of the H-CS group were higher in comparison to the CON group ($p < 0.05$). No significant difference was observed between the contents of propionate, iso-butyrate, valerate, iso-valerate, and caproate in feces among the three dietary treatments ($p \geq 0.05$).

Table 4. Effect of chitosan on fecal SCFAs in cats.

Items (mg/kg)	CON	L-CS	H-CS	SEM	*p*-Value
Acetate	3378 ^{b}	4079 ^{a}	4564 ^{a}	143.0	<0.01
Propionate	2703	2906	3179	209.6	0.665
Butyrate	1885 ^{b}	2298 ^{ab}	2645 ^{a}	100.4	<0.01
Iso-butyrate	217.6	226.1	254.1	19.04	0.729
Valerate	988.4	1037	1012	91.45	0.978
Iso-valerate	467.1	510.7	497.4	39.38	0.905
Caproate	69.81	70.34	77.97	10.72	0.945
Total SCFA	9710 ^{b}	11128 ^{ab}	12229 ^{a}	345.8	<0.01

Note: CON represents basic diet; L-CS, basic diet containing 500 mg/kg chitosan; H-CS, basic diet containing 2000 mg/kg chitosan; SEM, standard error of means, and SCFA, short-chain fatty acid. ^{a,b} Within a row, means without a common superscript differ at $p < 0.05$. Values are presented as means and pooled SEMs, $n = 10$.

3.4. Serum Anti-Oxidative Condition

The results for the antioxidant parameters of cats' serum are illustrated in Figure 1. The two chitosan-supplementation groups had markedly greater serum SOD concentrations than that of the CON group ($p < 0.05$). In addition, the serum SOD concentration in the L-CS group was significantly higher than the H-CS group (Figure 1A, $p < 0.05$). Furthermore, MDA was found to be significantly reduced in the L-CS and H-CS groups in comparison to the CON group (Figure 1B, $p < 0.05$).

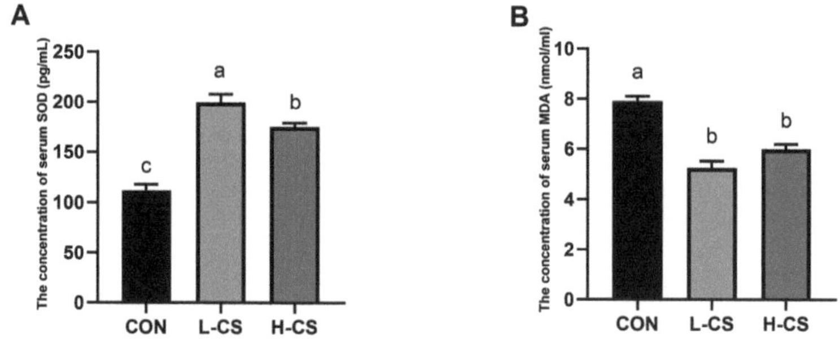

Figure 1. The effects of chitosan on anti-oxidative status in serum of cats: (**A**) Superoxide dismutase (SOD); (**B**) Malonaldehyde (MDA). Note: CON represents basic diet; L-CS, basic diet containing 500 mg/kg chitosan; H-CS, basic diet containing 2000 mg/kg chitosan; and SEM, standard error of means. Values are presented as means ± SEMs, $n = 10$. Different letters on the top of the column represent significant differences ($p < 0.05$).

3.5. Serum Intestinal Barrier Function Parameters

The parameters of serum intestinal barrier function in cats are illustrated in Figure 2. The serum levels of LPS (Figure 2A), DAO (Figure 2B) and D-LA (Figure 2C) were significantly lower in the L-CS and H-CS groups in comparison to the CON group ($p < 0.05$).

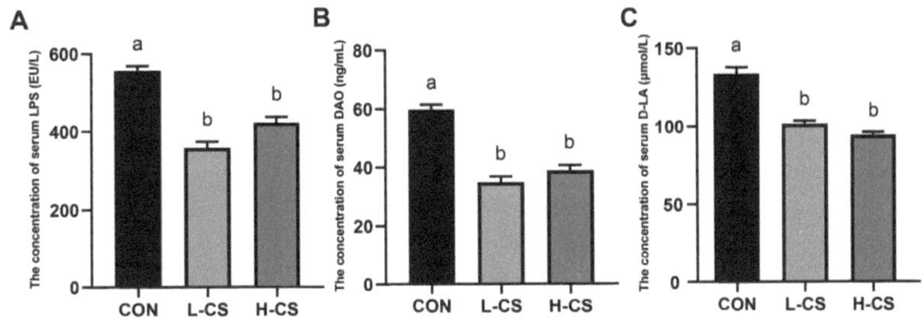

Figure 2. The effects of chitosan on functional parameters in the serum intestinal barrier: (**A**) Lipopolysaccharide (LPS); (**B**) Diamine oxidase (DAO); (**C**) D-lactate (D-LA). Note: CON represents basic diet; L-CS, basic diet containing 500 mg/kg chitosan; H-CS, basic diet containing 2000 mg/kg chitosan; and SEM, standard error of means. Values are presented as means ± SEMs, $n = 10$. Different letters on the top of the column represent significant differences ($p < 0.05$).

3.6. Serum Inflammatory Cytokines and Immunoglobulins

The effects of chitosan on inflammatory cytokines and immunoglobulins in the serum of cats are illustrated in Figure 3. The levels of serum IL-1β (Figure 3A) and IL-6 (Figure 3B) in the L-CS and H-CS groups were lower than in the CON group ($p < 0.05$). Meanwhile, the level of serum IL-10 (Figure 3C) in the H-CS group was higher in comparison to the CON group ($p < 0.05$). As the content of chitosan in the diet increased, the level of TNF-α (Figure 3D) in the serum of cats gradually decreased ($p < 0.05$). The level of serum IgA (Figure 3E) was the lowest in the CON group among the three groups ($p < 0.05$). No difference was observed for IgG (Figure 3F) among the three dietary treatments ($p < 0.05$).

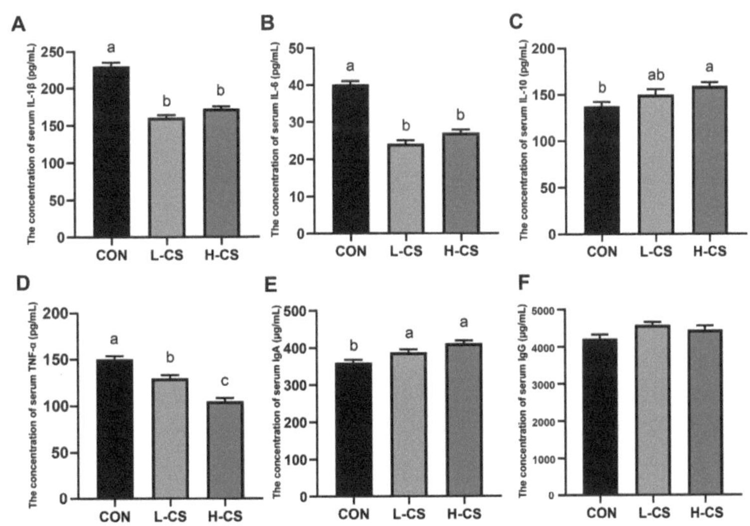

Figure 3. The effects of chitosan on the levels of inflammatory cytokines and immunoglobulins in the serum of cats: (**A**) Interleukin (IL)-1β; (**B**) IL-6; (**C**) IL-10; (**D**) Tumor necrosis factor (TNF)-α; (**E**) Immunoglobulin A (IgA); (**F**) Immunoglobulin G (IgG). Note: CON represents basic diet; L-CS, basic diet containing 500 mg/kg chitosan; H-CS, basic diet containing 2000 mg/kg chitosan; and SEM, standard error of means. Values are presented as means ± SEMs, $n = 10$. Different letters on the top of the column represent significant differences ($p < 0.05$).

3.7. Fecal Microbiota

The fecal microbiota composition of cats is illustrated in Figure 4. No differences were observed among the three groups in Chao (Figure 4A) and Simpson index (Figure 4B).

Furthermore, the beta diversity based on the PCoA analysis resulted in a significant separation of the fecal microbial communities among the three treatment groups (Figure 4C, $p < 0.05$). Meanwhile, the community bar plot at the phylum level revealed that Firmicutes, Actinobacteriota, Bacteroidota, Fusobacteriota, and Proteobacteria were the prevailing bacteria (Figure 4D). On the family level, the most abundant microbes were *Peptostreptococcaceae*, *Lachnospiraceae*, *Coriobacteriaceae*, *Prevotellaceae*, and *Streptococcaceae*. Interestingly, *Peptostreptococcaceae* were more prevalent in the L-CS and H-CS groups in comparison to the CON group, while *Lactobacillaceae* were the predominant bacteria in the CON group. The results of the differences in microbes at the genus level are illustrated in Figure 4F. The CON group demonstrated increased levels of *Lactobacillus* in comparison to the other two groups ($p < 0.05$). The abundance of *Bacillus* in fecal samples was significantly higher, while the abundance of *Alloprevotella* was significantly lower in the CON group in comparison to the L-CS and H-CS groups ($p < 0.05$). With increasing chitosan content in the diet, the abundance of unclassified_f_*Lachnospiraceae* in cat feces increased, whereas the abundance of *Subdoligranulum* decreased ($p < 0.05$). Meanwhile, *Allobaculum* was significantly higher in the H-CS group in comparison to the CON and L-CS groups ($p < 0.05$).

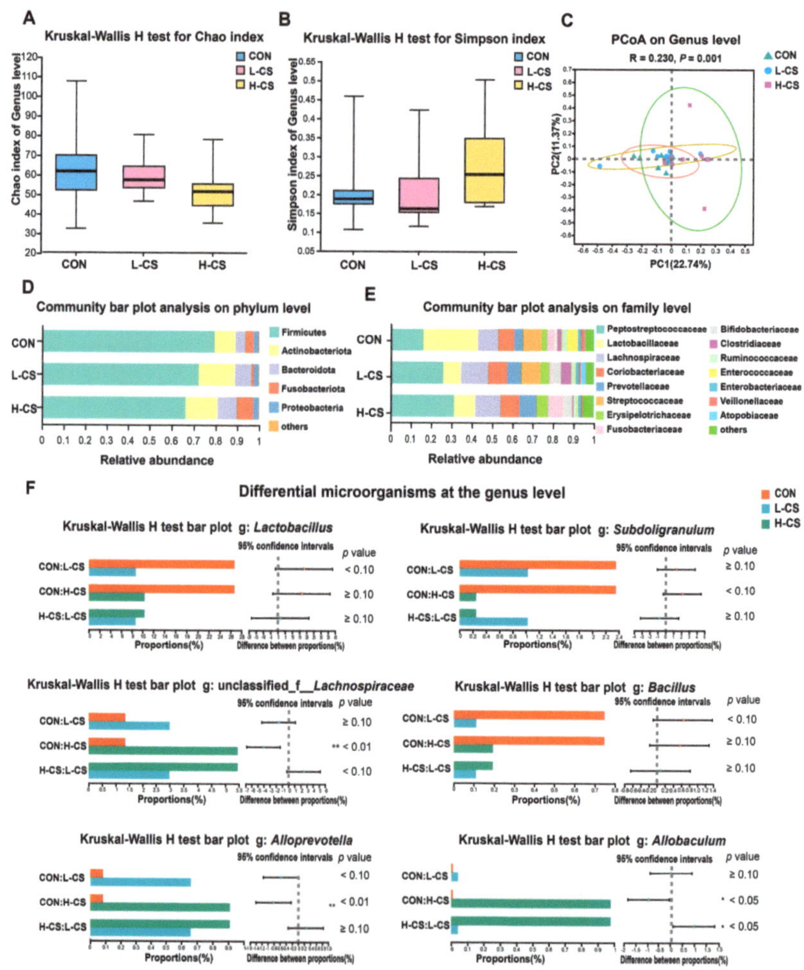

Figure 4. Description of the fecal microbiota composition of cats: (**A**) Comparison of alpha diversity Chao indexes in 3 dietary treatments; (**B**) Comparison of alpha diversity Simpson indexes in 3 dietary

treatments; (**C**) Principal coordinate analysis (PCoA) plots of bacterial communities based on genus level; (**D**, **E**) Community bar plot analysis on phylum and family levels; (**F**) Differential microbes at the genus level in the three groups. Note: CON represents basic diet; L-CS, basic diet containing 500 mg/kg chitosan; H-CS, basic diet containing 2000 mg/kg chitosan; and SEM, standard error of means. Values are presented as means ± SEMs, $n = 10$. Significances are presented as * $p < 0.05$ and ** $p < 0.01$.

4. Discussion

Chitosan is a complex of β-1,4-linked d-glucosamine oligosaccharides that are deacetylated derivatives of chitin and is derived from the cell walls of crustaceans, fungi, and plants [22]. Meanwhile, chitosan is considered to be the only natural polysaccharide with alkaline and cationic characteristics conferring water absorption, antibacterial, anti-inflammatory, antioxidant, and immunological properties [23]. Various studies have assessed the efficacy of chitosan in protecting the intestinal health of animals [24,25], and this major observation is consistent with our current findings. Our results revealed that chitosan exerted beneficial effects in reducing intestinal barrier dysfunction and improving intestinal problems associated with diarrhea in cats.

Diarrhea is considered a representative symptom of impaired intestinal health [26]. In this study, 30 cats with mild diarrhea symptoms were selected, and the addition of chitosan to the diet significantly reduced the fecal water content and alleviated the symptoms in cats. This might be a result of the fact that polar groups, such as hydroxyl and amino groups contained in chitosan, are highly hygroscopic and polymeric [23]. Moreover, as a dietary polysaccharide, we found that the addition of chitosan did not compromise the dietary palatability of cats, and even the ADFI was significantly higher in cats from the H-CS group in comparison to the CON group.

The integrity of the intestinal barrier functions and the homeostasis of the intestinal microecology are important to the health of the organism [27]. The intestinal epithelial barrier is a dynamic and permeable barrier that selectively absorbs nutrients but prevents harmful substances from entering the intestinal lumen [28]. LPS is a potent pro-inflammatory molecule, and when the gut microbiota is disrupted, an excessive LPS is secreted, which could damage the intestinal mucosal barrier and release toxic substances into the bloodstream, resulting in a systemic inflammatory response [29]. D-LA is a metabolite of bacterial metabolism, and the concentration of D-LA in the serum is associated with the degree of intestinal mucosal damage [30]. DAO activity and levels are high in intestinal epithelial cells, but DAO crosses the intestinal epithelial mucosa into the serum when intestinal permeability increases [31]. Therefore, elevated serum DAO levels are a biomarker of increased intestinal permeability. Excessive accumulation of D-LA might have toxic effects on the organism because the host lacks the necessary enzymes to metabolize it [32]. These markers of intestinal barrier function, including LPS, D-LA, and DAO, pass through the intestinal mucosa to enter the circulation in significant amounts during damage to the intestinal barrier and intestinal mucosa [33]. In an earlier study with mice, it was found that the administration of chitosan reduced the levels of serum DAO and D-LA in an LPS-challenge model [34], which was consistent with our findings. On day 60, there were significant decreases in serum LPS, D-LA, and DAO concentrations in L-CS and H-CS groups in comparison to the CON group. Our results demonstrated that chitosan alleviates intestinal epithelial cell damage and reduces intestinal mucosal permeability, thereby protecting the intestinal barrier function of cats.

Impaired intestinal barrier function results in increased intestinal permeability, which exacerbates pathogen invasion of the cells [35]. At this time, the excessive immune response to pathogens increases epithelial apoptosis and decreases connexin expression, aggravating the impairment of intestinal barrier function [35]. Cytokines secreted by immune cells regulate cell function and reflect immunological health and intestinal barrier damage [36]. The increase in pro-inflammatory cytokines and decrease in anti-inflammatory cytokines in peripheral circulation is one of the typical characteristics of intestinal inflammation [37].

TNF-α is a mediator of intestinal inflammation that stimulates the secretion of IL-1β and IL-6 produced by macrophages and dendritic cells in the lamina propria [38]. In contrast, IL-10 plays role in anti-inflammatory effects [39]. IL-10 primarily inhibits nuclear factor kappa-light-chain-enhancer of activated B cell target genes in macrophages, indirectly disrupting interferon regulatory factors in dendritic and mast cells [39]. Chitosan has been demonstrated to reduce the serum concentrations of TNF-α, IL-6, and IL-1β in weaned piglet models [40]. Our results also revealed that the addition of chitosan significantly increased IL-10 and decreased TNF-α, IL-6, and IL-1β in the serum of the L-CS and H-CS groups in comparison to the CON group. IgA could be excreted from the intestine to the blood during inflammatory flare-ups and plays a significant role in suppressing inflammation [41]. In our study, chitosan significantly increased the concentration of serum IgA in cats. Therefore, these results suggested that chitosan could alleviate inflammation in cats triggered by a compromised intestinal barrier.

The release of inflammatory cytokines induces a persistent overproduction of the reactive oxygen species, which results in oxidative stress in the body [42]. The concentrations of SOD and MAD are the prevalent sensitive indicators of antioxidative and pro-oxidative systems [43]. The compromised antioxidant capacity is generally reflected by the decreased SOD and increased MDA [44]. Moreover, chitosan alleviates oxidative stress in rats by reducing MDA levels in the serum [45]. The ability of chitosan to alleviate oxidative stress was also confirmed in another study on severe acute pancreatitis, in which the level of MDA decreased, and SOD increased in the ileum and pancreas of mice following supplementation with chitosan oligosaccharides [46]. Our results in cats also confirmed the effectiveness of the antioxidant properties of chitosan, which was indicated by an increase in SOD level and a decrease in MDA level in serum following chitosan feeding.

The gut microbiota plays an essential role in keeping felines healthy and preventing diseases [7]. The inflammation of the intestine or impaired barrier is connected to dysbiosis of microbiota and the overgrowth of pathogens [1]. Previous research studies have confirmed that the composition of the gut microbiota in cats is strongly correlated with gastrointestinal health problems, such as chronic enteritis and diarrhea [9]. Similar to a previous study [7], our findings revealed that Firmicutes, Actinobacteriota, and Bacteroidota are the most prevalent phyla in the feces of cats. Moreover, following the addition of chitosan, significant differences in microbial composition and structure were observed among the three treatments. A previous study revealed that the addition of chitosan lowered the abundance of *Lactobacillus* in weaned piglets [47]. In addition, during the in vitro fermentation of chitosan by the mouse gut microbiota, *Lactobacillus* was significantly reduced following 8 h chitosan treatment [17]. These results are consistent with our finding that the supplementation of chitosan in the diet reduced the abundance of *Lactobacillus* in the feces of cats. Notably, chitosan has been reported to possess an inhibitory effect on beneficial bacteria, including *Bifidobacterium,* in contrast to traditional prebiotic polysaccharides [17,48,49]. This is probably a result of the inhibitory effect of chitosan on Gram-positive bacteria [23]. The positively charged chitosan could interact electrostatically with negatively charged alginate in the peptidoglycan of Gram-positive bacteria, resulting in cell membrane disruption and leakage of intracellular components, causing bacterial death [50]. Additionally, in this finding, *Allobaculum* was significantly enriched in the H-CS group. *Allobaculum* is an efficient glucose metabolizer and producer of lactate and butyrate, which is also believed to be involved in inflammatory processes and could have a beneficial influence on the intestinal immune responses of the host [51]. In brief, our results found that chitosan affected the composition of intestinal microorganisms in cats, but the mechanism of chitosan action needs to be further investigated.

Intestinal microbes are able to ferment indigestible carbohydrates to produce SCFAs, primarily including acetate, propionate, and butyrate [11]. The beneficial functions of SCFAs, such as reducing the production of pro-inflammatory factors and enhancing the intestinal mucosal barrier, have been extensively reported [52]. Our study revealed that the total fecal SCFAs in the H-CS group was significantly higher in comparison to that in the CON group. Butyrate provides energy to intestinal epithelial cells through fatty acid oxidation, which regulates the health of the intestine and reduces the incidence of diarrhea and enhances the immunity of the host [53]. Furthermore, acetate stimulates an immune response against pathogens and protects intestinal function by increasing IgA secretion [54]. Our study revealed significantly higher levels of fecal acetate and serum IgA in the L-CS and H-CS groups in comparison to the CON group following the addition of chitosan. Meanwhile, the level of butyrate in the feces of cats gradually increased with the addition of increased chitosan, and the level of butyrate in the feces of the H-CS group was significantly higher in comparison to the CON group. *Allobaculum*, which has been reported to be an efficient butyrate producer, is significantly elevated in the H-CS group [55]. These results revealed that dietary chitosan supplementation impacts the SCFAs production-related bacteria, which ameliorated the intestinal barrier dysfunction and intestinal health of cats.

5. Conclusions

In conclusion, our study demonstrated that dietary chitosan supplementation improved mild diarrhea in cats, which may be attributed to the ameliorated inflammatory and redox status as well as the increase in SCFA and beneficial bacteria. Our study provides support for the incorporation of chitosan in pet foods to protect the intestinal health of pet cats.

Author Contributions: Conceptualization, R.M. and Y.W.; methodology, R.M. and Y.W.; software, M.Z. and H.W.; validation, M.Z., T.L. and P.L.; formal analysis, R.M.; investigation P.L.; resources, H.W.; data curation, R.M.; writing—original draft preparation, R.M., Y.W. and M.Z.; writing—review and editing, M.Z. and T.L.; visualization, H.W.; supervision, Y.W.; project administration, Y.W.; funding acquisition, Y.W. All authors have read and agreed to the published version of the manuscript.

Funding: This research received no external funding.

Institutional Review Board Statement: The animal study protocol was approved by the Care and Use of Experimental Animals Committee of the China Agricultural University (AW60203202-1-1).

Informed Consent Statement: Not applicable.

Data Availability Statement: For this study, data are available from the corresponding authors upon request. Data is not publicly available due to privacy.

Acknowledgments: All authors would thank the faculty and staff in the Ministry of Agriculture and Rural Affairs Feed Industry Centre (Beijing, China) for their support of this study.

Conflicts of Interest: The authors declare no conflict of interest.

Appendix A

Table A1. Fecal condition score.

Score	Standard
1	Dry, hard, easy to pick up, difficult to defecate, and forms a ball of dung
2	Forming, not hard, segmented, and easy to pick up
3	Wet, no section, and not easy to pick up
4	Wet, columnar, and difficult to pick up
5	Wet, heaped, and hard to pick up
6	Shapeless and heaped/shoal
7	Watery diarrhea

References

1. Hou, Q.; Huang, J.; Ayansola, H.; Masatoshi, H.; Zhang, B. Intestinal Stem Cells and Immune Cell Relationships: Potential Therapeutic Targets for Inflammatory Bowel Diseases. *Front. Immunol.* **2020**, *11*, 623691. [CrossRef] [PubMed]
2. Carding, S.R.; Davis, N.; Hoyles, L. Review article: The human intestinal virome in health and disease. *Aliment. Pharmacol. Ther.* **2017**, *46*, 800–815. [CrossRef] [PubMed]
3. Fan, Y.; Pedersen, O. Gut microbiota in human metabolic health and disease. *Nat. Rev. Microbiol* **2021**, *19*, 55–71. [CrossRef] [PubMed]
4. Sonnenburg, J.L.; Bäckhed, F. Diet-microbiota interactions as moderators of human metabolism. *Nature* **2016**, *535*, 56–64. [CrossRef]
5. Ley, R.E.; Hamady, M.; Lozupone, C.; Turnbaugh, P.J.; Ramey, R.R.; Bircher, J.S.; Schlegel, M.L.; Tucker, T.A.; Schrenzel, M.D.; Knight, R.; et al. Evolution of mammals and their gut microbes. *Science* **2008**, *320*, 1647–1651. [CrossRef]
6. Moon, C.D.; Young, W.; Maclean, P.H.; Cookson, A.L.; Bermingham, E.N. Metagenomic insights into the roles of Proteobacteria in the gastrointestinal microbiomes of healthy dogs and cats. *Microbiologyopen* **2018**, *7*, e00677. [CrossRef]
7. Alessandri, G.; Argentini, C.; Milani, C.; Turroni, F.; Cristina Ossiprandi, M.; van Sinderen, D.; Ventura, M. Catching a glimpse of the bacterial gut community of companion animals: A canine and feline perspective. *Microb. Biotechnol.* **2020**, *13*, 1708–1732. [CrossRef]
8. Rochus, K.; Janssens, G.P.; Hesta, M. Dietary fibre and the importance of the gut microbiota in feline nutrition: A review. *Nutr. Res. Rev.* **2014**, *27*, 295–307. [CrossRef]
9. Ziese, A.-L.; Suchodolski, J.S. Impact of Changes in Gastrointestinal Microbiota in Canine and Feline Digestive Diseases. *Vet. Clin. N. Am. Small Anim. Pract.* **2021**, *51*, 155–169. [CrossRef]
10. Kil, D.Y.; Swanson, K.S. Companion animals symposium: Role of microbes in canine and feline health. *J. Anim. Sci.* **2011**, *89*, 1498–1505. [CrossRef]
11. Martin-Gallausiaux, C.; Marinelli, L.; Blottière, H.M.; Larraufie, P.; Lapaque, N. SCFA: Mechanisms and functional importance in the gut. *Proc. Nutr. Soc.* **2021**, *80*, 37–49. [CrossRef] [PubMed]
12. Wang, G.; Huang, S.; Wang, Y.; Cai, S.; Yu, H.; Liu, H.; Zeng, X.; Zhang, G.; Qiao, S. Bridging intestinal immunity and gut microbiota by metabolites. *Cell Mol. Life Sci.* **2019**, *76*, 3917–3937. [CrossRef] [PubMed]
13. Cheung, R.C.; Ng, T.B.; Wong, J.H.; Chan, W.Y. Chitosan: An Update on Potential Biomedical and Pharmaceutical Applications. *Mar. Drugs* **2015**, *13*, 5156–5186. [CrossRef]
14. Zhou, D.Y.; Wu, Z.X.; Yin, F.W.; Song, S.; Li, A.; Zhu, B.W.; Yu, L.L. Chitosan and Derivatives: Bioactivities and Application in Foods. *Annu. Rev. Food Sci. Technol.* **2021**, *12*, 407–432. [CrossRef]
15. Tong, A.J.; Hu, R.K.; Wu, L.X.; Lv, X.C.; Li, X.; Zhao, L.N.; Liu, B. Ganoderma polysaccharide and chitosan synergistically ameliorate lipid metabolic disorders and modulate gut microbiota composition in high fat diet-fed golden hamsters. *J. Food Biochem.* **2020**, *44*, e13109. [CrossRef]
16. Liu, W.; Li, X.; Zhao, Z.; Pi, X.; Meng, Y.; Fei, D.; Liu, D.; Wang, X. Effect of chitooligosaccharides on human gut microbiota and antiglycation. *Carbohydr. Polym.* **2020**, *242*, 116413. [CrossRef]
17. Zhang, C.; Jiao, S.; Wang, Z.A.; Du, Y. Exploring Effects of Chitosan Oligosaccharides on Mice Gut Microbiota in in vitro Fermentation and Animal Model. *Front. Microbiol.* **2018**, *9*, 2388. [CrossRef] [PubMed]
18. Wu, X.; Kim, M.J.; Yang, H.J.; Park, S. Chitosan alleviated menopausal symptoms and modulated the gut microbiota in estrogen-deficient rats. *Eur. J. Nutr.* **2021**, *60*, 1907–1919. [CrossRef] [PubMed]
19. National Research Council. *Nutrient Requirements of Dogs and Cats*; The National Academies Press: Washington, DC, USA, 2006; p. 424.
20. Horwitz, W. (Ed.) *Official Methods of Analysis of AOAC International. Volume I, Agricultural Chemicals, Contaminants*; AOAC International: Gaithersburg, MD, USA, 2010.
21. Wu, Y.; Ma, N.; Song, P.; He, T.; Levesque, C.; Bai, Y.; Zhang, A.; Ma, X. Grape Seed Proanthocyanidin Affects Lipid Metabolism via Changing Gut Microflora and Enhancing Propionate Production in Weaned Pigs. *J. Nutr.* **2019**, *149*, 1523–1532. [CrossRef]
22. Thadathil, N.; Velappan, S.P. Recent developments in chitosanase research and its biotechnological applications: A review. *Food Chem.* **2014**, *150*, 392–399. [CrossRef]
23. Yan, D.; Li, Y.; Liu, Y.; Li, N.; Zhang, X.; Yan, C. Antimicrobial Properties of Chitosan and Chitosan Derivatives in the Treatment of Enteric Infections. *Molecules* **2021**, *26*, 7136. [CrossRef] [PubMed]
24. Huang, B.; Xiao, D.; Tan, B.; Xiao, H.; Wang, J.; Yin, J.; Duan, J.; Huang, R.; Yang, C.; Yin, Y. Chitosan Oligosaccharide Reduces Intestinal Inflammation That Involves Calcium-Sensing Receptor (CaSR) Activation in Lipopolysaccharide (LPS)-Challenged Piglets. *J. Agric. Food Chem.* **2016**, *64*, 245–252. [CrossRef] [PubMed]
25. Zhang, X.; Yang, H.; Zheng, J.; Jiang, N.; Sun, G.; Bao, X.; Lin, A.; Liu, H. Chitosan oligosaccharides attenuate loperamide-induced constipation through regulation of gut microbiota in mice. *Carbohydr. Polym.* **2021**, *253*, 117218. [CrossRef] [PubMed]
26. Li, Y.; Xia, S.; Jiang, X.; Feng, C.; Gong, S.; Ma, J.; Fang, Z.; Yin, J.; Yin, Y. Gut Microbiota and Diarrhea: An Updated Review. *Front. Cell Infect. Microbiol.* **2021**, *11*, 625210. [CrossRef] [PubMed]

27. Chelakkot, C.; Ghim, J.; Ryu, S.H. Mechanisms regulating intestinal barrier integrity and its pathological implications. *Exp. Mol. Med.* **2018**, *50*, 1–9. [CrossRef] [PubMed]
28. Luissint, A.C.; Parkos, C.A.; Nusrat, A. Inflammation and the Intestinal Barrier: Leukocyte-Epithelial Cell Interactions, Cell Junction Remodeling, and Mucosal Repair. *Gastroenterology* **2016**, *151*, 616–632. [CrossRef]
29. Mohammad, S.; Thiemermann, C. Role of Metabolic Endotoxemia in Systemic Inflammation and Potential Interventions. *Front. Immunol.* **2021**, *11*. [CrossRef]
30. Zheng, D.; Liao, H.; Chen, S.; Liu, X.; Mao, C.; Zhang, C.; Meng, M.; Wang, Z.; Wang, Y.; Jiang, Q.; et al. Elevated Levels of Circulating Biomarkers Related to Leaky Gut Syndrome and Bacterial Translocation Are Associated with Graves Disease. *Front. Endocrinol.* **2021**, *12*, 796212. [CrossRef]
31. Ji, M.; Huang, H.; Lan, X. Correlation between Intestinal Microflora in Irritable Bowel Syndrome and Severity. *Dis. Markers* **2022**, *2022*, 1031844. [CrossRef]
32. Vella, A.; Farrugia, G. D-lactic acidosis: Pathologic consequence of saprophytism. *Mayo. Clin. Proc.* **1998**, *73*, 451–456. [CrossRef]
33. Banerjee, A. Chapter 15—Gastrointestinal toxicity biomarkers. In *Biomarkers in Toxicology*; Gupta, R.C., Ed.; Academic Press: Boston, MA, USA, 2014; pp. 269–277.
34. Tao, W.; Wang, G.; Pei, X.; Sun, W.; Wang, M. Chitosan Oligosaccharide Attenuates Lipopolysaccharide-Induced Intestinal Barrier Dysfunction through Suppressing the Inflammatory Response and Oxidative Stress in Mice. *Antioxidants* **2022**, *11*, 1384. [CrossRef] [PubMed]
35. John, L.J.; Fromm, M.; Schulzke, J.D. Epithelial barriers in intestinal inflammation. *Antioxid. Redox Signal.* **2011**, *15*, 1255–1270. [CrossRef] [PubMed]
36. Neurath, M.F. Targeting immune cell circuits and trafficking in inflammatory bowel disease. *Nat. Immunol.* **2019**, *20*, 970–979. [CrossRef] [PubMed]
37. Turner, M.D.; Nedjai, B.; Hurst, T.; Pennington, D.J. Cytokines and chemokines: At the crossroads of cell signalling and inflammatory disease. *Biochim. Biophys. Acta* **2014**, *1843*, 2563–2582. [CrossRef]
38. Bamias, G.; Cominelli, F. Cytokines and intestinal inflammation. *Curr. Opin. Gastroenterol.* **2016**, *32*, 437–442. [CrossRef] [PubMed]
39. Saraiva, M.; Vieira, P.; OGarra, A. Biology and therapeutic potential of interleukin-10. *J. Exp. Med.* **2020**, *217*, 20190418. [CrossRef]
40. Na, K.; Wei, J.; Zhang, L.; Fang, Y.; Li, X.; Lu, S.; Guo, X. Effects of chitosan oligosaccharides (COS) and FMT from COS-dosed mice on intestinal barrier function and cell apoptosis. *Carbohydr. Polym.* **2022**, *297*, 120043. [CrossRef]
41. Rojas, O.L.; Pröbstel, A.K.; Porfilio, E.A.; Wang, A.A.; Charabati, M.; Sun, T.; Lee, D.S.W.; Galicia, G.; Ramaglia, V.; Ward, L.A.; et al. Recirculating Intestinal IgA-Producing Cells Regulate Neuroinflammation via IL-10. *Cell* **2019**, *176*, 610–624.E18. [CrossRef]
42. Strober, W.; Fuss, I.J. Proinflammatory cytokines in the pathogenesis of inflammatory bowel diseases. *Gastroenterology* **2011**, *140*, 1756–1767. [CrossRef]
43. Zheng, F.; Wang, Y.Z.; Wu, Y.X.; Zhang, M.Y.; Li, F.T.; He, Y.; Wen, L.K.; Yue, H. Effect of stabilization malvids anthocyanins on the gut microbiota in mice with oxidative stress. *J. Food Biochem.* **2021**, *45*, 4892–4902. [CrossRef]
44. Todorova, I.; Simeonova, G.; Kyuchukova, D.; Dinev, D.; Gadjeva, V. Reference values of oxidative stress parameters (MDA, SOD, CAT) in dogs and cats. *Comp. Clin. Pathol.* **2005**, *13*, 190–194. [CrossRef]
45. Li, X.; Wu, X.; Jia, S.; Zhang, J.; Dai, X.; Sun, Y. Effect of Chitosan Oligosaccharide on Learning and Memory Functions and Serum Antioxidant Status in a Rat Model of Amyloid-beta1-42-Induced Alzheimer's Disease. *Food Sci.* **2017**, *38*, 220–225.
46. Mei, Q.X.; Hu, J.H.; Huang, Z.H.; Fan, J.J.; Huang, C.L.; Lu, Y.Y.; Wang, X.P.; Zeng, Y. Pretreatment with chitosan oligosaccharides attenuate experimental severe acute pancreatitis via inhibiting oxidative stress and modulating intestinal homeostasis. *Acta Pharmacol. Sin.* **2021**, *42*, 942–953. [CrossRef] [PubMed]
47. Yang, C.M.; Ferket, P.R.; Hong, Q.H.; Zhou, J.; Cao, G.T.; Zhou, L.; Chen, A.G. Effect of chito-oligosaccharide on growth performance, intestinal barrier function, intestinal morphology and cecal microflora in weaned pigs. *J. Anim. Sci.* **2012**, *90*, 2671–2676. [CrossRef] [PubMed]
48. Yu, T.; Wang, Y.; Chen, S.; Hu, M.; Wang, Z.; Wu, G.; Ma, X.; Chen, Z.; Zheng, C. Low-Molecular-Weight Chitosan Supplementation Increases the Population of Prevotella in the Cecal Contents of Weanling Pigs. *Front. Microbiol.* **2017**, *8*, 2182. [CrossRef] [PubMed]
49. Mateos-Aparicio, I.; Mengíbar, M.; Heras, A. Effect of chito-oligosaccharides over human faecal microbiota during fermentation in batch cultures. *Carbohydr. Polym.* **2016**, *137*, 617–624. [CrossRef]
50. Matica, M.A.; Aachmann, F.L.; Tøndervik, A.; Sletta, H.; Ostafe, V. Chitosan as a Wound Dressing Starting Material: Antimicrobial Properties and Mode of Action. *Int. J. Mol. Sci.* **2019**, *20*, 5889. [CrossRef]
51. van Muijlwijk, G.H.; van Mierlo, G.; Jansen, P.; Vermeulen, M.; Bleumink-Pluym, N.M.C.; Palm, N.W.; van Putten, J.P.M.; de Zoete, M.R. Identification of Allobaculum mucolyticum as a novel human intestinal mucin degrader. *Gut. Microbes* **2021**, *13*, 1966278. [CrossRef]
52. Sivaprakasam, S.; Prasad, P.D.; Singh, N. Benefits of short-chain fatty acids and their receptors in inflammation and carcinogenesis. *Pharmacol. Ther.* **2016**, *164*, 144–151. [CrossRef]
53. Fu, X.; Liu, Z.; Zhu, C.; Mou, H.; Kong, Q. Nondigestible carbohydrates, butyrate, and butyrate-producing bacteria. *Crit. Rev. Food Sci. Nutr.* **2019**, *59*, S130–S152. [CrossRef]

54. Takeuchi, T.; Miyauchi, E.; Kanaya, T.; Kato, T.; Nakanishi, Y.; Watanabe, T.; Kitami, T.; Taida, T.; Sasaki, T.; Negishi, H.; et al. Acetate differentially regulates IgA reactivity to commensal bacteria. *Nature* **2021**, *595*, 560–564. [CrossRef] [PubMed]
55. Greetham, H.L.; Gibson, G.R.; Giffard, C.; Hippe, H.; Merkhoffer, B.; Steiner, U.; Falsen, E.; Collins, M.D. *Allobaculum stercoricanis* gen. nov., sp. nov., isolated from canine feces. *Anaerobe* **2004**, *10*, 301–307. [CrossRef] [PubMed]

metabolites MDPI

Article

Different Diet Energy Levels Alter Body Condition, Glucolipid Metabolism, Fecal Microbiota and Metabolites in Adult Beagle Dogs

Haoran Sun [1], Qiaoru Zhang [1,2], Chao Xu [1], Aipeng Mao [1,2], Hui Zhao [1], Miao Chen [1], Weili Sun [1,3], Guangyu Li [1,3,*] and Tietao Zhang [1,*]

1 Institute of Special Animal and Plant Sciences, Chinese Academy of Agricultural Sciences, Changchun 130112, China; solomonsunhr@outlook.com (H.S.); xuchao@caas.cn (C.X.); zhaohui02@caas.cn (H.Z.); sunweili@caas.cn (W.S.)
2 College of Animal Science and Technology, Hebei Normal University of Science and Technology, Qinhuangdao 066004, China
3 College of Animal Science and Technology, Qingdao Agricultural University, Qingdao 266000, China
* Correspondence: tcslgy@126.com (G.L.); zhangtietao@caas.cn (T.Z.)

Abstract: Diet energy is a key component of pet food, but it is usually ignored during pet food development and pet owners also have limited knowledge of its importance. This study aimed to explore the effect of diet energy on the body condition, glucolipid metabolism, fecal microbiota and metabolites of adult beagles and analyze the relation between diet and host and gut microbiota. Eighteen healthy adult neutered male beagles were selected and randomly divided into three groups. Diets were formulated with three metabolizable energy (ME) levels: the low-energy (Le) group consumed a diet of 13.88 MJ/kg ME; the medium-energy (Me) group consumed a diet of 15.04 MJ/kg ME; and the high-energy (He) group consumed a diet of 17.05 MJ/kg ME. Moreover, the protein content of all these three diets was 29%. The experiment lasted 10 weeks, with a two-week acclimation period and an eight-week test phase. Body weight, body condition score (BCS), muscle condition score (MCS) and body fat index (BFI) decreased in the Le group, and the changes in these factors in the Le group were significantly higher than in the other groups ($p < 0.05$). The serum glucose and lipid levels of the Le and He groups changed over time ($p < 0.05$), but those of the Me group were stable ($p > 0.05$). The fecal pH of the Le and He groups decreased at the end of the trial ($p < 0.05$) and we found that the profiles of short-chain fatty acids (SCFAs) and bile acids (BAs) changed greatly, especially secondary BAs ($p < 0.05$). As SCFAs and secondary BAs are metabolites of the gut microbiota, the fecal microbiota was also measured. Fecal 16S rRNA gene sequencing found that the Me group had higher α-diversity indices ($p < 0.05$). The Me group had notably higher levels of gut probiotics, such as *Faecalibacterium prausnitzii*, *Bacteroides plebeius* and *Blautia producta* ($p < 0.05$). The diet–host–fecal microbiota interactions were determined by network analysis, and fecal metabolites may help to determine the best physical condition of dogs, assisting pet food development. Overall, feeding dogs low- or high-energy diets was harmful for glucostasis and promoted the relative abundance of pathogenic bacteria in the gut, while a medium-energy diet maintained an ideal body condition. We concluded that dogs that are fed a low-energy diet for an extended period may become lean and lose muscle mass, but diets with low energy levels and 29% protein may not supply enough protein for dogs losing weight.

Keywords: diet energy; pet food; beagle dog; obesity; glucolipid metabolism; fecal microbiota; metabolomics; SCFAs; bile acids

Citation: Sun, H.; Zhang, Q.; Xu, C.; Mao, A.; Zhao, H.; Chen, M.; Sun, W.; Li, G.; Zhang, T. Different Diet Energy Levels Alter Body Condition, Glucolipid Metabolism, Fecal Microbiota and Metabolites in Adult Beagle Dogs. *Metabolites* **2023**, *13*, 554. https://doi.org/10.3390/metabo13040554

Academic Editors: Lian Li, Baichuan Deng and Giulia Pignataro

Received: 9 March 2023
Revised: 31 March 2023
Accepted: 7 April 2023
Published: 13 April 2023

1. Introduction

Obesity is the most common nutritional disease in domestic dogs and cats [1–3] and there are a variety of commercial veterinary diets in the pet food market and animal practice targeting this issue. The energy levels and nutritional compositions of veterinary diets for weight loss vary considerably. The metabolizable energy (ME) level ranges from 11.41 to 16.22 MJ/kg for dry diets and from 2.23 to 4.19 MJ/kg for canned diets [4]. The protein content ranges from 24.5 to 33.82 g/MJ ME, fat from 6.88 to 10.70 g/MJ ME [5] and dry mater (DM)-based total dietary fiber from 7.7 to 42.6% [6]. Selecting an effective weight loss diet for a pet is therefore difficult for pet owners and veterinarians [7]. The core of a therapeutic weight management diet is to restrict energy intake and to avoid potential nutritional deficiencies that may occur with energy restriction [8,9]. The diversity of the nutritional composition of weight loss diets may result from the limited knowledge of the interactions between diet energy and body metabolism [10–12].

Recent research on canine obesity has focused on validating the results of weight loss diets and the metabolism characteristics of high-fat-induced obesity. In two studies, obese or overweight dogs with restricted energy intake showed decreased body weight and body condition scores (BCS) over two months [2,13]. Methods of energy restriction include changing the macronutrient composition, such as high-protein high-fiber diets [14–17] or high-protein medium-carbohydrate diets [18], and the use of nutritional additives [19,20] or high fiber or polysaccharides [21–24]. During high-fat-diet-induced canine obesity modeling, research showed changes in glucose homeostasis, the gut microbiota and the metabolome [25–29].

Carbohydrates and fat are two of the main energy substances that are important in losing weight, and there has been some research in recent years on the effects of energy substances on glucolipid metabolism and the fecal microbiota [30,31]. The interactions between diet, host and fecal microbiota are also poorly explored in canine nutrition research. If the diet energy level changes, then the host and gut microbiota metabolism will also change and have an interaction. This study therefore aimed to investigate the long-term effects of the diet energy level on body condition, glucolipid metabolism, fecal microbiota and metabolites in beagles, reveal its possible health effects and interactions and lay a foundation for developing canine weight loss diets.

2. Materials and Methods

The experimental procedures for animal trials were approved by the Animal Ethics Committee of the Chinese Academy of Agricultural Sciences and performed according to the guidelines for animal experiments set by the National Institute of Animal Health.

2.1. Animals and Experimental Design

The experiment lasted 10 weeks, with a two-week acclimation period and an eight-week test phase, and was conducted at the Companion Animal Test Base, in Zuojia, Jilin, China (125°19′50″ N 43°49′46″ E; elevation 313 m). The average temperature was 21.6 °C, ranging from 12 °C to 29 °C, and the average humidity was 63%, ranging from 44% to 77% in the test phase.

Eighteen healthy adult neutered male beagles aged 1.33 ± 0.47 years, with mean body weight of 12.52 ± 2.09 kg and body condition scores (BCS) of 3.89 ± 0.99, were randomly assigned to three groups, based on the test diet. Individual groups were maintained in the same semi-closed paddock (5 × 7 m), and every dog was placed in an individual 1.2 × 1 × 1 m cage when feeding. All dogs were vaccinated and dewormed before the study.

The diets were prepared as dry kibble and were formulated to meet the nutrient requirements of adult dogs at maintenance [32], with three different energy levels, shown in Table 1. All dogs were fed the same medium-energy diet with 15.04 MJ/kg ME in the two-week acclimation period to remove the effects of the previous dog food and then randomly allotted to one of the three dry diet groups: the low-energy diet (Le) group, with dietary ME of 13.88 MJ/kg; the medium-energy diet (Me) group, with dietary ME of 15.04 MJ/kg; and the high-energy (He) group, with dietary ME of 17.05 MJ/kg. The three diets were designed with different levels of fat and nitrogen-free extracts (NEF), but the same content of protein.

Table 1. Ingredients and chemical compositions of experimental diets.

Item	Treatment		
	Le	Me	He
Ingredient		% as-is	
Corn	39.2	31.8	21.0
Corn gluten meal	7.0	8.1	9.2
Chicken fat	-	6.3	16.0
Broken rice	11	11	11
Beef meal	6.5	6.5	6.5
Bean pulp	6	6	6
Wheat shorts	5	5	5
Beet pulp	5	5	5
Poultry meal	5	5	5
Beef and bone meal	3	3	3
Vitamin and mineral premix	3	3	3
Liquid palatant	3	3	3
Hydrolyzed spray-dried chicken blood cells	2	2	2
Beer yeast powder	2	2	2
$CaHPO_4$	1.3	1.3	1.3
Chicken liver meal	1	1	1
Analyzed composition			
Dry matter (DM), %	93.85	93.89	94.25
		% DM	
Crude protein	29.99	28.93	29.06
Crude fat	4.69	10.22	19.74
Nitrogen-free extracts [1]	53.38	48.91	39.40
Crude fiber	4.18	4.10	4.55
Ash	7.76	7.83	7.26
Calcium	1.04	1.08	1.02
Phosphorus	0.81	0.78	0.80
Lysine	1.29	1.18	1.21
GE, MJ/kg [2]	18.28	19.46	21.23
ME, MJ/kg [2]	13.88	15.04	17.05

[1] Nitrogen-free extracts = 100 − (ash + crude protein + crude fat + crude fiber). [2] Gross energy (GE) was measured by bomb calorimetry; metabolizable energy (ME) = 8.5 kcal ME/g fat + 3.5 kcal ME/g protein + 3.5 kcal ME/g nitrogen-free extracts; 1 kcal = 4.184 kJ.

All dogs were individually fed 32 g/BW kg/d twice a day at 9 a.m. and 4 p.m. and had free access to fresh water. We performed multiple measurements: fasting serum biochemistry analysis; fasting weighing; BCS (using a nine-point system, where 1–3 indicates a less than ideal body condition, 4–5 indicates an ideal body condition and 6–9 indicates an overly ideal body condition) [33]; muscle condition score (MCS) (using a four-point system, where a score of four means normal muscle mass, three indicates mild muscle loss, two is moderate muscle loss and one is severe muscle loss) [34]; body fat index (BFI) (using a BFI risk chart, where a score of 20 (16–25% body fat) = low risk, 30 (26–35% body fat) = moderate risk, 40 (36–45% body fat) = high risk, 50 (46–55% body fat) = serious risk, 60 (56–65% body fat) = severe risk, 70 (66–75% body fat) = extreme risk) [35]; and fecal pH. On day zero of the trial (T0) and on the last day of the trial (T8), all these measurements were analyzed, and fresh feces were collected from the rectum.

Fecal samples were kept at −80 °C. During the trial, weighing was carried out once every two weeks and serum biochemistry analysis once every four weeks.

2.2. Serum Biochemistry Analysis

Blood samples were taken on T0, T4 and T8 after overnight fasting. After centrifugation, the supernatants were collected as serum samples to measure glucose (GLU), total cholesterol (CHO), triglyceride (TG), low-density lipoprotein cholesterol (LDL-C) and high-density lipoprotein cholesterol (HDL-C) using an automatic Vitalab Selectra E biochemical analyzer (Vital Scientific, Spankeren, The Netherlands).

2.3. Fecal pH Analysis

On T0 and T8, fresh fecal pH was immediately measured after 10% fecal suspension (*w*/*v*) in ultrapure water with a PHS-3C pH meter (INESA Scientific Instrument Co., Ltd., Shanghai, China).

2.4. Content Analysis of SCFAs

Fecal samples collected at T0 and T8 were homogenated for 1 min with 500 μL of water and 100 mg of glass beads and then centrifuged at 4 °C for 10 min at 12,000 rpm. Then, 200 μL supernatant was extracted with 100 μL of 15% phosphoric acid and 20 μL of 375 μg/mL 4-methyl valeric acid solution as the internal standard and 280 μL ether. The samples were then centrifuged at 4 °C for 10 min at 12,000 rpm after vortexing for 1 min, and the supernatant was transferred into the vial before gas chromatography–mass spectrometer (GC-MS) analysis [36–38]. SCFA content was analyzed at Shanghai Personal Biotechnology Co., Ltd. (Shanghai, China).

All target standards were purchased from Sigma-Aldrich (St. Louis, MO, USA). The GC analysis was performed on a trace 1310 gas chromatograph (Thermo Fisher Scientific, Waltham, MA, USA). The GC was fitted with a HP-Innowax 30 m × 0.25 mm ID × 0.25 μm capillary column (Agilent, Santa Clara, CA, USA) and helium was used as the carrier gas at 1 mL/min. The injection was performed in split mode at 10:1, with an injection volume of 1 μL and an injector temperature of 250 °C. The temperatures of the ion source and MS transfer line were 300 °C and 250 °C, respectively. The column temperature was programmed to increase from an initial temperature of 90 °C, followed by an increase to 120 °C at 10 °C per min, then to 150 °C at 5 °C per min and finally to 250 °C at 25 °C per min, which was maintained for 2 min. Mass spectrometric detection of metabolites was performed on an ISQ 7000 (Thermo Fisher Scientific, Waltham, MA, USA) in electron impact ionization mode. Single ion monitoring (SIM) mode was used with the electron energy of 70 eV. Six SCFAs were measured, including acetic acid, propionic acid, isobutyric acid, butyric acid, isovaleric acid and valeric acid.

2.5. Content Analysis of BAs

The protocol for quantifying bile acids was adapted and modified from the methods previously described by Bhargava et al. [39]. Fifty mg samples were extracted in 400 μL of methanol at −20 °C, with 100 mg of glass beads, vortexed for 60 s. They were then placed into a tissue grinder at 55 Hz for 1 min and this was repeated at least twice. The sample was next sonicated for 30 min at room temperature, centrifuged at 12,000 rpm and 4 °C for 10 min, and 300 μL of the supernatant was mixed with 600 μL of water and vortexed for 30 s. Then, 10 μL supernatant was added to 30% methanol and diluted 10 times. The supernatant was filtered through a 0.22 μm membrane and the filtrate was added to the LC-MS bottle. BA content was analyzed at Shanghai Personal Biotechnology Co., Ltd. (Shanghai, China).

The target standards were purchased from Sigma-Aldrich (Shanghai, China), Macklin (Shanghai, China), Shyuanye (Shanghai, China), Toronto Research Chemicals (Toronto, ON, Canada), Zzsrm (Shanghai, China) and the National Institutes for Food and Drug Control (Beijing, China). The supernatant was analyzed by gas chromatography and mass spectrometry according to the methods described by Hu et al. [40] and Yang et al. [41]. A 2.1 × 100 mm, 1.7 μm Acquity UPLC® BEH C18 column (Waters, Milford, MA, USA) was used, the injection volume was 5 μL, the column temperature was 40 °C and the mobile phase was A-0.01% formic acid water and B-acetonitrile. The gradient elution conditions were 0–9 min, 30% B, 9–14 min, 30–36% B, at a flow rate of 0.3 mL/min, 14–18 min, 36–38% B; 18–24 min, 38% to 50% B; 24–32 min, 50–75% B; 32 to 33 min, 75–90% B; 33–35.5 min and 90–30% B at a flow rate of 0.25 mL/min. The mass spectrophotometer used an electrospray ionization (ESI) source, in negative ionization mode. The ion source temperature was 500 °C, the ion source voltage was −4500 V, the collision gas was 6 psi, the curtain gas was 30 psi and the atomizing gas and auxiliary gas were both 50 psi. Scans were performed using multiple reaction monitoring (MRM). Thirty-nine bile acids were measured, including deoxycholic acid (DCA), ursocholic acid (UCA), allocholic acid (ACA), glycohyodeoxycholic acid (GHDCA), chenodeoxycholic acid-3-β-D-glucuronide (CDCA-G), dehydrocholic acid (DHCA), taurodeoxycholic acid sodium salt (TDCA), taurochenodeoxycholic acid (TCDCA), lithocholic acid (LCA), chenodeoxycholic acid (CDCA), ursodeoxycholic acid (UDCA), cholic acid (CA), glycochenodeoxycholic acid sodium salt (GCDCA), glycodeoxycholic acid sodium salt, (GDCA), sodium glycocholate hydrate (GCA), taurocholic acid sodium salt (TCA), taurohyodeoxycholic acid sodium salt (THDCA), allolithocholic acid (isoLCA), isolithocholic acid (alloLCA), 23-nordeoxycholic acid (NorDCA), 6-ketolithocholic acid acetate, (6-ketoLCA), 12-ketolithocholic acid (12-ketoLCA), 7-ketolithocholic acid (7-ketoLCA), 3β-ursodeoxycholic acid (β-UDCA), hyodeoxycholic acid, (HDCA), norcholic acid (NorCA), 7,12-diketolithocholic acid (7,12-diketoLCA), 6,7-diketolithocholic acid (6,7-diketoLCA), α-muricholic acid (α-MCA), β-muricholic acid (β-MCA), 3β-cholic acid (β-CA), glycolithocholic acid sodium salt (GLCA), glycoursodeoxycholic acid (GUDCA), lithocholic acid 3-sulfate sodium salt (LCA-3S), taurolithocholic acid sodium salt (TLCA), tauro-α-muricholic acid sodium salt (T-α-MCA), taurohyocholic acid sodium salt (THCA), tauro-β-muricholic acid sodium salt (T-β-MCA) and tauroursodeoxycholic acid (TUDCA).

2.6. Microbial Analysis

2.6.1. DNA Extraction, Amplification and Sequencing

A 0.2 g fecal sample collected at T0 and T8 was stored in dry ice and sent to Shanghai Personal Biotechnology Co., Ltd. for 16S rRNA detection of the intestinal microbiome. Total genomic DNA samples were extracted using the M5635-02 soil DNA kit (Omega Bio-Tek, Norcross, GA, USA), following the manufacturer's instructions, and stored at −20 °C before further analysis. The quantity and quality of extracted DNA were measured using a NanoDrop NC2000 spectrophotometer (Thermo Fisher Scientific, Waltham, MA, USA) and agarose gel electrophoresis, respectively.

Polymerase chain reaction (PCR) amplification of the bacterial 16S rRNA genes V3–V4 region was performed using the primers 5′-AYTGGGYDTAAAGNG-3′ (520F) and 5′-TACNVGGGTATCTAATCC-3′ (802R). Paired-end sequencing was used in the Illumina NovaSeq platform with the NovaSeq 6000 SP Reagent Kit (San Diego, CA, USA) at Shanghai Personal Biotechnology Co., Ltd. (Shanghai, China).

2.6.2. Bioinformatics Analysis

Microbiome bioinformatics was performed with QIIME2(2019.4) [42] with slight modifications according to the official tutorials https://docs.qiime2.org/2019.4/tutorials/ (accessed on 1 March 2023). Sequences were then quality-filtered, denoised, merged and chimera-removed using the DADA2 plugin [43]. Non-singleton amplicon sequence variants (ASVs) were aligned with MAFFT [44] and used to construct a phylogeny with FastTree2 [45]. Alpha-diversity and beta-diversity metrics were estimated using the diversity plugin with samples rarefied to 97% sequences per sample. Taxonomy was assigned to ASVs using the classify-sklearn naïve Bayes taxonomy classifier in the feature-classifier plugin [46] against the SILVA Release 132 Database [47].

Sequence data analyses were mainly performed using QIIME2 (2019.4) and R packages (v3.2.0). A heat map showed the top 20 genera in terms of relative abundance, and it was constructed using the R language and PheatMap package. A Venn diagram was generated to visualize the shared and unique ASVs among samples or groups using the R package VennDiagram, based on the occurrence of ASVs across groups regardless of their relative abundance [48]. Linear discriminant analysis effect size (LEfSe) was performed to detect differentially abundant taxa across groups using the default parameters [49]. Random forest analysis was applied to discriminate the samples from different groups using QIIME2 with default settings, and the target variable was diet [50,51]. Microbial functions were predicted using the phylogenetic investigation of communities by reconstruction of unobserved states (PICRUSt2) [52] based on the Kyoto Encyclopedia of Genes and Genomes (KEGG) at https://www.kegg.jp (accessed on 1 March 2023).

2.7. Statistical Analysis

Statistical analysis was performed using GraphPad Prism 9.5 (GraphPad, San Diego, CA). Data are presented as the mean ± SEM (standard error of mean). Statistical software SPSS 19.0 was used for one-way or two-way with time and treatment interaction analysis of variance (ANOVA) and the Turkey postmortem test was used to compare the differences between the groups with the three diet energy levels. The fixed variable was diet and the random variable was dog. $p < 0.05$ was considered significant.

3. Results

3.1. Changes in Body Weight and Body Condition of Beagles with Different Diet Energy Levels

During the eight-week experiment, the body weight (BW) of dogs was as shown in Figure 1A and Table 2. As the BW of the Le group decreased rapidly, the average daily gain (ADG) was −44.05 g/d, and the other two groups were significantly higher than Le ($p < 0.05$) and over zero, as seen in Table 3. The BCS, MCS and BFI of the Le group at T8 were significantly lower than those of the Me and He groups ($p < 0.05$), and changes in body condition items showed a linear relation with diet energy ($p < 0.05$), as seen in Figure 1B–D and Table 3.

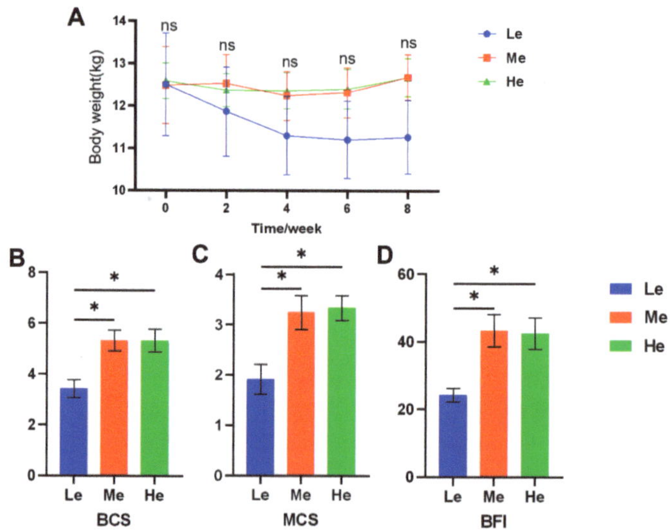

Figure 1. Effects of diet energy level on body weight in adult beagles (**A**); body condition score (BCS), muscle condition score (MCS) and body fat index (BFI) at the end of the experiment (**B**–**D**). Data are shown as mean ± SEM. * $p < 0.05$.

Table 2. Effects of diet energy levels on body weight (kg).

Time	Le	Me	He	SEM	*p*-Value
T0	12.50	12.48	12.58	0.49	0.997
T2	11.87	12.53	12.37	0.42	0.815
T4	11.30	12.24	12.36	0.39	0.500
T6	11.20	12.32	12.40	0.39	0.409
T8	11.27	12.68	12.68	0.38	0.233

Table 3. Changes in body weight and body condition of adult beagles.

Item [1]	Energy Levels			SEM	*p*-Value		
	Le	Me	He		Energy	Liner	Quadratic
ADG (g/d)	-44.05^{a}	7.14^{b}	3.27^{b}	8.68	0.016	0.015	0.084
ΔBCS [2]	-0.25^{a}	1.00^{b}	1.67^{b}	1.20	0.014	0.004	0.565
ΔMCS [2]	-0.58^{a}	0.25^{ab}	0.83^{b}	0.93	0.024	0.007	0.757
ΔBFI [2]	-0.83^{a}	6.67^{ab}	12.50^{b}	8.43	0.017	0.005	0.816

[1] ADG = average daily gain; BCS = body condition score, MCS = muscle condition score and BFI = body fat index. [2] Changes in BCS, MCS and BFI between T8 and T0. ^{a,b} Within a row, means with different superscript letters were significantly different ($p < 0.05$).

3.2. Changes in Blood Glucose and Blood Lipid Levels of Beagles with Different Diet Energy Levels

As shown in Table 4, the CHO content of the He group and the TG content of the Le group in serum increased significantly ($p < 0.05$) and the serum CHO of eight beagles in the Me and He groups was more than the normal reference range in the eight-week trial, compared with only one beagle in the Le group.

Table 4. Effects of diet energy levels on serum glucose and lipid levels.

Item		GLU [1]	CHO [1]	TG [1]	LDL-C [1]	HDL-C [1]	LDL-C% [1]	HDL-C% [1]
Unit		mmol/L	mmol/L	mmol/L	mmol/L	mmol/L	%	%
Reference Values [53]		3.61–6.55	3.50–6.99	0.20–1.30	-	-	-	-
T0	Le	4.89^{AB}	5.44	0.62^{A}	0.39	4.18	7.13^{A}	77.03^{B}
	Me	4.96	5.84	0.67	0.43	4.36	7.15^{A}	75.80
	He	5.34^{A}	5.44^{A}	0.62	0.38^{A}	4.28	6.82^{A}	79.27^{B}
T4	Le	5.59^{B}	6.09	0.70^{AB}	0.54	4.13	8.88^{AB}	68.34^{A}
	Me	5.64	7.01	0.69	0.66	4.60	9.28^{B}	66.40
	He	6.41^{B}	6.97^{B}	0.84	0.62^{B}	4.82	8.89^{B}	69.45^{A}
T8	Le	4.45^{A}	5.43	0.88^{B}	0.56	3.74^{a}	10.50^{B}	69.28^{A}
	Me	4.69	6.85	0.65	0.73	4.28^{ab}	10.17^{B}	65.02
	He	4.95^{A}	6.70^{B}	0.72	0.62^{B}	4.82^{b}	9.21^{B}	72.58^{AB}
SEM		0.11	0.18	0.03	0.03	0.08	0.28	1.05
p-value	Time	<0.001	0.030	0.142	0.001	0.392	<0.001	<0.001
	Energy	0.058	0.085	0.591	0.317	0.005	0.662	0.192
	Time × Energy	0.840	0.755	0.217	0.935	0.286	0.915	0.923

[1] GLU = glucose; CHO = total cholesterol; TG = triglyceride; LDL-C = low-density lipoprotein cholesterol; HDL-C = high-density lipoprotein cholesterol; LDL-C% = the ratio of LDL-C in CHO; HDL-C% = the ratio of HDL-C in CHO. ^{A,B} Within a column, means with different superscript capital letters were significantly different ($p < 0.05$) for the time factor in the same group. ^{a,b} Within a column, means with different superscript small letters were significantly different ($p < 0.05$) for the diet energy factor at the same time.

The LDL-C content and the ratio in CHO rose significantly with time ($p < 0.05$), especially in the He group. The serum HDL-C content at T8 increased as the diet energy levels were increased ($p < 0.05$), and the ratio of HDL-C in CHO increased over time ($p < 0.05$).

The GLU content changed greatly over time in the Le and He groups ($p < 0.05$). In general, the diet energy levels affected the glucose and lipid metabolism of the trial beagles, with higher diet energy levels harming the serum lipid levels and medium diet energy levels assisting glucostasis.

3.3. Changes in Fecal pH, SCFAs and BA Levels of Beagles with Different Diet Energy Levels

Feces became more acidic when beagles were fed a lower- or higher-energy diet, as the fecal pH of the Le and He groups significantly decreased by T8 ($p < 0.05$, Figure 2A). The content of SCFAs and BAs in beagle feces was analyzed and showed that different diet energy levels changed the metabolite profiles of fecal SCFAs and BAs, as shown in Figure 3A,B.

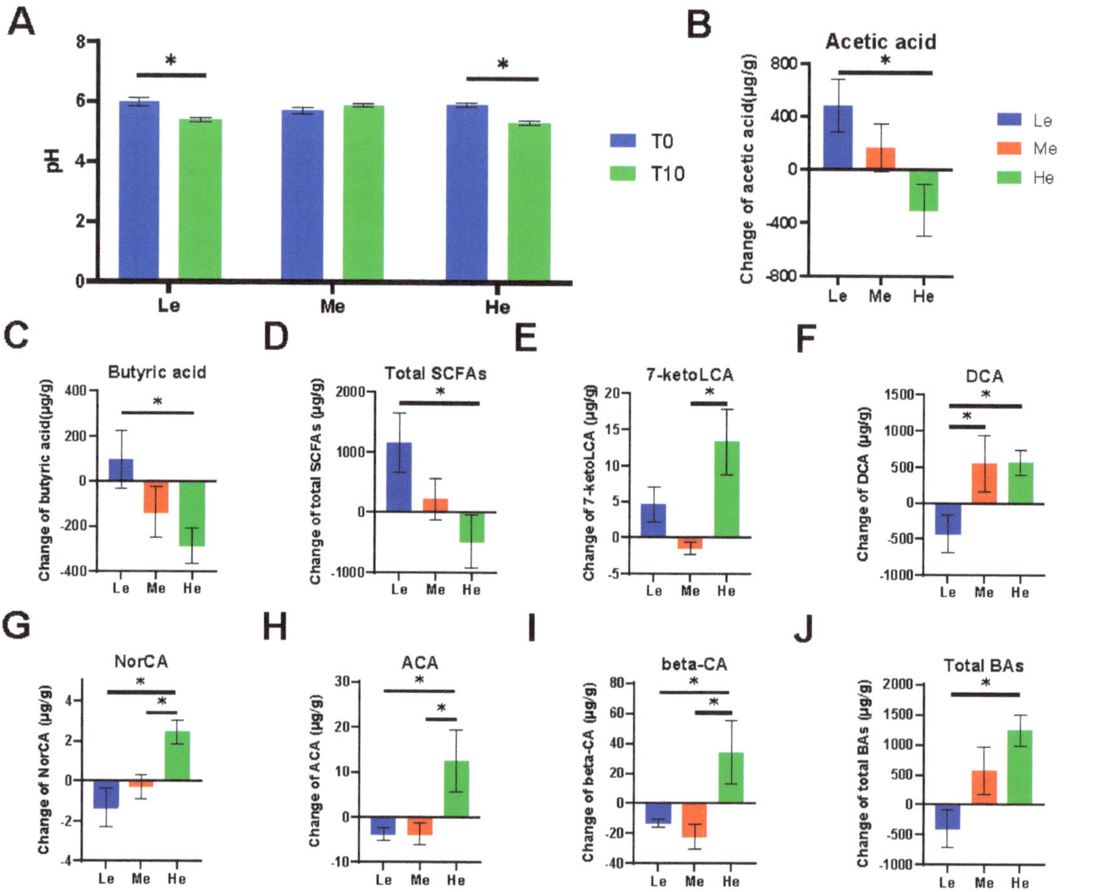

Figure 2. Effects of diet energy levels on fecal pH of beagles (**A**); changes in fecal acetic acid, butyric acid and total short-chain fatty acids (SCFAs) between T0 and T8 (**B–D**); changes in fecal 7-ketolithocholic acid (7-ketoLCA), deoxycholic acid (DCA), norcholic acid (NorCA), allocholic acid (ACA), 3β-cholic acid (β-CA) and total bile acids (BAs) between T0 and T8 (**E–J**). Data are shown as the mean ± SEM. * $p < 0.05$.

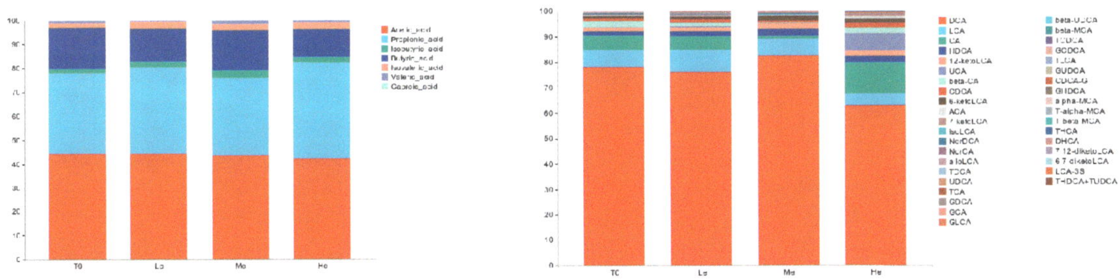

(A) (B)

Figure 3. Effects of diet energy levels on fecal SCFA (**A**) and BA (**B**) proportions in adult beagles.

SCFAs are major end products of bacterial carbohydrate fermentation in the intestinal tracts of dogs [54]. Fecal total SCFAs increased the most in the Le group, which had the most carbohydrates, but decreased in the He group, which had the least carbohydrates ($p < 0.05$), as shown in Figure 2D. The same trend occurred for linear chain SCFAs, such as acetic acid and butyric acid ($p < 0.05$), as shown in Figure 2B,C.

Bile acids have lipid-digestive functions [55]. The changes in fecal total BAs were significantly higher in the He group compared with the Le group ($p < 0.05$) between T8 and T0, and the content of total BAs decreased in the Le group, as seen in Figure 2J. Changes in five secondary BAs in the He group, including 7-ketoLCA, DCA, NorCA, ACA and β-CA, were the highest between the three groups ($p < 0.05$), as shown in Figure 2E–I. However, changes in primary BAs, such as CDCA, CA, GCDCA, GCA and TCDCA, were not altered by diet energy levels ($p > 0.05$).

Using all the quantitative analysis data of six SCFAs and 39 BAs at T8 to construct volcano plots, it was found that 6-ketoLCA, 12-ketoLCA, DCA and HDCA were downregulated in the Le group, while 7-ketoLCA, β-CA, CA, CDCA, UCA and ACA were upregulated and NorDCA downregulated in the He group when compared with the Me group ($-1 \geq \log_2$ fold change ≥ 1 and $p < 0.05$), as seen in Figure 4A,B.

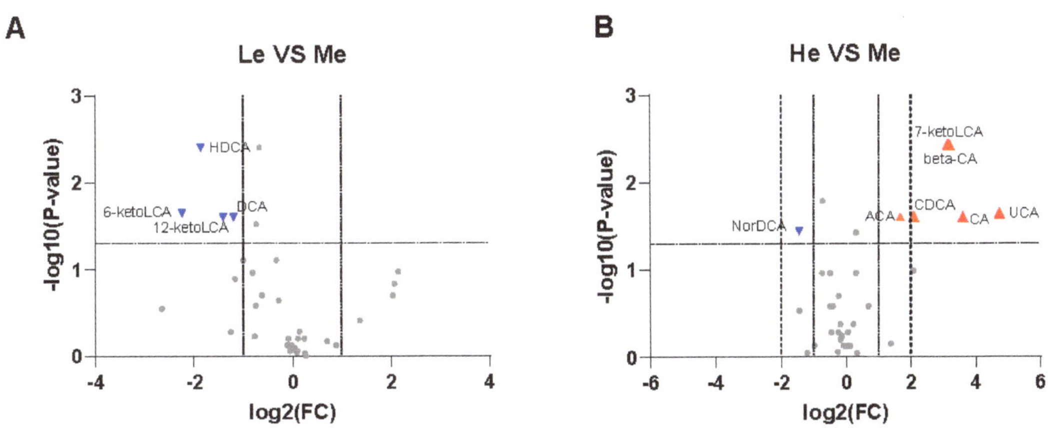

Figure 4. Volcano plots for Le vs. Me (**A**) and He vs. Me (**B**). Blue triangles represent significantly downregulated metabolites (log2 fold change < -1 and $p < 0.05$) whereas red triangles represent significantly upregulated metabolites (log2 fold change > 1 and $p < 0.05$).

3.4. Changes in the Structure and Composition of Fecal Microbiota of Beagles with Different Diet Energy Levels

As the gut microbiota ferment SCFAs and modify primary BAs to secondary BAs, the structure and composition of the fecal microbiota from the rectum were analyzed. As shown in Figure 5A, the time factor had no significant effects on the Chao1, Shannon, Simpson and Goods coverage indexes of fecal microbiota α-diversity when dogs were fed the same energy diet ($p > 0.05$). However, dogs in the Le and He groups showed lower levels of fecal microbiota α-diversity compared with the T0 group ($p < 0.05$). As seen from the results of the PCoA analysis of β-diversity in Figure 5B, the distribution ranges of the fecal microbiota in the Le and He groups were similar, but they were different from the T0 group. The PCoA of Bray–Curtis distances showed that the diet energy level could influence the gut microbiota composition and structure in dogs.

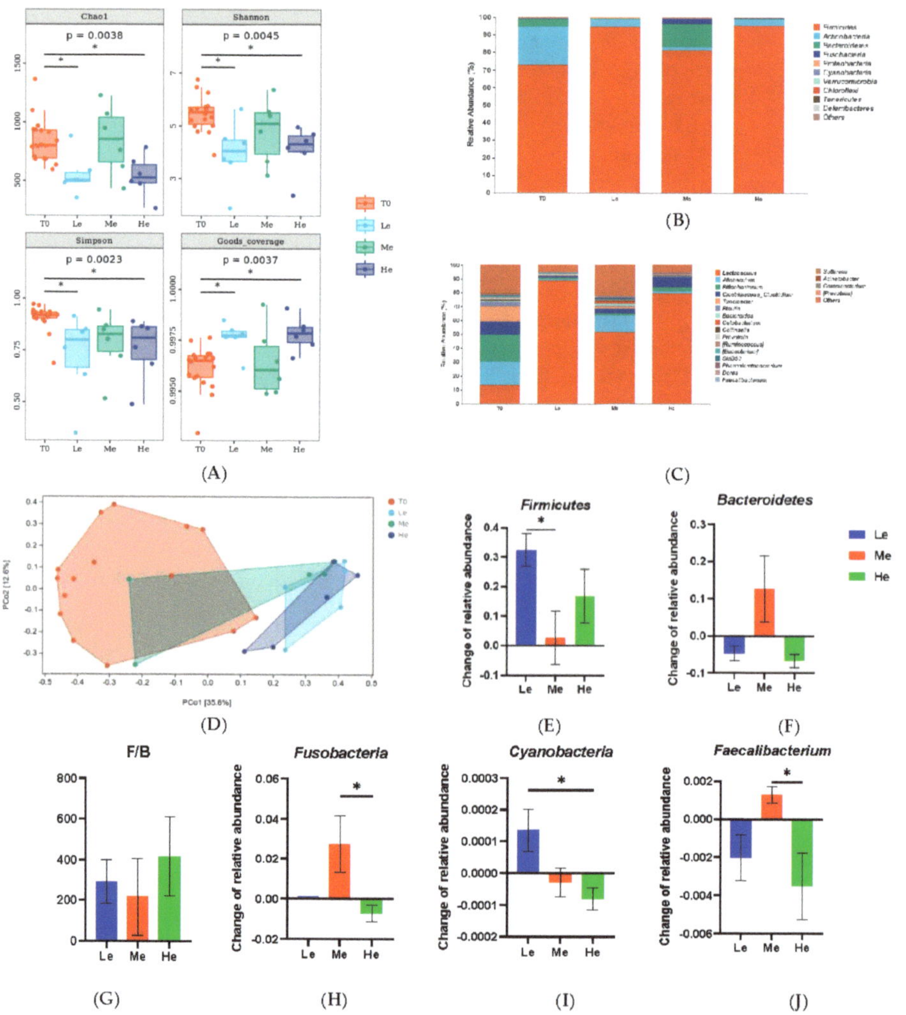

Figure 5. α-Diversity indices (Chao1, Shannon, Simpson and Goods coverage) of fecal microbial communities of the T0, Le, Me and He groups (**A**). The taxonomic composition of fecal microbiota at the phylum (**B**) and genus (**C**) levels of the T0, Le, Me and He groups. Principal coordinate analysis (PCoA) of β-diversity of the T0, Le, Me and He groups (**D**). Species with major differences at the phylum (**E–I**) and genus (**J**) levels. Data are shown as mean ± SEM. * $p < 0.05$.

As shown in Figure 5C,D, the profiles of the phylum and genus levels in the four groups were different. At the phylum level, as shown in Figure 5C,E–I, the changes in the relative abundances of *Firmicutes* between T8 and T0 in the Le group were significantly higher than in the Me group ($p < 0.05$). The changes in *Bacteroidetes* were higher and the ratio of *Firmicutes*/*Bacteroidetes* (F/B) was lower in the Me group than that in the other groups ($p > 0.05$), but it did not show a significant difference because of the great intra-group difference. *Fusobacteria* and *Cyanobacteria* of the He group were also different from those of the other groups ($p < 0.05$). At the genus level, as seen in Figure 4D, *Faecalibacterium* had the most obvious change in abundance. The relative abundance of *Faecalibacterium* increased in the Me group while decreasing in the Le and He groups ($p < 0.05$).

3.5. Effects of Diet Energy Level on Fecal Microbiota and Metabolic Pathways

To further study the effects of diet energy factors on fecal microbiota without the time factor, the species differences and marker species between the three groups at T8 were analyzed. As shown in Figure 6A, 1801 amplicon sequence variants (ASVs) were obtained in the Le group, 3091 ASVs were obtained in the Me group and 1743 ASVs were obtained in the He group. Compared with the Me group, the numbers of ASVs in the Le and He groups decreased by 41.73% and 43.61%, respectively. The number of ASVs shared among the three groups was 512, the proportion of unique ASVs in the Le and He groups decreased and the numbers of unique ASVs in the two groups changed from 68.62% in the Me group to 49.98% and 51.29%, respectively.

To select the differential bacteria, differential abundant taxa were confirmed by LEfSe analysis, where LDA exceeded 2.0. At the phylum level, *Firmicutes* was enriched in the Le group and *Bacteroidetes* and *Fusobacteria* were enriched in the Me group. At the family level, *Enterococcaceae* was enriched in the He group, *ACK M1* was enriched in the Le group and *S24 7*, *Fusobacteriaceae*, *Bacteroidaceae* and *Ruminococcaceae* were enriched in the Me group.

At the genus level, *Megamonas*, *Dolichospermum*, *Peptostreptococcus* and *Eubacterium* were enriched in the He group and *Bacteroides*, *Rothia* and *Faecalibacterium* were enriched in the Me group, as shown in Figure 6B,C. Heat maps at the species level further revealed the influences of the diet energy level on the fecal microbiota structure, as in Figure 6D. Compared with the Me group, the relative abundances of *Lactobacillus helveticus*, *Bifidobacterium pseudolongum*, *Lactobacillus vaginalis* and *Clostridium spiroforme* in both the Le and ME groups, and *Clostridium hiranonis*, *Lactobacillus coleohominis*, *Clostridium perfringens*, *Eubacterium biforme*, *Collinsella stercoris* and *Ruminococcus gnavus* in the He group, were all significantly increased. The relative abundances of *Blautia producta*, *Bacteroides plebeius*, *Faecalibacterium prausnitzii*, *Lactobacillus pontis*, *Prevotella copri*, *Cetobacterium somerae* and *Lactobacillus salivarius* in both the Le and ME groups, and *Ruminococcus torques*, *Bifidobacterium animalis* and *Clostridium celatum* in the He group, were low.

To select the differential bacteria, differential abundant taxa were confirmed by LEfSe analysis, where LDA exceeded 2.0. At the phylum level, *Firmicutes* was enriched in the Le group and *Bacteroidetes* and *Fusobacteria* were enriched in the Me group. At the family level, *Enterococcaceae* was enriched in the He group, *ACK M1* was enriched in the Le group and *S24 7*, *Fusobacteriaceae*, *Bacteroidaceae* and *Ruminococcaceae* were enriched in the Me group.

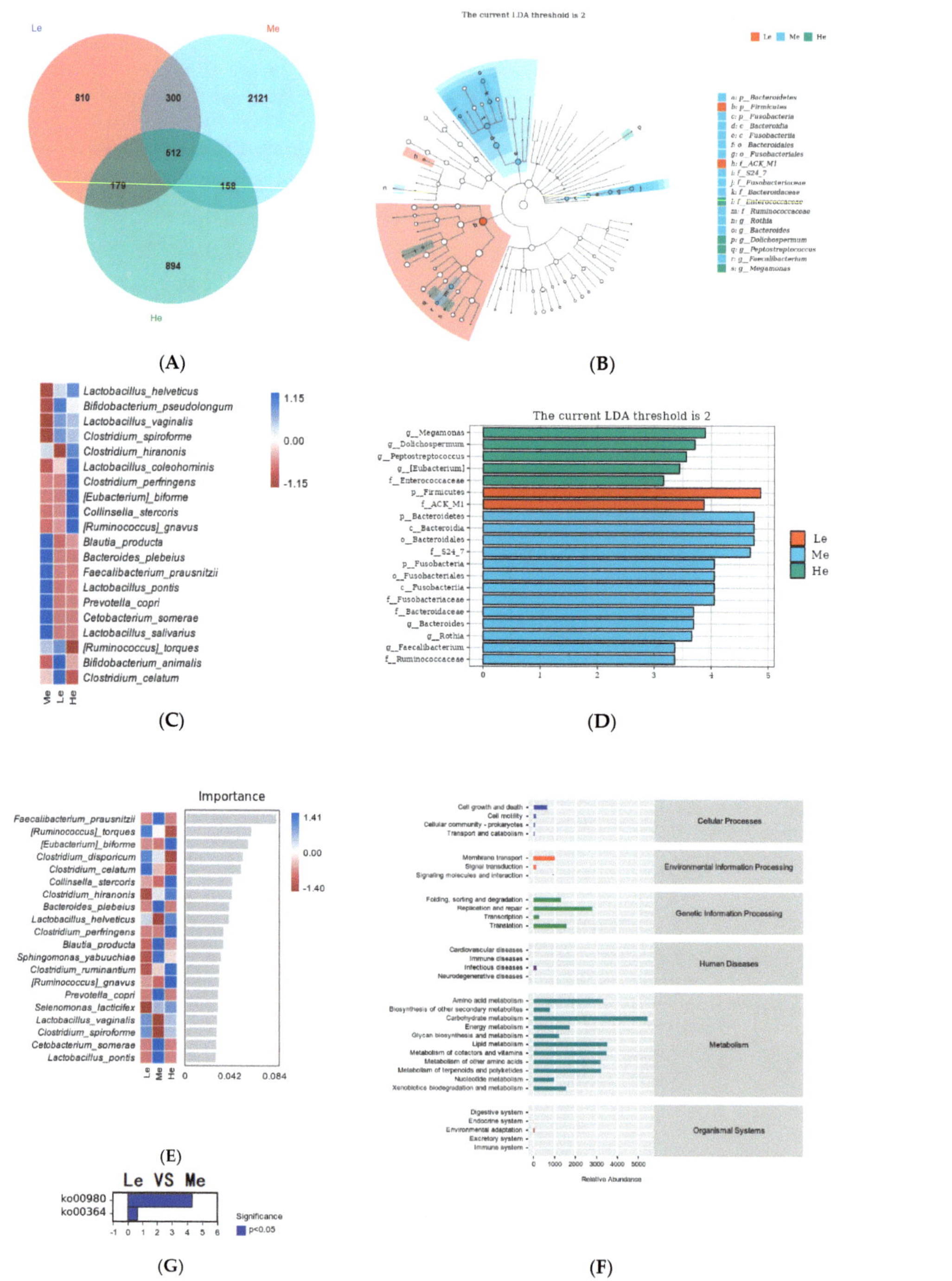

Figure 6. Venn diagram of the Le, Me and He groups (**A**). LEfSe analysis identified fecal bacterial biomarkers of the Le, Me and He groups (**B**,**C**). Heat map of species composition (**D**). Random forest model of species level (**E**). Metabolic pathways analysis (KEGG database) of the Le, Me and He groups (**F**). The upregulated metabolic pathways of the Le group compared with the Me group (**G**).

At the genus level, *Megamonas, Dolichospermum, Peptostreptococcus* and *Eubacterium* were enriched in the He group and *Bacteroides, Rothia* and *Faecalibacterium* were enriched in the Me group, as shown in Figure 6B,C. Heat maps at the species level further revealed the influences of the diet energy level on the fecal microbiota structure, as in Figure 6D. Compared with the Me group, the relative abundances of *Lactobacillus helveticus, Bifidobacterium pseudolongum, Lactobacillus vaginalis* and *Clostridium spiroforme* in both the Le and ME groups, and *Clostridium hiranonis, Lactobacillus coleohominis, Clostridium perfringens, Eubacterium biforme, Collinsella stercoris* and *Ruminococcus gnavus* in the He group, were all significantly increased. The relative abundances of *Blautia producta, Bacteroides plebeius, Faecalibacterium prausnitzii, Lactobacillus pontis, Prevotella copri, Cetobacterium somerae* and *Lactobacillus salivarius* in both the Le and ME groups, and *Ruminococcus torques, Bifidobacterium animalis* and *Clostridium celatum* in the He group, were low.

Random forest analysis can identify the complex nonlinear dependence between variables to enable the more effective and accurate classification of intestinal flora samples from each group [56]. As shown in Figure 6E, the relative abundances of *Eubacterium biforme, Collinsella stercoris, Clostridium hiranonis, Lactobacillus helveticus, Clostridium perfringens, Clostridium ruminantium* and *Ruminococcus gnavus* significantly increased and became the dominant strains in the He group. The relative abundances of *Ruminococcus torques, Clostridium disporicum, Clostridium celatum, Lactobacillus vaginalis* and *Clostridium spiroforme* significantly increased and became the dominant strains of the Le group. Most of the strains above are harmful bacteria, and the relative abundances of a variety of probiotics, such as *Faecalibacterium prausnitzii, Bacteroides plebeius, Blautia producta, Prevotella copri, Cetobacterium somerae* and *Lactobacillus pontis,* increased in the Me group. The above results show that adequate diet energy levels promoted the growth of probiotics and prevented the growth of harmful bacteria.

After examining the structural characteristics, the Picrust2 software was used to observe the influence of the fecal microbiota on metabolic pathways. As shown in Figure 6F, metabolic pathways in each group were concentrated in the aspects of metabolism and genetic information processing, among which the metabolism-related pathways accounted for the highest proportion of the total. The identification of bacterial functions between the Le and Me groups revealed that two distinctive pathways, including the metabolism of xenobiotics by cytochrome P450 (ko00980) and fluorobenzoate degradation (ko00364), were significantly represented, as shown in Figure 6G. There were no differential metabolic pathways between the He and Me groups.

3.6. Network Relation between Diet, Host Microbiota and Fecal Microbiota

To comprehensively evaluate our data and support a system-level understanding of the relation between diet, host and fecal microbiota, an integrative analysis of data from four dimensions was employed. We used four items in diet nutrition content, eight body condition indexes and nine serum biochemistry and fecal pH indexes in host phenotypic data, 10 phylum-level taxa and 20 genus-level taxa in fecal microbiota and eight fecal SCFAs, five primary BAs and 34 secondary BAs in fecal metabolites.

The Le, Me and He groups at T8 with 18 beagles were chosen for an association analysis. The Pearson correlation between all the data was evaluated. Data with strong correlation ($-0.6 > r > 0.6$ and $p < 0.05$) were used to construct an integrated visualization network containing eight circles. As shown in Figure 7, nine taxa were in the circle of the phylum level of the fecal microbiota, nineteen taxa were in the circle of the genus level of the fecal microbiota and twenty-five secondary BAs were in the fecal BAs circle. The diet nutrition content items, host phenotypic items, primary BAs and fecal SCFAs were all selected in the network and the results showed that secondary BAs, such as TDCA, isoLCA, LCA and GDCA et al., were strongly correlated with fecal microbiota, especially *Firmicutes, Fusobacteria, Bacteroidetes* and *Actinobacteria* at the phylum level, and *Cetobacterium, Bacteroides, Sutterella, Phascolarctobacterium, Turcibacter, Faecalibacterium, Bifidobacterium* and *Dorea* at the genus level. The taxa of these fecal microbiota all showed a positive correlation with secondary BAs, except *Firmicutes* and *Lactobacillus,* which were negatively correlated with BAs. It should be noted that primary BAs were mainly associated with five secondary BAs, including NorCA, bata-MCA, ACA, β-CA and 7-ketoLDA,d and were only correlated with *Bifidobacterium* at the genus level.

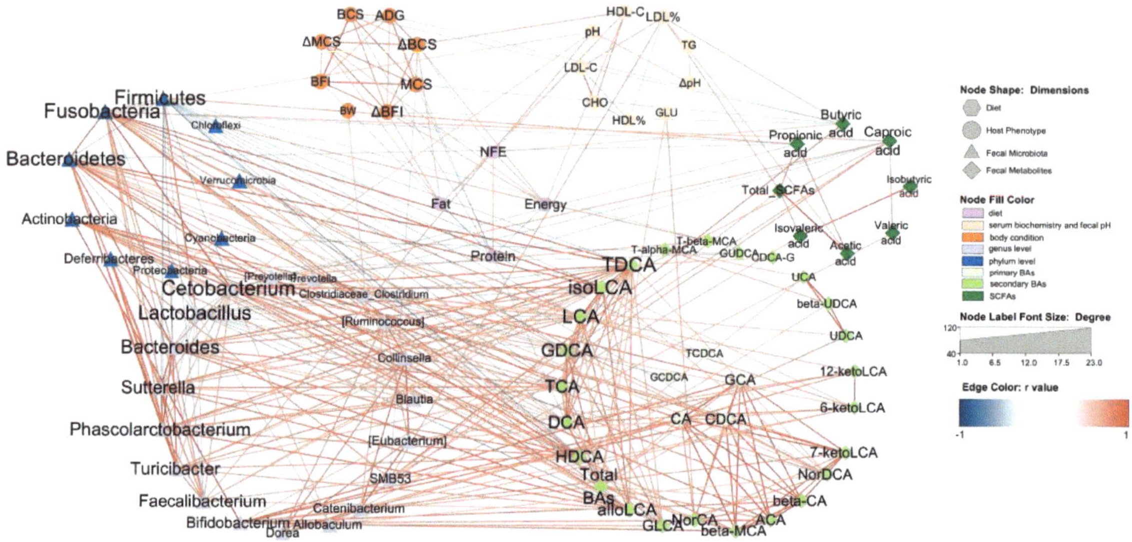

Figure 7. Network relation between diet, host and fecal microbiota. The network illustrates the interaction among diet, host and fecal microbiota. The hexagon nodes indicate the diet factor, the ellipse nodes indicate the host phenotype, the triangle nodes indicate the fecal microbiota, and the diamond nodes indicate fecal metabolites. The red edges indicate a positive correlation, and the blue edges indicate a negative correlation. The edges are the correlations with a Pearson's correlation coefficient < -0.6 (or >0.6) and $p < 0.05$.

Different SCFAs showed different functions in the network. Butyric acid was negatively correlated with body condition indexes and positively correlated with the NFE content in the diet. Caproic acid was the only SCFA associated with BAs, such as GDCA and isoLCA. Most SCFAs are associated with *Fusobabacteria, Firmicutes, SMB53, Allobaculum* and LDL%. On the host phenotypic dimension, fecal pH showed a positive correlation with *Fusobacteria, Faecalibacterium* and *Cetobacterium.* LDL% was positively correlated with *SMB53,* butyric acid, acetic acid and total SCFAs. The GLU was the only serum biochemistry index associated with BAs. Diet dimension is mainly associated with body condition indexes and HDL-C, and it had no edge with the fecal microbiota dimension. Butyric acid, caproic acid and DCA were the only three fecal metabolites correlated with diet nutrition.

4. Discussion

Animals need energy for all their life activities, and energy comes from fat, protein and carbohydrates in the diet [57]. Changes in the composition of these components may alter the diet energy level and also affect the metabolism of the host, and fecal microbiota and related studies have been reported in humans [58,59], mice [60], pigs [61] and cattle [62]. In the current pet food market, brands and consumers tend to pay attention to the content and source of protein in pet food and usually ignore the energy content. Excessive or insufficient energy intake may have adverse effects on animals [63]. Research on the nutritional regulation of obesity in dogs and cats is increasing [64,65], but the question of how to ensure healthy weight loss in animals while controlling energy is currently an important topic in dog and cat nutrition research [1].

This study investigated the effects of diet energy factors on body condition, glycolipid metabolism, fecal microbiota and metabolites in adult beagles. It found that changing the energy level of the diet, while keeping the protein level constant, significantly affected the body condition of adult beagles. At the end of the eight-week trial, weight loss was observed in the Le group, whose dietary ME was 13.88 MJ/kg, while the Me (ME = 15.04 MJ/kg) and He (ME = 17.05 MJ/kg) groups did not show significant changes.

During feeding, it was found that dogs in the Le group could consume almost all of the 32 g/BW kg/d food, while dogs in the Me and He groups had varying degrees of leftovers, which may have been related to the fact that dogs can adjust their food intake according to their energy intake. A similar finding was obtained by Xue et al. [14], where adult beagle dogs lost 15 to 25% of their body weight after eight weeks of consuming a diet with ME of 13.54 or 12.93 MJ/kg diet, while no significant changes were observed after 24 weeks of consuming a dietary ME of 15.07 MJ/kg diet.

It is the nature of dogs to eat for energy. When the energy intake of a canine is satisfied, its appetite and consequently food intake decreases, which is regulated by the body's hormones and the nervous system [66]. In a study by Hall et al. [67], it was found that dogs preferred balanced food with 25.2% protein, 15.8% fat and 44.9% carbohydrates over high-fat food with 24.6% protein, 28.4% fat and 44.9% carbohydrates, when the two diets had similar protein levels. The study also found that healthy adult dogs chose to consume most of their energy from fat (41%) and carbohydrates (36%) and there was a negative relation between fat and carbohydrate intake ($r = -0.87$). In view of the effects of the macronutrient composition of the diet on palatability and the appetite regulation mechanism of canines, it can be explained why there were no significant changes in body weight in the Me and He groups over eight weeks under adequate food conditions, and the body weights of the two groups were similar.

Compared with the other two groups, the BCS, MCS and BFI of the Le group showed significant decreases. The results of the Le group were mainly due to the imbalance of fat and carbohydrates and the lack of fat. The fat content of the Le diet was only 4.69%; it was lower than the NRC recommendation for adult dogs. The BCS is a common clinical scoring system to assess the body condition of dogs and cats, which can be used to determine the degree of obesity scoring. In this trial, the BCS of adult beagles increased with an adequate energy supply, while the Le group showed a decrease in BCS and entered a leaner body condition after being fed a low-energy diet for eight weeks. Many studies have shown that lowering the diet energy level by adjusting the macronutrient composition may reduce body weight and body condition [11,13,15,19,68]. These results were associated with inadequate energy intake, resulting in an inability to ensure the maintenance of energy requirements. The use of MCS for the clinical assessment of the nutritional status of dogs and cats is also an important indicator, where overweight animals may have significant muscle loss and animals with lower BCS may also have good muscle condition [69,70]. The BFI is used in the clinical assessment of the body fat percentage (BF%) by palpation to assess the risk of obesity [69]. In this trial, both the Me and He groups maintained good muscle mass and BF%, but dogs in the Le group showed a significant decrease in muscle mass and BF% after the eight-week trial. In combination with the BCS results, this suggests that if the energy supply is inadequate, the dogs will consume fat and protein to ensure maintenance energy requirements used for the most basic vital and voluntary activities. In one weight loss program study using a diet with ME = 12.39 MJ/kg, the protein content needed to be increased to 29% to ensure that no significant decrease in lean body mass or significant downregulation of fat mass and BF% occurred during the six-month weight loss period [13], but these results showed that dietary protein of 29% may be not enough for canine weight loss. Phungviwatnikul et al. [15] found a significant decrease in total body mass and lean mass in dogs fed a 42% protein, high-fiber diet with an ME of 11.76 MJ/kg for six weeks, suggesting that the content of dietary protein should be noted in the development of dog prescription diets for weight loss to avoid lean weight loss by reducing the diet energy level.

From the results of the network diagram, food factors were also associated with glycolipid metabolism, in addition to being more associated with body condition indexes. In this trial, the serum GLU of the Le and He groups was the highest at T4 and changed greatly over time, while the Me group was relatively stable. Changes in environmental temperature can cause changes in glucose metabolism in the body, as shown by changes in blood glucose [71], and diets with appropriate energy levels reduce this effect [72,73]. For serum biochemical indexes related to lipid metabolism, both the Le and He groups showed changes related to temporal factors, while the Me group was stable. The content of LDL-C in serum shows a significant positive correlation with the incidence of coronary heart disease, and it is an important index to evaluate the risk factors of individual coronary heart disease [74]. In canine obesity modeling and weight loss trials, high-fat diets were found to lead to a significant increase in serum CHO levels, while low-energy diets led to a decrease in CHO [13–15,75].

By analyzing the pH of the fresh feces of dogs, it was shown that the long-term consumption of either low- or high-energy diets caused fecal acidification, which was related to the metabolites in the feces and the composition of the gut microbiota, in several studies on canine nutrition [50–52,76–78]. Based on the above, the total SCFAs in the feces were measured in this study and were increased significantly in the Le group, especially for acetic acid and butyric acid. SCFAs are the end-products of carbohydrate metabolism by the gut flora, and the Le group had the highest carbohydrate content in the diet, explaining the above results. However, a significant decrease in SCFAs was also found in the He group compared to the beginning of the trial, which may be related to the lower carbohydrate and higher fat content in the diet. It has been shown that SCFAs can improve the inflammatory response of the body, induced by a high-fat diet [79,80], where butyric acid alleviates both oxidative stress and inflammation [81]. In contrast, SCFA deficiency leads to increased intestinal permeability, which triggers an inflammatory cascade response and induces the development of inflammatory diseases [82]. In addition, a sustained increase in glucagon-like peptide-1 (GLP-1) and PYY induced by propionate affects appetite regulation circuits in the brain and inhibits food intake [83,84]. In a study of pigs, both fecal acetic acid and butyric acid content were correlated with finishing weight [85].

Bile acids are involved in diet lipid absorption and utilization [86], and bile acids were analyzed that were closely related to lipid digestion and absorption in the intestine. The results showed that the profiles of BAs differed significantly between groups, and that the highest rise in total BAs was observed in the He group. Moreover, 7-ketolithocholic acid (7-ketoLCA), deoxycholic acid (DCA), norcholic acid (NorCA), allocholic acid (ACA) and 3β-cholic acid (β-CA) also showed similar trends with total BAs. Bile acids can be divided into two major groups according to the secretion pathway: one is primary bile acids directly secreted by the host, while the other is secondary bile acids after the modification of primary bile acids by intestinal flora. The bile acids, such as DCA mentioned above, are precisely secondary bile acids, which suggests that the digestion and utilization process of high-fat diets altered the metabolic activity of intestinal microbiota, reflected by similar results in humans [87,88]. A study in mice demonstrated that high-fat diets induced an increase in fecal DCA content and low-fat diets decreased the total BA content [89]. They also found that DCA disrupts intestinal mucosal barrier function by interfering with aryl hydrocarbon receptor (AHR) signaling in intestinal stem cells. The marker metabolites of bile acids in the feces of adult beagle dogs fed different energy level diets were significantly different. Decreased microbial secondary BA metabolites of DCA and HDCA were revealed in the feces of the low-energy group, and research showed that they were found responsible for stimulating 5-hydroxy tryptamine (5-HT) levels [90], involved in the regulation of appetite, weight and behavior [91]. Pan et al. [92] found that HDCA-treated mice exhibited reduced fat content and the metabolite HDCA was found to be significantly increased in the feces of uncoupling protein 1 (UCP1) knock-in pigs and had a negative relationship with backfat thickness. It is hypothesized that the decrease in fecal HDCA may be associated with energy deficiency in dog diets.

Changes in diet energy levels caused changes in intestinal metabolites, and there was an opposite trend in the changes in SCFAs and BAs between the Le and He groups. The reason may be the differences in energy levels in this experiment, which stemmed from the different carbohydrate and fat ratios, and the different intakes of carbohydrates and fat changed the structure and function of the intestinal flora—the network analysis confirmed this idea. The fecal microbiota compositions of the three groups were analyzed and it was found that changes in diet energy levels changed the flora structure. The change in diet energy level caused a decrease in microbiota diversity. In terms of flora composition, both temporal and energy level factors influenced the profiles of the genus and phylum levels and the number of ASVs. Similar findings have been obtained in canine studies on the relationship between food and the gut microbiota [93–95] and related studies have also demonstrated that *Fusobacterium* and *Clostridium* appear to be very strongly associated with the administration of raw-meat-based diets to cats and dogs [96].

The fecal microbiota results showed that either elevated or decreased diet energy levels inhibited the growth of probiotics and increased the growth of harmful bacteria. The relative abundances of *Faecalibacterium prausnitzii*, which is considered as one of the most important bacterial indicators of a healthy gut [97], significantly decreased in the Le and He groups. *Ruminococcus torques*, linked with gastrointestinal disease [98], was increased in the Le group. *Clostridium* was mainly associated with the Le and He groups and most of the associated strains were pathogenic or harmful [99,100]. *Bacteroides plebeius* improves muscle wasting in the chronic kidney [101], *Blautia producta* alleviates lipopolysaccharide-induced acute liver injury [102] and ameliorates high-fat-diet-induced hyperlipidemia [103], and *Cetobacterium somerae* and its metabolites are beneficial for liver health [104,105]. These probiotics were all enriched in the Me group. In terms of microbial function, a reduction in diet energy levels promoted the metabolism of xenobiotics by cytochrome P450 and fluorobenzoate degradation function. It has been found that the metabolism of xenobiotics via the cytochrome P450 pathway can be activated by increased energy restriction [106]. Fluorobenzoate degradation may be the key metabolic pathway involved in myasthenia gravis [107] and may be related to Crohn's disease [108]. These facts suggest a potential risk associated with the prolonged feeding of low-fat, high-carbohydrate diets to adult beagles.

Diet–host–fecal microbiota interactions exist, and this relationship was also evident in the network analysis in this paper. Diets modify the body condition of dogs through their effect on body metabolism and act as a link between the host and the gut microbiota through secondary bile acids [109,110]. This suggests that further research is needed on the mechanism by which diet energy levels affect canine metabolism, and it suggests that attention should be paid to the role of fecal metabolites as biomarkers in pet food development to better monitor food–pet interactions.

5. Conclusions

We found that different diet energy levels altered the body condition, the glucolipid metabolism and the structure of fecal SCFAs, BAs and microbiota in adult beagles because of the interaction between diet, host and gut microbiota. Feeding dogs with a medium-energy-level diet with an ME of 15.04 MJ/kg was suitable for maintaining an ideal body condition and stable glucolipid metabolism and promoting the relative abundance of probiotics in the gut. Feeding low- (ME = 13.88 MJ/kg) or high-energy (ME = 17.05 MJ/kg) diets was harmful for glucostasis, decreased the diversity of the intestinal microbiota and promoted the relative abundance of pathogenic bacteria in the gut. Dogs fed a low-energy diet for an extended time may become lean and lose muscle mass, but a diet with low energy and 29% protein may not supply enough protein for dogs losing weight.

Author Contributions: Conceptualization, H.S. and G.L.; methodology, H.S.; software, H.S.; validation, H.S., Q.Z., A.M. and M.C.; formal analysis, H.S.; resources, C.X., H.Z. and W.S.; data curation, H.S.; writing—original draft preparation, H.S.; writing—review and editing, T.Z. and H.S.; visualization, H.S.; supervision, T.Z.; project administration, H.S.; funding acquisition, T.Z and G.L. All authors have read and agreed to the published version of the manuscript.

Funding: This research was funded by the Agricultural Science and Technology Innovation Program of China (Grant No. CAAS-ASTIP-2021-ISAPS).

Institutional Review Board Statement: The animal experiment was conducted in accordance with the Regulations of the People's Republic of China on the Control of Laboratory Animals and approved by the Laboratory Animal Management and Ethics Committee of the Institute of Special Animal and Plant Sciences, Chinese Academy of Agricultural Sciences (No. ISAPSAEC-2021-004D).

Informed Consent Statement: Not applicable.

Data Availability Statement: The data presented in this study are available in the article.

Conflicts of Interest: The authors declare no conflict of interest.

References

1. Marchi, P.H.; Vendramini, T.H.A.; Perini, M.P.; Zafalon, R.V.A.; Amaral, A.R.; Ochamotto, V.A.; Da Silveira, J.C.; Dagli, M.L.Z.; Brunetto, M.A. Obesity, Inflammation, and Cancer in Dogs: Review and Perspectives. *Front. Vet. Sci.* **2022**, *9*, 1004122. [CrossRef] [PubMed]
2. Partington, C.; Hodgkiss-Geere, H.; Woods, G.R.T.; Dukes-McEwan, J.; Flanagan, J.; Biourge, V.; German, A.J. The Effect of Obesity and Subsequent Weight Reduction on Cardiac Structure and Function in Dogs. *BMC Vet. Res.* **2022**, *18*, 351. [CrossRef]
3. Ahuja, R.P.; Fletcher, J.M.; Granger, L.A.; Liu, C.-C.; Miessler, B.; Mitchell, M.A. Changes in Glucose Tolerance and Insulin Secretion in a Cohort of Cats with Chronic Obesity. *Can. J. Vet. Res.* **2022**, *86*, 181–187. [PubMed]
4. Linder, D.E.; Freeman, L.M. Evaluation of Calorie Density and Feeding Directions for Commercially Available Diets Designed for Weight Loss in Dogs and Cats. *J. Am. Vet. Med. Assoc.* **2010**, *236*, 74–77. [CrossRef] [PubMed]
5. Olivindo, R.F.G.; Zafalon, R.V.A.; Teixeira, F.A.; Vendramini, T.H.A.; Pedrinelli, V.; Brunetto, M.A. Evaluation of the Nutrients Supplied by Veterinary Diets Commercialized in Brazil for Obese Dogs Undergoing a Weight Loss Program. *J. Anim. Physiol. Anim. Nutr.* **2022**, *106*, 355–367. [CrossRef]
6. Farcas, A.K.; Larsen, J.A.; Owens, T.J.; Nelson, R.W.; Kass, P.H.; Fascetti, A.J. Evaluation of Total Dietary Fiber Concentration and Composition of Commercial Diets Used for Management of Diabetes Mellitus, Obesity, and Dietary Fat-Responsive Disease in Dogs. *J. Am. Vet. Med. Assoc.* **2015**, *247*, 501–507. [CrossRef]
7. Bland, I.M.; Guthrie-Jones, A.; Taylor, R.D.; Hill, J. Dog Obesity: Veterinary Practices' and Owners' Opinions on Cause and Management. *Prev. Vet. Med.* **2010**, *94*, 310–315. [CrossRef]
8. Shepherd, M. Canine and Feline Obesity Management. *Vet. Clin. N. Am. Small Anim. Pract.* **2021**, *51*, 653–667. [CrossRef]
9. Grant, C.E.; Shoveller, A.K.; Blois, S.; Bakovic, M.; Monteith, G.; Verbrugghe, A. Dietary Intake of Amino Acids and Vitamins Compared to NRC Requirements in Obese Cats Undergoing Energy Restriction for Weight Loss. *BMC Vet. Res.* **2020**, *16*, 426. [CrossRef]
10. Asaro, N.J.; Seymour, D.J.; Mansilla, W.D.; Cant, J.P.; Zijlstra, R.T.; Berendt, K.D.; Brewer, J.; Shoveller, A.K. Modelling Net Energy of Commercial Cat Diets. *PLoS ONE* **2019**, *14*, e0218173. [CrossRef]
11. Keller, E.; Sagols, E.; Flanagan, J.; Biourge, V.; German, A.J. Use of Reduced-Energy Content Maintenance Diets for Modest Weight Reduction in Overweight Cats and Dogs. *Res. Vet. Sci.* **2020**, *131*, 194–205. [CrossRef]
12. Serisier, S.; Weber, M.; Feugier, A.; Fardet, M.-O.; Garnier, F.; Biourge, V.; German, A.J. Maintenance Energy Requirements in Miniature Colony Dogs. *J. Anim. Physiol. Anim. Nutr.* **2013**, *97* (Suppl. S1), 60–67. [CrossRef]
13. Piantedosi, D.; Palatucci, A.T.; Giovazzino, A.; Ruggiero, G.; Rubino, V.; Musco, N.; Carriero, F.; Farina, F.; Attia, Y.A.E.W.; Terrazzano, G.; et al. Effect of a Weight Loss Program on Biochemical and Immunological Profile, Serum Leptin Levels, and Cardiovascular Parameters in Obese Dogs. *Front. Vet. Sci.* **2020**, *7*, 398. [CrossRef]
14. Xue, J.; Lu, Y.; Zou, T.; Shi, W.; Wang, S.; Cheng, X.; Wan, J.; Chen, Y.; Wang, M.; Wang, Q.; et al. A Protein- and Fiber-Rich Diet with Astaxanthin Alleviates High-Fat Diet-Induced Obesity in Beagles. *Front. Nutr.* **2022**, *9*, 1019615. [CrossRef]
15. Phungviwatnikul, T.; Lee, A.H.; Belchik, S.E.; Suchodolski, J.S.; Swanson, K.S. Weight Loss and High-Protein, High-Fiber Diet Consumption Impact Blood Metabolite Profiles, Body Composition, Voluntary Physical Activity, Fecal Microbiota, and Fecal Metabolites of Adult Dogs. *J. Anim. Sci.* **2022**, *100*, skab379. [CrossRef]
16. Weber, M.; Bissot, T.; Servet, E.; Sergheraert, R.; Biourge, V.; German, A.J. A High-Protein, High-Fiber Diet Designed for Weight Loss Improves Satiety in Dogs. *J. Vet. Intern. Med.* **2007**, *21*, 1203–1208. [CrossRef]

17. Bermudez Sanchez, S.; Pilla, R.; Sarawichitr, B.; Gramenzi, A.; Marsilio, F.; Steiner, J.M.; Lidbury, J.A.; Woods, G.R.T.; Suchodolski, J.S.; German, A.J. Untargeted Fecal Metabolome Analysis in Obese Dogs after Weight Loss Achieved by Feeding a High-Fiber-High-Protein Diet. *Metabolomics* **2021**, *17*, 66. [CrossRef]
18. André, A.; Leriche, I.; Chaix, G.; Thorin, C.; Burger, M.; Nguyen, P. Recovery of Insulin Sensitivity and Optimal Body Composition after Rapid Weight Loss in Obese Dogs Fed a High-Protein Medium-Carbohydrate Diet. *J. Anim. Physiol. Anim. Nutr.* **2017**, *101* (Suppl. S1), 21–30. [CrossRef]
19. Murphy, M.; Bartges, J.W.; Zemel, M.B.; Kirk, C.A.; Witzel-Rollins, A. Effect of a Leucine/Pyridoxine Nutraceutical on Caloric Intake and Body Composition of Obese Dogs Losing Weight. *Front. Vet. Sci.* **2020**, *7*, 555. [CrossRef]
20. Floerchinger, A.M.; Jackson, M.I.; Jewell, D.E.; MacLeay, J.M.; Hahn, K.A.; Paetau-Robinson, I. Effect of Feeding a Weight Loss Food beyond a Caloric Restriction Period on Body Composition and Resistance to Weight Gain in Cats. *J. Am. Vet. Med. Assoc.* **2015**, *247*, 365–374. [CrossRef]
21. Sabchuk, T.T.; Scheraiber, M.; Zanatta, C.P.; Maiorka, A.; Félix, A.P.; Oliveira, S.G. Body Condition of Dogs Fed Diets Containing Soya Hulls. *J. Nutr. Sci.* **2014**, *3*, e42. [CrossRef] [PubMed]
22. McKnight, L.L.; Eyre, R.; Gooding, M.A.; Davenport, G.M.; Shoveller, A.K. Dietary Mannoheptulose Increases Fasting Serum Glucagon like Peptide-1 and Post-Prandial Serum Ghrelin Concentrations in Adult Beagle Dogs. *Animals* **2015**, *5*, 442–454. [CrossRef] [PubMed]
23. Alexander, C.; Cross, T.-W.L.; Devendran, S.; Neumer, F.; Theis, S.; Ridlon, J.M.; Suchodolski, J.S.; de Godoy, M.R.C.; Swanson, K.S. Effects of Prebiotic Inulin-Type Fructans on Blood Metabolite and Hormone Concentrations and Faecal Microbiota and Metabolites in Overweight Dogs. *Br. J. Nutr.* **2018**, *120*, 711–720. [CrossRef]
24. Ferreira, C.S.; Vendramini, T.H.A.; Amaral, A.R.; Rentas, M.F.; Ernandes, M.C.; da Silva, F.L.; Oba, P.M.; de Oliveira Roberti Filho, F.; Brunetto, M.A. Metabolic Variables of Obese Dogs with Insulin Resistance Supplemented with Yeast Beta-Glucan. *BMC Vet. Res.* **2022**, *18*, 14. [CrossRef] [PubMed]
25. Apper, E.; Privet, L.; Taminiau, B.; Le Bourgot, C.; Svilar, L.; Martin, J.-C.; Diez, M. Relationships between Gut Microbiota, Metabolome, Body Weight, and Glucose Homeostasis of Obese Dogs Fed with Diets Differing in Prebiotic and Protein Content. *Microorganisms* **2020**, *8*, 513. [CrossRef]
26. Söder, J.; Höglund, K.; Dicksved, J.; Hagman, R.; Eriksson Röhnisch, H.; Moazzami, A.A.; Wernersson, S. Plasma Metabolomics Reveals Lower Carnitine Concentrations in Overweight Labrador Retriever Dogs. *Acta Vet. Scand.* **2019**, *61*, 10. [CrossRef]
27. Kilburn, L.R.; Koester, L.R.; Schmitz-Esser, S.; Serão, N.V.L.; Rossoni Serão, M.C. High-Fat Diets Led to OTU-Level Shifts in Fecal Samples of Healthy Adult Dogs. *Front. Microbiol.* **2020**, *11*, 564160. [CrossRef]
28. Kilburn, L.R.; Allenspach, K.; Jergens, A.E.; Bourgois-Mochel, A.; Mochel, J.P.; Serao, M.C.R. Apparent Total Tract Digestibility, Fecal Characteristics, and Blood Parameters of Healthy Adult Dogs Fed High-Fat Diets. *J. Anim. Sci.* **2020**, *98*, skaa043. [CrossRef]
29. Moinard, A.; Payen, C.; Ouguerram, K.; André, A.; Hernandez, J.; Drut, A.; Biourge, V.C.; Suchodolski, J.S.; Flanagan, J.; Nguyen, P.; et al. Effects of High-Fat Diet at Two Energetic Levels on Fecal Microbiota, Colonic Barrier, and Metabolic Parameters in Dogs. *Front. Vet. Sci.* **2020**, *7*, 566282. [CrossRef]
30. Lyu, Y.; Liu, D.; Nguyen, P.; Peters, I.; Heilmann, R.M.; Fievez, V.; Hemeryck, L.Y.; Hesta, M. Differences in Metabolic Profiles of Healthy Dogs Fed a High-Fat vs. a High-Starch Diet. *Front. Vet. Sci.* **2022**, *9*, 801863. [CrossRef]
31. Schauf, S.; de la Fuente, G.; Newbold, C.J.; Salas-Mani, A.; Torre, C.; Abecia, L.; Castrillo, C. Effect of Dietary Fat to Starch Content on Fecal Microbiota Composition and Activity in Dogs. *J. Anim. Sci.* **2018**, *96*, 3684–3698. [CrossRef]
32. Nutritional Research Council. *Nutrient Requirements of Dogs and Cats*; The National Academies Press: Washington, DC, USA, 2006.
33. Laflamme, D.R.P.C. Development and Validation of a Body Condition Score System for Dogs. *Canine Pract.* **1997**, *22*, 10–15.
34. Freeman, L.M.; Michel, K.E.; Zanghi, B.M.; Vester Boler, B.M.; Fages, J. Evaluation of the Use of Muscle Condition Score and Ultrasonographic Measurements for Assessment of Muscle Mass in Dogs. *Am. J. Vet. Res.* **2019**, *80*, 595–600. [CrossRef]
35. Witzel, A.L.; Kirk, C.A.; Henry, G.A.; Toll, P.W.; Brejda, J.J.; Paetau-Robinson, I. Use of a Novel Morphometric Method and Body Fat Index System for Estimation of Body Composition in Overweight and Obese Dogs. *J. Am. Vet. Med. Assoc.* **2014**, *244*, 1279–1284. [CrossRef]
36. Han, X.; Guo, J.; You, Y.; Yin, M.; Ren, C.; Zhan, J.; Huang, W. A Fast and Accurate Way to Determine Short Chain Fatty Acids in Mouse Feces Based on GC-MS. *J. Chromatogr. B Analyt. Technol. Biomed. Life Sci.* **2018**, *1099*, 73–82. [CrossRef]
37. Zhang, S.; Wang, H.; Zhu, M.-J. A Sensitive GC/MS Detection Method for Analyzing Microbial Metabolites Short Chain Fatty Acids in Fecal and Serum Samples. *Talanta* **2019**, *196*, 249–254. [CrossRef]
38. Hsu, Y.-L.; Chen, C.-C.; Lin, Y.-T.; Wu, W.-K.; Chang, L.-C.; Lai, C.-H.; Wu, M.-S.; Kuo, C.-H. Evaluation and Optimization of Sample Handling Methods for Quantification of Short-Chain Fatty Acids in Human Fecal Samples by GC-MS. *J. Proteome Res.* **2019**, *18*, 1948–1957. [CrossRef]
39. Bhargava, P.; Smith, M.D.; Mische, L.; Harrington, E.; Fitzgerald, K.C.; Martin, K.; Kim, S.; Reyes, A.A.; Gonzalez-Cardona, J.; Volsko, C.; et al. Bile Acid Metabolism Is Altered in Multiple Sclerosis and Supplementation Ameliorates Neuroinflammation. *J. Clin. Investig.* **2020**, *130*, 3467–3482. [CrossRef]

40. Hu, T.; An, Z.; Shi, C.; Li, P.; Liu, L. A Sensitive and Efficient Method for Simultaneous Profiling of Bile Acids and Fatty Acids by UPLC-MS/MS. *J. Pharm. Biomed. Anal.* **2020**, *178*, 112815. [CrossRef]
41. Yang, T.; Shu, T.; Liu, G.; Mei, H.; Zhu, X.; Huang, X.; Zhang, L.; Jiang, Z. Quantitative Profiling of 19 Bile Acids in Rat Plasma, Liver, Bile and Different Intestinal Section Contents to Investigate Bile Acid Homeostasis and the Application of Temporal Variation of Endogenous Bile Acids. *J. Steroid Biochem. Mol. Biol.* **2017**, *172*, 69–78. [CrossRef]
42. Bolyen, E.; Rideout, J.R.; Dillon, M.R.; Bokulich, N.A.; Abnet, C.C.; Al-Ghalith, G.A.; Alexander, H.; Alm, E.J.; Arumugam, M.; Asnicar, F.; et al. Reproducible, Interactive, Scalable and Extensible Microbiome Data Science Using QIIME 2. *Nat. Biotechnol.* **2019**, *37*, 852–857. [CrossRef] [PubMed]
43. Callahan, B.J.; McMurdie, P.J.; Rosen, M.J.; Han, A.W.; Johnson, A.J.A.; Holmes, S.P. DADA2: High-Resolution Sample Inference from Illumina Amplicon Data. *Nat. Methods* **2016**, *13*, 581–583. [CrossRef] [PubMed]
44. Katoh, K.; Misawa, K.; Kuma, K.; Miyata, T. MAFFT: A Novel Method for Rapid Multiple Sequence Alignment Based on Fast Fourier Transform. *Nucleic Acids Res.* **2002**, *30*, 3059–3066. [CrossRef] [PubMed]
45. Price, M.N.; Dehal, P.S.; Arkin, A.P. FastTree: Computing Large Minimum Evolution Trees with Profiles Instead of a Distance Matrix. *Mol. Biol. Evol.* **2009**, *26*, 1641–1650. [CrossRef] [PubMed]
46. Bokulich, N.A.; Kaehler, B.D.; Rideout, J.R.; Dillon, M.; Bolyen, E.; Knight, R.; Huttley, G.A.; Gregory Caporaso, J. Optimizing Taxonomic Classification of Marker-Gene Amplicon Sequences with QIIME 2′s Q2-Feature-Classifier Plugin. *Microbiome* **2018**, *6*, 90. [CrossRef]
47. Kõljalg, U.; Nilsson, R.H.; Abarenkov, K.; Tedersoo, L.; Taylor, A.F.S.; Bahram, M.; Bates, S.T.; Bruns, T.D.; Bengtsson-Palme, J.; Callaghan, T.M.; et al. Towards a Unified Paradigm for Sequence-Based Identification of Fungi. *Mol. Ecol.* **2013**, *22*, 5271–5277. [CrossRef]
48. Zaura, E.; Keijser, B.J.F.; Huse, S.M.; Crielaard, W. Defining the Healthy "Core Microbiome" of Oral Microbial Communities. *BMC Microbiol.* **2009**, *9*, 259. [CrossRef]
49. Segata, N.; Izard, J.; Waldron, L.; Gevers, D.; Miropolsky, L.; Garrett, W.S.; Huttenhower, C. Metagenomic Biomarker Discovery and Explanation. *Genome Biol.* **2011**, *12*, R60. [CrossRef]
50. Liaw, A.; Wiener, M. Classification and Regression by RandomForest. *Forest* **2001**, *23*, 18–22.
51. Breiman, L. Random Forests. *Mach. Learn.* **2001**, *45*, 5–32. [CrossRef]
52. Douglas, G.; Maffei, V.; Zaneveld, J.; Yurgel, S.; Brown, J.; Taylor, C.; Huttenhower, C.; Langille, M. PICRUSt2: An Improved and Extensible Approach for Metagenome Inference. *BioRxiv* **2019**, 672295. [CrossRef]
53. Kaneko, J.; Harvey, W.J.; Bruss, M.L. *Clinical Biochemistry of Domestic Animals*; Academic Press: Cambridge, MA, USA, 2008.
54. Minamoto, Y.; Minamoto, T.; Isaiah, A.; Sattasathuchana, P.; Buono, A.; Rangachari, V.R.; McNeely, I.H.; Lidbury, J.; Steiner, J.M.; Suchodolski, J.S. Fecal Short-Chain Fatty Acid Concentrations and Dysbiosis in Dogs with Chronic Enteropathy. *J. Vet. Intern. Med.* **2019**, *33*, 1608–1618. [CrossRef]
55. Vítek, L.; Haluzík, M. The Role of Bile Acids in Metabolic Regulation. *J. Endocrinol.* **2016**, *228*, R85–R96. [CrossRef]
56. Thompson, J.; Johansen, R.; Dunbar, J.; Munsky, B. Machine Learning to Predict Microbial Community Functions: An Analysis of Dissolved Organic Carbon from Litter Decomposition. *PLoS ONE* **2019**, *14*, e0215502. [CrossRef]
57. Jobgen, W.S.; Fried, S.K.; Fu, W.J.; Meininger, C.J.; Wu, G. Regulatory Role for the Arginine-Nitric Oxide Pathway in Metabolism of Energy Substrates. *J. Nutr. Biochem.* **2006**, *17*, 571–588. [CrossRef]
58. Bolesławska, I.; Kowalówka, M.; Dobrzyńska, M.; Karaźniewicz-Łada, M.; Przysławski, J. Differences in the Concentration of Vitamin D Metabolites in Plasma Due to the Low-Carbohydrate-High-Fat Diet and the Eastern European Diet—A Pilot Study. *Nutrients* **2021**, *13*, 2774. [CrossRef]
59. Bütikofer, U.; Burnand, D.; Portmann, R.; Blaser, C.; Schwander, F.; Kopf-Bolanz, K.A.; Laederach, K.; Badertscher, R.; Walther, B.; Vergères, G. Serum Metabolites Responding in a Dose-Dependent Manner to the Intake of a High-Fat Meal in Normal Weight Healthy Men Are Associated with Obesity. *Metabolites* **2021**, *11*, 392. [CrossRef]
60. Yang, J.; Wei, H.; Zhou, Y.; Szeto, C.-H.; Li, C.; Lin, Y.; Coker, O.O.; Lau, H.C.H.; Chan, A.W.H.; Sung, J.J.Y.; et al. High-Fat Diet Promotes Colorectal Tumorigenesis Through Modulating Gut Microbiota and Metabolites. *Gastroenterology* **2022**, *162*, 135–149.e2. [CrossRef]
61. Liu, Y.; Kong, X.; Jiang, G.; Tan, B.; Deng, J.; Yang, X.; Li, F.; Xiong, X.; Yin, Y. Effects of Dietary Protein/Energy Ratio on Growth Performance, Carcass Trait, Meat Quality, and Plasma Metabolites in Pigs of Different Genotypes. *J. Anim. Sci. Biotechnol.* **2015**, *6*, 36. [CrossRef]
62. Liu, H.; Zhou, J.; Degen, A.; Liu, H.; Cao, X.; Hao, L.; Shang, Z.; Ran, T.; Long, R. A Comparison of Average Daily Gain, Apparent Digestibilities, Energy Balance, Rumen Fermentation Parameters, and Serum Metabolites between Yaks (*Bos grunniens*) and Qaidam Cattle (*Bos taurus*) Consuming Diets Differing in Energy Level. *Anim. Nutr.* **2023**, *12*, 77–86. [CrossRef]
63. Loftus, J.P.; Wakshlag, J.J. Canine and Feline Obesity: A Review of Pathophysiology, Epidemiology, and Clinical Management. *Vet. Med. Res. Rep.* **2015**, *6*, 49–60. [CrossRef]
64. Kleinert, M.; Clemmensen, C.; Hofmann, S.M.; Moore, M.C.; Renner, S.; Woods, S.C.; Huypens, P.; Beckers, J.; de Angelis, M.H.; Schürmann, A.; et al. Animal Models of Obesity and Diabetes Mellitus. *Nat. Rev. Endocrinol.* **2018**, *14*, 140–162. [CrossRef] [PubMed]

65. Garcia-Mazcorro, J.F.; Minamoto, Y.; Kawas, J.R.; Suchodolski, J.S.; de Vos, W.M. *Akkermansia* and Microbial Degradation of Mucus in Cats and Dogs: Implications to the Growing Worldwide Epidemic of Pet Obesity. *Vet. Sci.* **2020**, *7*, 44. [CrossRef] [PubMed]
66. Kim, K.-S.; Seeley, R.J.; Sandoval, D.A. Signalling from the Periphery to the Brain That Regulates Energy Homeostasis. *Nat. Rev. Neurosci.* **2018**, *19*, 185–196. [CrossRef]
67. Hall, J.A.; Vondran, J.C.; Vanchina, M.A.; Jewell, D.E. When Fed Foods with Similar Palatability, Healthy Adult Dogs and Cats Choose Different Macronutrient Compositions. *J. Exp. Biol.* **2018**, *221*, jeb.173450. [CrossRef]
68. de Marchi, P.N.; Cardoso, M.J.L.; Fagnani, R.; Calesso, J.R.; Melussi, M.; Machado, L.H.d.A. Metabolic Parameters in Obese Dogs Undergoing to the Diet with Calorie Restriction. *Open J. Vet. Med.* **2018**, *8*, 25–35. [CrossRef]
69. Cline, M.G.; Burns, K.M.; Coe, J.B.; Downing, R.; Durzi, T.; Murphy, M.; Parker, V. 2021 AAHA Nutrition and Weight Management Guidelines for Dogs and Cats. *J. Am. Anim. Hosp. Assoc.* **2021**, *57*, 153–178. [CrossRef]
70. Freeman, L.; Becvarova, I.; Cave, N.; MacKay, C.; Nguyen, P.; Rama, B.; Takashima, G.; Tiffin, R.; Tsjimoto, H.; WSAVA Nutritional Assessment Guidelines Task Force Members. WSAVA Nutritional Assessment Guidelines. *J. Small Anim. Pract.* **2011**, *52*, 385–396. [CrossRef]
71. Baumgard, L.H.; Rhoads, R.P. Effects of Heat Stress on Postabsorptive Metabolism and Energetics. *Annu. Rev. Anim. Biosci.* **2013**, *1*, 311–337. [CrossRef]
72. Oh, S.; Hosseindoust, A.; Ha, S.; Moturi, J.; Mun, J.; Tajudeen, H.; Kim, J. Metabolic Responses of Dietary Fiber during Heat Stress: Effects on Reproductive Performance and Stress Level of Gestating Sows. *Metabolites* **2022**, *12*, 280. [CrossRef]
73. Kim, K.; Choi, Y.; Hosseindoust, A.; Kim, M.; Hwang, S.; Bu, M.; Lee, J.; Kim, Y.; Chae, B. Evaluation of High Nutrient Diets and Additional Dextrose on Reproductive Performance and Litter Performance of Heat-stressed Lactating Sows. *Anim. Sci. J.* **2019**, *90*, 1212–1219. [CrossRef]
74. Khaki, Z.; Nooshirvani, P.; Shirani, D.; Masoudifard, M. Diagnostic Value of Atrial Natriuretic Peptide (ANP), B-Type Natriuretic Peptide (BNP) and Their Correlation with Lipoproteins in Dogs with Myxomatous Mitral Valve Disease. *BMC Vet. Res.* **2022**, *18*, 448. [CrossRef]
75. Park, M.; Kim, K.H.; Jaiswal, V.; Choi, J.; Chun, J.L.; Seo, K.M.; Lee, M.-J.; Lee, H.-J. Effect of Black Ginseng and Silkworm Supplementation on Obesity, the Transcriptome, and the Gut Microbiome of Diet-Induced Overweight Dogs. *Sci. Rep.* **2021**, *11*, 16334. [CrossRef]
76. do Nascimento, R.C.; Souza, C.M.M.; Bastos, T.S.; Kaelle, G.C.B.; de Oliveira, S.G.; Félix, A.P. Effects of an Herbal Source of Choline on Diet Digestibility and Palatability, Blood Lipid Profile, Liver Morphology, and Cardiac Function in Dogs. *Animals* **2022**, *12*, 2658. [CrossRef]
77. Phungviwatnikul, T.; Alexander, C.; Do, S.; He, F.; Suchodolski, J.; Godoy, M.; Swanson, K. Effects of Dietary Macronutrient Profile on Apparent Total Tract Macronutrient Digestibility and Fecal Microbiota, Fermentative Metabolites, and Bile Acids of Female Dogs after Spay Surgery. *J. Anim. Sci.* **2021**, *99*, skab225. [CrossRef]
78. Geary, E.L.; Oba, P.M.; Applegate, C.C.; Clark, L.V.; Fields, C.J.; Swanson, K.S. Effects of a Mildly Cooked Human-Grade Dog Diet on Gene Expression, Skin and Coat Health Measures, and Fecal Microbiota of Healthy Adult Dogs. *J. Anim. Sci.* **2022**, *100*, skac265. [CrossRef]
79. Wang, L.; Ren, B.; Hui, Y.; Chu, C.; Zhao, Z.; Zhang, Y.; Zhao, B.; Shi, R.; Ren, J.; Dai, X.; et al. Methionine Restriction Regulates Cognitive Function in High-Fat Diet-Fed Mice: Roles of Diurnal Rhythms of SCFAs Producing- and Inflammation-Related Microbes. *Mol. Nutr. Food Res.* **2020**, *64*, e2000190. [CrossRef]
80. Hou, Q.; Huang, J.; Zhao, L.; Pan, X.; Liao, C.; Jiang, Q.; Lei, J.; Guo, F.; Cui, J.; Guo, Y.; et al. Dietary Genistein Increases Microbiota-Derived Short Chain Fatty Acid Levels, Modulates Homeostasis of the Aging Gut, and Extends Healthspan and Lifespan. *Pharmacol. Res.* **2023**, *188*, 106676. [CrossRef]
81. Silva, Y.P.; Bernardi, A.; Frozza, R.L. The Role of Short-Chain Fatty Acids From Gut Microbiota in Gut-Brain Communication. *Front. Endocrinol.* **2020**, *11*, 25. [CrossRef]
82. Wu, T.; Guo, X.; Zhang, M.; Yang, L.; Liu, R.; Yin, J. Anthocyanins in Black Rice, Soybean and Purple Corn Increase Fecal Butyric Acid and Prevent Liver Inflammation in High Fat Diet-Induced Obese Mice. *Food Funct.* **2017**, *8*, 3178–3186. [CrossRef]
83. Chambers, E.S.; Viardot, A.; Psichas, A.; Morrison, D.J.; Murphy, K.G.; Zac-Varghese, S.E.K.; MacDougall, K.; Preston, T.; Tedford, C.; Finlayson, G.S.; et al. Effects of Targeted Delivery of Propionate to the Human Colon on Appetite Regulation, Body Weight Maintenance and Adiposity in Overweight Adults. *Gut* **2015**, *64*, 1744–1754. [CrossRef]
84. Byrne, C.S.; Chambers, E.S.; Alhabeeb, H.; Chhina, N.; Morrison, D.J.; Preston, T.; Tedford, C.; Fitzpatrick, J.; Irani, C.; Busza, A.; et al. Increased Colonic Propionate Reduces Anticipatory Reward Responses in the Human Striatum to High-Energy Foods. *Am. J. Clin. Nutr.* **2016**, *104*, 5–14. [CrossRef] [PubMed]
85. Lan, Q.; Lian, Y.; Peng, P.; Yang, L.; Zhao, H.; Huang, P.; Ma, H.; Wei, H.; Yin, Y.; Liu, M. Association of Gut Microbiota and SCFAs with Finishing Weight of Diannan Small Ear Pigs. *Front. Microbiol.* **2023**, *14*, 1117965. [CrossRef] [PubMed]
86. Manjarín, R.; Dillard, K.; Coffin, M.; Hernandez, G.V.; Smith, V.A.; Noland-Lidell, T.; Gehani, T.R.; Smart, H.J.; Wheeler, K.; Sprayberry, K.A.; et al. Dietary Fat Composition Shapes Bile Acid Metabolism and Severity of Liver Injury in a Pig Model of Pediatric NAFLD. *Am. J. Physiol. Endocrinol. Metab.* **2022**, *323*, E187–E206. [CrossRef] [PubMed]

87. Breuninger, T.A.; Wawro, N.; Freuer, D.; Reitmeier, S.; Artati, A.; Grallert, H.; Adamski, J.; Meisinger, C.; Peters, A.; Haller, D.; et al. Fecal Bile Acids and Neutral Sterols Are Associated with Latent Microbial Subgroups in the Human Gut. *Metabolites* **2022**, *12*, 846. [CrossRef]
88. Visekruna, A.; Luu, M. The Role of Short-Chain Fatty Acids and Bile Acids in Intestinal and Liver Function, Inflammation, and Carcinogenesis. *Front. Cell Dev. Biol.* **2021**, *9*, 703218. [CrossRef]
89. Liu, L.; Xu, J.; Xu, X.; Mao, T.; Niu, W.; Wu, X.; Lu, L.; Zhou, H. Intestinal Stem Cells Damaged by Deoxycholic Acid via AHR Pathway Contributes to Mucosal Barrier Dysfunction in High-Fat Feeding Mice. *Int. J. Mol. Sci.* **2022**, *23*, 15578. [CrossRef]
90. Xu, Y.; Wang, J.; Wu, X.; Jing, H.; Zhang, S.; Hu, Z.; Rao, L.; Chang, Q.; Wang, L.; Zhang, Z. Gut Microbiota Alteration after Cholecystectomy Contributes to Post-Cholecystectomy Diarrhea via Bile Acids Stimulating Colonic Serotonin. *Gut Microbes* **2023**, *15*, 2168101. [CrossRef]
91. He, Y.; Brouwers, B.; Liu, H.; Liu, H.; Lawler, K.; Mendes de Oliveira, E.; Lee, D.-K.; Yang, Y.; Cox, A.R.; Keogh, J.M.; et al. Human Loss-of-Function Variants in the Serotonin 2C Receptor Associated with Obesity and Maladaptive Behavior. *Nat. Med.* **2022**, *28*, 2537–2546. [CrossRef]
92. Pan, J.; Chui, L.; Liu, T.; Zheng, Q.; Liu, X.; Liu, L.; Zhao, Y.; Zhang, L.; Song, M.; Han, J.; et al. Fecal Microbiota Was Reshaped in UCP1 Knock-In Pigs via the Adipose-Liver-Gut Axis and Contributed to Less Fat Deposition. *Microbiol. Spectr.* **2023**, *11*, e0354022. [CrossRef]
93. Zhang, L.; Yang, K.; Jian, S.; Xin, Z.; Wen, C.; Zhang, L.; Huang, J.; Deng, B.; Deng, J. Effects of Softening Dry Food with Water on Stress Response, Intestinal Microbiome, and Metabolic Profile in Beagle Dogs. *Metabolites* **2022**, *12*, 1124. [CrossRef]
94. Xu, J.; Becker, A.A.M.J.; Luo, Y.; Zhang, W.; Ge, B.; Leng, C.; Wang, G.; Ding, L.; Wang, J.; Fu, X.; et al. The Fecal Microbiota of Dogs Switching to a Raw Diet Only Partially Converges to That of Wolves. *Front. Microbiol.* **2021**, *12*, 701439. [CrossRef]
95. Yang, K.; Jian, S.; Guo, D.; Wen, C.; Xin, Z.; Zhang, L.; Kuang, T.; Wen, J.; Yin, Y.; Deng, B. Fecal Microbiota and Metabolomics Revealed the Effect of Long-Term Consumption of Gallic Acid on Canine Lipid Metabolism and Gut Health. *Food Chem. X* **2022**, *15*, 100377. [CrossRef]
96. Butowski, C.F.; Moon, C.D.; Thomas, D.G.; Young, W.; Bermingham, E.N. The Effects of Raw-Meat Diets on the Gastrointestinal Microbiota of the Cat and Dog: A Review. *N. Z. Vet. J.* **2022**, *70*, 1–9. [CrossRef]
97. Munukka, E.; Rintala, A.; Toivonen, R.; Nylund, M.; Yang, B.; Takanen, A.; Hänninen, A.; Vuopio, J.; Huovinen, P.; Jalkanen, S.; et al. Faecalibacterium Prausnitzii Treatment Improves Hepatic Health and Reduces Adipose Tissue Inflammation in High-Fat Fed Mice. *ISME J.* **2017**, *11*, 1667–1679. [CrossRef]
98. Wang, L.; Christophersen, C.T.; Sorich, M.J.; Gerber, J.P.; Angley, M.T.; Conlon, M.A. Increased Abundance of Sutterella Spp. and Ruminococcus Torques in Feces of Children with Autism Spectrum Disorder. *Mol. Autism* **2013**, *4*, 42. [CrossRef]
99. Tang, Z.; Li, X.; Wang, X.; Zhang, C.; Zou, L.; Ren, H.; Liu, W. Characterization and Genomic Analysis of a Novel Lytic Phage DCp1 against Clostridium Perfringens Biofilms. *Int. J. Mol. Sci.* **2023**, *24*, 4191. [CrossRef]
100. Paul, G.C.; Friend, D.G. Clostridial Enterotoxemia and Coccidiosis in Weanling Cottontail Rabbits (*Sylvilagus audubonii, Sylvilagus floridanus, Sylvilagus nuttallii*) from Colorado, USA. *J. Wildl. Dis.* **2019**, *55*, 189–195. [CrossRef]
101. Pei, T.; Zhu, D.; Yang, S.; Hu, R.; Wang, F.; Zhang, J.; Yan, S.; Ju, L.; He, Z.; Han, Z.; et al. Bacteroides Plebeius Improves Muscle Wasting in Chronic Kidney Disease by Modulating the Gut-Renal Muscle Axis. *J. Cell. Mol. Med.* **2022**, *26*, 6066–6078. [CrossRef]
102. Mao, B.; Guo, W.; Liu, X.; Cui, S.; Zhang, Q.; Zhao, J.; Tang, X.; Zhang, H. Potential Probiotic Properties of Blautia Producta Against Lipopolysaccharide-Induced Acute Liver Injury. *Probiotics Antimicrob. Proteins* **2023**. [CrossRef]
103. Yang, Y.-N.; Wang, Q.-C.; Xu, W.; Yu, J.; Zhang, H.; Wu, C. The Berberine-Enriched Gut Commensal Blautia Producta Ameliorates High-Fat Diet (HFD)-Induced Hyperlipidemia and Stimulates Liver LDLR Expression. *Biomed. Pharmacother.* **2022**, *155*, 113749. [CrossRef] [PubMed]
104. Wang, A.; Zhang, Z.; Ding, Q.; Yang, Y.; Bindelle, J.; Ran, C.; Zhou, Z. Intestinal *Cetobacterium* and Acetate Modify Glucose Homeostasis via Parasympathetic Activation in Zebrafish. *Gut Microbes* **2021**, *13*, 1–15. [CrossRef] [PubMed]
105. Xie, M.; Hao, Q.; Xia, R.; Olsen, R.E.; Ringø, E.; Yang, Y.; Zhang, Z.; Ran, C.; Zhou, Z. Nuclease-Treated Stabilized Fermentation Product of Cetobacterium Somerae Improves Growth, Non-Specific Immunity, and Liver Health of Zebrafish (Danio Rerio). *Front. Nutr.* **2022**, *9*, 918327. [CrossRef] [PubMed]
106. Wang, L.; Derous, D.; Huang, X.; Mitchell, S.E.; Douglas, A.; Lusseau, D.; Wang, Y.; Speakman, J.R. The Effects of Graded Levels of Calorie Restriction: XIX. Impact of Graded Calorie Restriction on Protein Expression in the Liver. *J. Gerontol. Ser. A* **2023**, glad017. [CrossRef]
107. Zhao, M.; Liu, L.; Liu, F.; Liu, L.; Liu, Z.; Gao, Y.; Cao, J. Traditional Chinese Medicine Improves Myasthenia Gravis by Regulating the Symbiotic Homeostasis of the Intestinal Microbiota and Host. *Front. Microbiol.* **2022**, *13*, 1082565. [CrossRef]
108. Montassier, E.; Gastinne, T.; Vangay, P.; Al-Ghalith, G.A.; Bruley des Varannes, S.; Massart, S.; Moreau, P.; Potel, G.; de La Cochetière, M.F.; Batard, E.; et al. Chemotherapy-Driven Dysbiosis in the Intestinal Microbiome. *Aliment. Pharmacol. Ther.* **2015**, *42*, 515–528. [CrossRef]

109. He, Z.; Ma, Y.; Yang, S.; Zhang, S.; Liu, S.; Xiao, J.; Wang, Y.; Wang, W.; Yang, H.; Li, S.; et al. Gut Microbiota-Derived Ursodeoxycholic Acid from Neonatal Dairy Calves Improves Intestinal Homeostasis and Colitis to Attenuate Extended-Spectrum β-Lactamase-Producing Enteroaggregative Escherichia Coli Infection. *Microbiome* **2022**, *10*, 79. [CrossRef]
110. Chen, Y.; Shen, X.; Ma, T.; Yu, X.; Kwok, L.-Y.; Li, Y.; Sun, Z.; Li, D.; Zhang, H. Adjunctive Probio-X Treatment Enhances the Therapeutic Effect of a Conventional Drug in Managing Type 2 Diabetes Mellitus by Promoting Short-Chain Fatty Acid-Producing Bacteria and Bile Acid Pathways. *mSystems* **2023**, *8*, e0130022. [CrossRef]

metabolites

MDPI

Article

Comparison of the Effects of Enzymolysis Seaweed Powder and *Saccharomyces boulardii* on Intestinal Health and Microbiota Composition in Kittens

Mingrui Zhang [1], Ruixia Mo [1], Mingtan Li [2], Yuankai Qu [2], Haotian Wang [1], Tianyi Liu [1], Pan Liu [1] and Yi Wu [1,*]

[1] State Key Laboratory of Animal Nutrition, College of Animal Science and Technology, China Agricultural University, Beijing 100193, China

[2] Shidai Marine Biotechnology Co., Ltd., Weihai 264319, China

* Correspondence: wuyi2013@cau.edu.cn; Tel.: +86-13611301596

Abstract: Kittens are prone to intestinal health problems as their intestines are not completely developed. Seaweed is rich in plant polysaccharides and bioactive substances that are highly beneficial to gut health. However, the effects of seaweed on cat gut health have not been assessed. This study compared the effects of dietary supplementation with enzymolysis seaweed powder and *Saccharomyces boulardii* on the intestinal health of kittens. In total, 30 Ragdoll kittens (age: 6 months; weight: 1.50 ± 0.29 kg) were assigned to three treatment groups for a 4-week feeding trial. The dietary treatment given was as follows: (1) basal diet (CON); (2) CON + enzymolysis seaweed powder (20 g/kg of feed) mixed evenly with the diet (SE); and (3) CON + *Saccharomyces boulardii* (2×10^{10} CFU/kg of feed) mixed evenly with the diet (SB). Compared with the CON and SB groups, dietary supplementation with the enzymolysis seaweed powder improved the immune and antioxidant capacity and also reduced the intestinal permeability and inflammation levels of kittens. The relative abundance of Bacteroidetes, Lachnospiraceae, Prevotellaceae, and *Faecalibacterium* in the SE group was higher than those in the CON and SB groups ($p \leq 0.05$), while the relative abundance of Desulfobacterota, Sutterellaceae, and *Erysipelatoclostridium* in the SB group was lower than that in the SE group ($p \leq 0.05$). Moreover, enzymolysis seaweed powder did not alter the level of intestinal SCFAs in kittens. Conclusively, supplementing kitten diet with enzymolysis seaweed powder can promote intestinal health by enhancing the gut barrier function and optimizing the microbiota composition. Our findings provide new perspectives on the application of enzymolysis seaweed powder.

Keywords: enzymolysis seaweed; *Saccharomyces boulardii*; kitten; intestinal health; gut microbiota

Citation: Zhang, M.; Mo, R.; Li, M.; Qu, Y.; Wang, H.; Liu, T.; Liu, P.; Wu, Y. Comparison of the Effects of Enzymolysis Seaweed Powder and *Saccharomyces boulardii* on Intestinal Health and Microbiota Composition in Kittens. *Metabolites* **2023**, *13*, 637. https://doi.org/10.3390/metabo13050637

Academic Editor: Bruno Stefanon

Received: 22 February 2023
Revised: 19 March 2023
Accepted: 5 May 2023
Published: 8 May 2023

1. Introduction

Gut health determines the whole health of the animal because the gut is not only a vital organ for digestion and nutrient absorption but also an immune organ [1]. Kittens have underdeveloped gut; therefore, they are more prone to intestinal health problems and exhibit symptoms of vomiting, diarrhea, reflux, and weight loss [2]. Notably, intestinal health in early life is critical to health and disease development in later life [3]. The colonization of gut microbes in infancy may define the lifelong composition of the microbiome [4]. Moreover, gut microbes influence nutritional intake, regulate intestinal homeostasis and physiological function, and drive immune responses in pets, thereby affecting their health [5]. Nutritional interventions in the diet can rapidly and repeatedly modify the composition and function of the gut microbiota [6]. Numerous studies have indicated that altering the dietary composition of cats can significantly influence the gut microbiome within 1 to 2 weeks [7,8]. There were significant differences in the composition of the intestinal microbiota between young and adult cats, and a marked increase in the structural and functional diversity of microbes was reported in adult cats [9]. Therefore, optimizing the intestinal microbiota structure through dietary interventions may be an essential means of improving the gut health of kittens. However, very few studies have used kittens as test subjects, with little data available for reference. Current studies have generally been conducted on adult cats.

Prebiotic supplementation is currently among the significant methods for improving the intestinal health of pets. Seaweed is rich in various bioactive compounds with anticancer, antioxidant, anti-inflammatory, and antibacterial properties [10–12]. More importantly, most of these active substances can resist gastric acid and host digestion, making seaweed a promising substrate for prebiotics [13]. Seaweed and its extracts can enhance nutrient digestibility and immune system function, thereby improving the growth and production performance of economic animals [14–16]. Furthermore, studies have shown that polysaccharides in seaweed can scavenge free radicals and enhance the endogenous antioxidant system of animals, thereby treating various diseases [17]. Meanwhile, seaweed polysaccharides are responsible for anticancer effects through inducing apoptosis of cancer cells, inhibiting lysosome-dependent autophagy, and suppressing cancer cell invasion and metastasis [18]. Conversely, research on the impacts of seaweed on pets is limited. A study showed that dogs fed a seaweed-supplemented diet exhibited no change in nutrient digestibility, fecal microbiota, and the metabolome [19]. Moreover, studies on the effects of seaweed on cat intestinal health have not yet been conducted.

Probiotics alter the host gut microbiota through different mechanisms, such as by changing the abundance of pathogenic bacteria, promoting microbiota interactions to stimulate the growth of resident bacteria, or indirectly influencing the microbiota structure through metabolite secretion [20]. Notably, probiotics are effective only if they tolerate stomach acid and bile salts, and adhere to and colonize the intestinal lumen. Otherwise, they will be excreted as passers-by bacteria. Factors such as resident microbiota, host characteristics, and diet determine the colonization success of probiotics [21]. In addition, some studies have questioned whether probiotics are safe for use in pets [22]. Therefore, additional studies are required to track the benefits of probiotics as well as the frequency and severity of their adverse outcomes in pets over time. Accumulating evidence suggests that *Saccharomyces boulardii* (*S. boulardii*) can inhibit the colonization of pathogenic microorganisms, improve intestinal barrier function, and regulate immunity [23]. This microbe is widely used as a commercial probiotic in pets. *S. boulardii* supplementation can improve the intestinal status, reduce stress, and act as an effective adjunct to the treatment of chronic enteropathies in dogs [24,25]. However, studies on the application of *S. boulardii* in cats are few, and its effects on the intestinal health of cats and the underlying action mechanisms remain unclear.

Ragdoll kittens were selected in this study because they have delicate intestines and a higher incidence of intestinal diseases. The study compared the effects of the enzymolysis seaweed powder and *S. boulardii* on intestinal health in kittens, particularly focusing on the gut barrier function and microbiota composition.

2. Materials and Methods

Experimental protocols were approved by the Institutional Animal Care and Use Committee of China Agricultural University (AW31402202-1-6).

2.1. Animals and Experimental Treatments

Before the start of the research, a medical examination was performed on a colony of Ragdoll kittens, which included blood and serum analyses, urinalysis, appetite, body condition, fecal score, and parasite. Thirty (half male and half female) Ragdoll kittens (6-month old; weighed 1.50 ± 0.29 kg) with mild diarrhea were selected. The body condition score of all kittens was between 4 and 5 points (on a 9-point scale) [26], and their fecal score was between 3.5 and 4.5 points (on a 5-point scale) [27]. None of the kittens had immune-mediated diseases, allergies, or other conditions causing chronic gastrointestinal dysfunction. For 3 months before the trial, the kittens were not given any function-related drugs, and they were kept in the same feeding environment with the same basal diet. Thirty cats in good health were fully involved in the trial.

Thirty kittens were randomly assigned to three treatment groups with 10 replicates per group, including 5 males and 5 females. The dietary treatment given was as follows: (1) basal diet (CON); (2) CON + enzymolysis seaweed powder (20 g/kg of feed, SE); and (3) CON + *S. boulardii* (2×10^{10} CFU/k g of feed, SB). The basal diet was a dry extruded diet, which met the nutritional requirements recommended by the NRC (2006), and its raw material composition and nutritional level were close to those of commercially available cat food (Appendix A Table A1). Enzymolysis seaweed and *S. boulardii* were both powders, and evenly sprinkled and attached to the surface of the dry extruded diet. The enzymolysis seaweed powder (Shidai Marine Biotechnology Co., Ltd., Weihai, China) was obtained by hydrolyzing a *Laminaria* spp. with complex enzymes, containing 7.8% laminarin and 69% ash. *S. boulardii* (Levucell SB) was obtained from the Lallemand Animal Nutrition Group (Montreal, Canada). The experiment lasted for 28 days. The feed consumption of each kitten was recorded daily. Body weights (BWs) and fecal scores in each treatment were recorded on d 0, 14, and 28 to evaluate growth status and fecal quality, respectively.

2.2. Sample Collection

On d 28, 1 mL of blood was drawn from the forelimb head vein of each kitten into a vacutainer containing heparin sodium, centrifuged at 3000 r/min for 15 min at 4 °C to separate plasma, and stored at −20 °C for short-term preservation and further analysis. On d 28, a sterilized tray was placed under each cage to hold kitten feces; 5 g of fresh feces from each kitten was collected separately in sterile tubes with sterilized forceps, avoiding contact with other surfaces or materials. The collected fecal samples were preserved at −80 °C for microbiota and SCFA analyses.

2.3. Plasma Parameters Measurement

The plasma concentrations of immunoglobulins (IgA and IgG), inflammatory factors interleukin-1β (IL-1β), IL-6, IL-10, and tumor necrosis factor-α (TNF-α) were analyzed using commercially available ELISA kits following the manufacturer's instructions. Plasma superoxide dismutase (SOD) and malondialdehyde (MDA) levels that determine the antioxidant capacity were also assessed using ELISA kits according to the manufacturer's instructions. The concentrations of d-lactate (D-LA), lipopolysaccharide (LPS), diamine oxidase (DAO), and intestinal fatty acid-binding protein (iFABP) were determined using ELISA kits following the manufacturer's protocols. All commercial kits were sourced from Shanghai Enzyme-linked Biotechnology Co., Ltd. (Shanghai, China).

2.4. Fecal Microbiota Analysis

The QIAamp Fast DNA Stool Mini Kit (Qiagen, Germany) was used for the extraction of total bacterial DNA in kitten feces. The V3–V4 region of the 16S rRNA was amplified using universal primers 338F (5′-ACTCCTACGGGAGGCAGCAG-3′) and 806R (5′-GGACTACHVGGGTWTCTAAT-3′), subsequently pooled into equimolar amounts, and sequenced on the Illumina MiSeq platform to generate paired-end reads of 300 base pairs (bp). The fastp software (version 0.19.6) was used for quality control of the raw sequences of 16s rRNA, and FLASH software (version 1.2.11) was used for splicing. Then the optimized sequences were clustered into operational taxonomic units (OTUs) using UPARSE software (Version 11) with 97% sequence similarity level. The taxonomy of each OTU representative sequence was analyzed by the RDP Classifier (Version 2.2) against the Silva 138 database using a confidence threshold of 0.7. Alpha diversity indices (the Shannon index and Simpson index) were calculated with Mothur (Version 1.30.2). Beta diversity was analyzed using Bray Curtis distance and visualized by principal component analysis (PCA). The community species composition of each sample was determined at each taxonomic level. The Kruskal–Wallis H test was used to detect significant differences in abundance among the groups. Raw data were uploaded to the NCBI SRA Database with accession number SRP422749.

2.5. Determination of Fecal SCFAs

The SCFA concentrations in the feces were determined through GC-MS (Shanghai Major Biomedical Technology Co., Ltd., Shanghai, China). Briefly, methanol and the standard of each SCFA were used to prepare SCFA standard working solutions at different concentrations (10, 50, 100, 200, 300, 400, and 500 μg/mL) for GC-MS detection and analysis to obtain the standard curves of each SCFA. After that, 100 mg of feces was weighed, added to 450 μL methanol and 50 μL 2-ethylbutyric acid (1000 μg/mL) for 30 min of sonication in an ice bath, and then centrifuged at 13,000× *g* for 15 min at 4 °C to separate the supernatant. Next, 50 mg of anhydrous sodium sulfate was added to the supernatant, and the supernatant was centrifuged at 13,000× *g* for 15 min at 4 °C after the vortex. The supernatant solution was collected and analyzed using the 8890B-5977B GC/MSD (Agilent Technologies Inc., Santa Clara, CA, USA).

2.6. Statistical Analysis

Data analysis was performed using IBM SPSS Statistics 26 (Chicago, IL, USA) and GraphPad Prism (version 8.3.0, San Diego, CA, USA). Microbial sequencing data were graphed using the R tool. The bar graphs of the bacterial community were prepared using the R ggplot package, and heatmaps were constructed using the R vegan package. The abundance of microbiota and fecal scores were compared using the Kruskal–Wallis H test. Other data were compared through one-way ANOVA using the Tukey post-hoc test. A score of $p \leq 0.05$ was considered statistically significant. Data are expressed as the mean ± SEM.

3. Results

3.1. Growth Performance

Table 1 presents the effects of different dietary treatments on the growth performance of kittens. No significant differences in the average daily feed intake and the BW of kittens were observed on d 0, 14, and 28 among the three treatments ($p > 0.05$). However, compared with kittens that consumed CON and SB diets, the ADG of kittens that consumed the SE diet significantly increased during d 15 to 28 ($p \leq 0.05$). Additionally, on d 0 to 28, ADG in the SE group was greater than that in the CON group ($p \leq 0.05$). However, ADG did not differ significantly between the SE and SB groups or the SB and CON groups ($p > 0.05$).

Table 1. Effects of enzymolysis seaweed powder and *S. boulardii* on ADG and ADFI in kittens.

Items	CON	SE	SB	SEM	*p*-Value
BW, kg					
d 0	1.50	1.51	1.50	0.13	0.99
d 14	1.69	1.72	1.75	0.15	0.93
d 28	1.91	2.10	2.03	0.16	0.47
ADG, g/d					
d 0 to 14	13.6	15.3	17.9	2.93	0.35
d 15 to 28	15.6^{b}	27.5^{a}	19.6^{b}	3.85	0.02
d 0 to 28	14.6^{b}	21.4^{a}	18.7^{ab}	2.20	0.02
ADFI, g/d					
d 0 to 14	90.0	90.0	93.2	2.47	0.33
d 15 to 28	65.4	63.8	62.2	6.24	0.88
d 0 to 28	90.0	88.9	90.5	1.78	0.64

^{a,b} Within a row, means without a common superscript differ at $p \leq 0.05$. CON, basal diet; SE, basal diet + enzymolysis seaweed powder (20 g/kg of feed); SB, basal diet + *S. boulardii* (2×10^{10} CFU/kg of feed). BW, body weight; ADG, average daily gain; ADFI, average daily feed intake.

3.2. Fecal Score

Table 2 presents the effects of different dietary treatments on fecal scores. Because of our screening of experimental kittens, no significant difference was observed in the fecal scores among the three groups on d 0. On d 14, the fecal score of kittens in the SE group was lower than that in the CON group ($p \leq 0.05$). However, the score did not differ significantly between the SE and SB groups or the SB and CON groups ($p > 0.05$), but, compared with kittens that consumed the CON diet, the fecal score of kittens that consumed with SE and SB diets was significantly decreased on d 28 ($p \leq 0.05$).

Table 2. Effect of enzymolysis seaweed powder and *S. boulardii* on fecal scores in kittens.

Items	CON	SE	SB	SEM	*p*-Value
d 0	4.00	4.05	4.10	0.07	0.86
d 14	3.65 ^{b}	2.75 ^{a}	3.55 ^{ab}	0.15	0.03
d 28	3.05 ^{b}	2.20 ^{a}	2.15 ^{a}	0.10	<0.01

^{a,b} Within a row, means without a common superscript differ at $p \leq 0.05$. CON, basal diet; SE, basal diet + enzymolysis seaweed powder (20 g/kg of feed); SB, basal diet + *S. boulardii* (2×10^{10} CFU/kg of feed).

3.3. Immunoglobulin and Antioxidant Parameters

Table 3 presents the immunoglobulin and antioxidant parameters of kittens subjected to different treatments. The IgA and IgG levels were significantly higher in the SE group than those in the CON and SB groups ($p \leq 0.05$) but significantly lower in the CON group than those in the SB group ($p \leq 0.05$). Compared with kittens that consumed CON and SB diets, the SOD concentration significantly increased and the MDA concentration significantly decreased in kittens that consumed the SE diet ($p \leq 0.05$). The SB group had higher ($p \leq 0.05$) SOD concentration and lower ($p \leq 0.05$) MDA concentration compared with the CON group.

Table 3. Effects of enzymolysis seaweed powder and *S. boulardii* on immunoglobulin and antioxidant parameters in kittens.

Items	CON	SE	SB	SEM	*p*-Value
IgA, μg/mL	177.4 ^{c}	461.5 ^{a}	387.9 ^{b}	10.6	<0.01
IgG, μg/mL	2520 ^{c}	5091 ^{a}	4359 ^{b}	138	<0.01
SOD, pg/mL	92.57 ^{c}	220.6 ^{a}	178.9 ^{b}	5.48	<0.01
MDA, nmol/mL	8.602 ^{a}	4.572 ^{c}	5.954 ^{b}	0.19	<0.01

$^{a–c}$ Within a row, means without a common superscript differ at $p \leq 0.05$. CON, basal diet; SE, basal diet + enzymolysis seaweed powder (20 g/kg of feed); SB, basal diet + *S. boulardii* (2×10^{10} CFU/kg of feed). IgA, immunoglobulin A; IgG, immunoglobulin G; SOD, superoxide dismutase; MDA, malondialdehyde.

3.4. Inflammatory Factors

As shown in Table 4, the SE group exhibited increased serum IL-10 levels compared with the CON and SB groups ($p \leq 0.05$). Meanwhile, compared with the CON and SB groups, IL-1β, IL-6 and TNF-α levels decreased in the SE group ($p \leq 0.05$). Moreover, serum IL-10 levels increased but IL-1β, IL-6 and TNF-α levels decreased in the SB group compared with those in the CON group ($p \leq 0.05$).

Table 4. Effects of enzymolysis seaweed powder and *S. boulardii* on plasma inflammatory parameters in kittens.

Items	CON	SE	SB	SEM	*p*-Value
IL-1β, pg/mL	216.3 ^{a}	149.9 ^{c}	167.4 ^{b}	3.78	<0.01
IL-6, pg/mL	38.91 ^{a}	20.41 ^{c}	26.17 ^{b}	1.16	<0.01
IL-10, pg/mL	102.0 ^{c}	219.2 ^{a}	184.3 ^{b}	6.12	<0.01
TNF-α, pg/mL	140.7 ^{a}	88.11 ^{c}	105.0 ^{b}	2.75	<0.01

$^{a–c}$ Within a row, means without a common superscript differ at $p \le 0.05$. CON, basal diet; SE, basal diet + enzymolysis seaweed powder (20 g/kg of feed); SB, basal diet + *S. boulardii* (2×10^{10} CFU/kg of feed). IL-1β, interleukin-1β; IL-6, interleukin-6; IL-10, interleukin-10; TNF-α, tumor necrosis factor-α.

3.5. Gut Barrier Function Parameters

The gut barrier function parameters of kittens in the different dietary treatment groups are listed in Table 5. The D-LA, LPS, DAO, and iFABP levels were decreased in the SE group compared with those in the CON and SB groups ($p \le 0.05$), and were decreased in the SB group compared with those in the CON group ($p \le 0.05$).

Table 5. Effects of enzymolysis seaweed powder and *S. boulardii* on plasma intestinal barrier function parameters in kittens.

Items	CON	SE	SB	SEM	*p*-Value
D-LA, μmol/L	145.3 ^{a}	98.54 ^{c}	113.8 ^{b}	2.44	<0.01
LPS, EU/L	873.6 ^{a}	363.3 ^{c}	496.8 ^{b}	24.6	<0.01
DAO, ng/mL	56.06 ^{a}	32.32 ^{c}	41.57 ^{b}	0.82	<0.01
iFABP, pg/mL	1842 ^{a}	1130 ^{c}	1392 ^{b}	32.2	<0.01

$^{a–c}$ Within a row, means without a common superscript differ at $p \le 0.05$. CON, basal diet; SE, basal diet + enzymolysis seaweed powder (20 g/kg of feed); SB, basal diet + *S. boulardii* (2×10^{10} CFU/kg of feed). D-LA, D-lactate; LPS, lipopolysaccharide; DAO, Diamine oxidase; iFABP, intestinal fatty acid-binding protein.

3.6. Fecal Microbiota Composition

Figure 1 presents changes in the microbiota composition in kitten feces after different treatments. PCA demonstrated distinct clustering of fecal microbial communities for each group ($p \le 0.05$; Table A1). Compared with the SB group, the SE group had a higher Shannon index and a lower Simpson index ($p \le 0.05$; Figure 1B). This indicated that SE consumption increased microbial richness and diversity. Furthermore, we identified fecal bacteria at the phylum, family, and genus levels, exhibiting differences in the abundance across groups. At the phylum level (Figure 1C), the abundance of Bacteroidetes in the fecal microbiota of the SE group was higher than that of the CON and SB groups ($p \le 0.05$). The abundance of Desulfobacterota was significantly lower in the SB group than that in the CON and SE groups ($p \le 0.05$). At the family level (Figure 1D), the abundance of Lachnospiraceae, Prevotellaceae, and Bacteroidaceae was higher in the SE group than that in the CON and SB groups ($p \le 0.05$). The abundance of the Muribaculaceae was significantly increased in the SE group compared with the SB group ($p \le 0.05$). The fecal microbiota of the SB group had lower abundance of Sutterellaceae and Desulfovibrionaceae compared with that of the SE group ($p \le 0.05$). At the genus level (Figure 1E), the abundance of *Faecalibacterium*, *Bacteroides*, and *Eubacterium_hallii_group* increased in the SE group compared with those in the CON and SB groups ($p \le 0.05$). The abundance of *Erysipelatoclostridium* and *Flavonifractor* in the fecal microbiota of the SB group was lower than that of the SE group ($p \le 0.05$); however, it did not differ significantly between the CON and SB groups ($p > 0.05$). The relative abundance of *Sutterella* was lower in the SB group than that in the CON and SE groups ($p \le 0.05$).

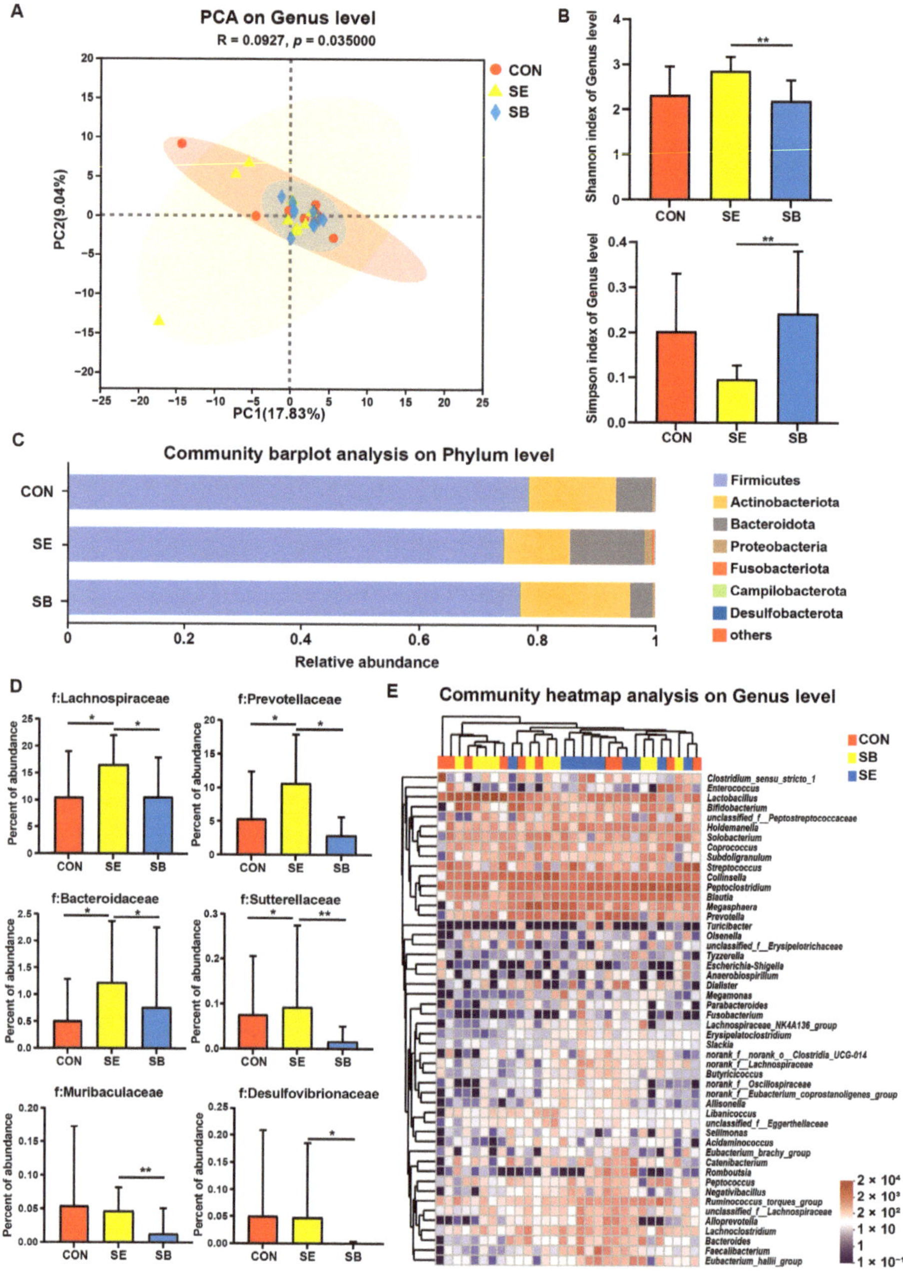

Figure 1. Changes in the microbial structure in feces of kittens subjected to different treatments. (**A**) Principal component analysis at the genus level. (**B**) The α-diversity analysis using the Shannon and Simpson indices. (**C**) The relative abundance of fecal bacteria at the phylum level. (**D**) The relative abundance of Lachnospiraceae, Prevotellaceae, Bacteroidaceae, Sutterellaceae, Muribaculaceae, and Desulfovibrionaceae. (**E**) The relative abundance of fecal bacteria at the genus level. CON, basal diet; SE, basal diet + enzymolysis seaweed powder (20 g/kg of feed); SB, basal diet + *S. boulardii* (2×10^{10} CFU/kg of feed). ** $p < 0.01$, * $p < 0.05$.

3.7. Fecal SCFAs

As shown in Table 6, there were no significant differences in fecal concentrations of acetate, propionate, butyrate, isobutyrate, valerate, and isovalerate among the three treatment groups ($p > 0.05$).

Table 6. Effects of enzymolysis seaweed powder and *S. boulardii* on SCFA concentrations in kittens.

Items, mg/kg	CON	SE	SB	SEM	*p*-Value
Acetate	3280	3609	3634	324.3	0.49
Propionate	2792	2782	2494	423.5	0.73
Butyrate	2247	2049	2158	360.3	0.86
Isobutyrate	280.5	240.2	238.9	51.71	0.66
Valerate	1027	1099	1063	194.1	0.93
Isovalerate	461.5	528.7	445.4	96.11	0.66

CON, basal diet; SE, basal diet + enzymolysis seaweed powder (20 g/kg of feed); SB, basal diet + *S. boulardii* (2×10^{10} CFU/kg of feed).

4. Discussion

The intestinal development of young animals is not sound, and they lack a stable intestinal microbiota, which makes them prone to intestinal microecology disorders. It is well known that dietary supplementation with probiotics and prebiotics is beneficial to animal intestines and, therefore, related products have been used in pet food. In vivo studies have focused on revealing the action mechanisms of specific dietary components. However, because of the large individual differences, the number of in vivo trials conducted on pets is very low, and adult animals are commonly used for experiments. In this study, it will be of special significance to take the lead in using kittens with larger scale and better sample uniformity. Numerous studies have shown that enzymolysis seaweed powder has a crucial application value in the breeding of economic animals and positively influences the production performance and body health of livestock and poultry [14,28,29]. Unfortunately, no application research has been conducted on enzymolysis seaweed powder in cats. Therefore, to provide data supporting the application of enzymolysis seaweed powder in pet food, this study explored the effects of this seaweed powder on immunity, antioxidant properties, intestinal permeability, and microbiota composition of kittens compared with the common commercial *S. boulardii*.

Immunoglobulin is a globulin with antibody activity and is widely found in mammalian serum, interstitial fluid, and exocrine fluid. The plasma immunoglobulin level can indirectly reflect the animal's ability to resist exogenous stimuli and pathogens. It is a critical parameter to evaluate the immunity of animals. As the major component of serum immunoglobulin, IgG accounts for 75–80% of total serum immunoglobulin. It is mainly synthesized and secreted by plasma cells in the spleen and lymph nodes [30]. IgG plays a pivotal role in combating infections in young animals [31]. IgA is crucial for local anti-infection action of the body's mucosa. The blocking of IgA synthesis increases susceptibility to local infections with microorganisms [32]. Studies have shown that dietary supplementation with seaweed by-products increased IgA concentration in the serum of chicks [33]. Serum IgG concentration in rats supplemented with heat-treated dried brown seaweed increased significantly after 16 weeks [34]. These results are in line with our findings, which showed that direct dietary supplementation with enzymolysis seaweed powder increased IgA and IgG levels. In addition, supplementation with seaweed extract increased IgA and IgG levels in maternal blood and colostrum, further enhancing the circulating IgG concentration and the percentage of leukocytes and lymphocytes phagocytosing Escherichia coli in lactating juvenile animals. This suggested that enzymolysis seaweed powder could improve the immune function by increasing immunoglobulin levels in young animals [35,36].

A stable reactive oxygen species (ROS) is essential for maintaining a normal physiological function. Under normal circumstances, the free radical content is at an equilibrium level. These free radicals are involved in cell signal transduction, synthesis and metabolism of substances, and energy production in the body. However, oxidative stress can trigger disease development when an imbalance between ROS production and antioxidant networks occurs [37]. Owing to their high sensitivity and activity, cellular stress levels may be higher in young animals [38]. SOD scavenges superoxide free radicals, and MDA is a lipid peroxidation product generated in the metabolism of oxygen free radicals in the organism. In this study, compared with the basal diet and the diet supplemented with *S. boulardii*, the addition of enzymolysis seaweed powder to the diet significantly increased the SOD level and reduced the MDA level in kittens. Consistent with our findings, a study showed that seaweed polysaccharides may increase SOD, CAT, and GSH levels and decrease the MDA level by regulating the NRF2 signaling pathway [39]. In addition, seaweed polysaccharides exert a beneficial effect on acetaminophen-induced acute liver injury in rats by elevating GSH, GSH-Px, and SOD expression and decreasing the MDA concentration [40]. Notably, supplementation of seaweed-derived polysaccharides increased the SOD level of weaned piglets at 24 days of age, along with changes in intestinal microbiota and increase in acetic acid and butyric acid concentrations [41].

Studies have confirmed the role of gut microbes in contributing to growth and development, maintaining health, and regulating the occurrence and treatment of various diseases in pets [42,43]. Moreover, the intestinal microbes and their functional products affect host health [44,45]. Cats are obligate carnivores and need a protein-rich diet. Therefore, the gut microbiota composition of cats is different from those of humans and many other mammals. However, the amount of microbiome research conducted in cats is relatively small compared with that in humans and other mammals. *Firmicutes*, *Bacteroidetes*, *Actinobacteria*, *Proteobacteria*, and *Fusobacteria* were the predominant phyla in previous feline microbiota reports that used different sequencing technologies [7,46,47], and these results are consistent with those of our study. Aging in the gut is related to changes in microbial composition, developing from colonization in early life to relative stability in adulthood and then decreasing diversity in old age [7,9,48]. Bacteroides were less represented in the kitten intestines but significantly increased in the adult cats, probably balanced by a reduced abundance of *Firmicutes* and *Actinobacteria* [7,49]. In our study, enzymolysis seaweed powder increased the Shannon index, decreased the Simpson index, and increased the abundance of *Bacteroidetes*, indicating that enzymolysis seaweed powder is beneficial for the stability and maturity of kitten intestinal microbiota. The abundance of Prevotellaceae and Bacteroidaceae (members of the phylum Bacteroidetes) as well as Lachnospiraceae and *Faecalibacterium* (members of the phylum Firmicutes) was significantly reduced in dogs with chronic enteropathies [50–53]. These gut bacteria are known to produce SCFAs. In our study, enzymolysis seaweed powder added to the basal diet increased the abundance of these bacteria. In addition, Muribaculaceae, much like Bacteroides, ferment plant polysaccharides to produce propionate [54]. Consistent with the study findings, dietary addition with brown seaweed elevated the relative abundance of Muribaculaceae in the gut of high-fat- and high-sugar-diet-induced obese rats [55]. Although *Eubacterium hallii* cannot degrade complex polysaccharides and oligosaccharides, it can produce butyrate by using fermentation metabolites such as acetate and lactic acid [56]. Supplementing enzymolysis seaweed powder significantly increased the abundance of *Eubacterium_hallii_group*, suggesting that SCFA generation changes to some extent. Thus, according to these results, enzymolysis seaweed powder may promote intestinal health by increasing the abundance of beneficial bacteria. Moreover, the abundance of *Sutterella* significantly increased in dogs with acute diarrhea [51]. Meanwhile, the abundance of *Desulfovibrio* increased markedly in the intestines of cats with inflammatory bowel disease [57]. *Erysipelatoclostridium* and its related metabolite, ptilosteroid A, have been considered markers of radiation-induced intestinal damage [58]. In this study, dietary addition with *S. boulardii* markedly reduced the relative abundance of *Desulfovibrionaceae*, *Erysipelatoclostridium*, and *Sutterella*, suggesting that *S. boulardii* improves the gut microbiota structure in kittens by lowering harmful bacteria.

The SCFAs are the main metabolites of intestinal microorganisms and serve as a source of energy for gut epithelial cells. They have various biological functions, such as enhancing the gut mucosal barrier function and participating in intestinal immune regulation [59]. Acetic acid, propionic acid, and butyric acid are the most abundant SCFAs in the animal's digestive tract. Similar to the case in other species, probiotics and prebiotics can affect intestinal SCFA production in felines by regulating the intestinal microbiota structure, especially by increasing the amount and activity of SCFA-producing bacteria [1]. The acetate and propionate contents of pre-digested red seaweed increased significantly after fermentation in a colonic model in vitro [60]. Furthermore, dietary supplementation with 400 mg/kg seaweed-derived polysaccharides significantly increased the cecal concentrations of acetate and butyrate in weaned piglets [41]. However, dietary addition with seaweed had no effect on the SCFA composition and concentration in dog feces [19], which is consistent with our findings in kittens. In addition, we did not find an effect of S. boulardii on SCFAs in kitten feces. Similarly, a study reported no significant effect of *S. boulardii* on SCFAs in the feces of breeding dogs [25]. Conversely, a recent study found that the addition of *S. boulardii* and *Pediococcus acidilactici* to feline diets increased the concentration of total SCFAs and butyric acid in cat feces, which may be beneficial for intestinal health [61]. Additionally, humanized mice supplemented with *S. boulardii* can increase the colonic contents of acetate, propionate, and butyrate, thereby alleviating colitis induced by dextran sulfate sodium [62]. As obligate carnivores, cats possess a distinctive intestinal structure and digestive system that enable them to efficiently digest and absorb high-protein and high-fat diets. Nevertheless, their capacity for fiber digestion is relatively poor [63,64]. Although some in vitro studies have shown that cats may increase the production of SCFAs by digesting fiber [65,66], current commercial cat foods are rich in protein and fat while limiting the use of fiber-rich ingredients. Similarly, only sweet potatoes, potatoes, and rice in this experimental diet could provide cats with a small amount of fiber. Most SCFAs are bacterial metabolites produced by specific anaerobic bacteria of the colon after the fermentation of dietary fiber. It is possible that the low fiber intake of experimental cats resulted in a lack of fermentable substrates for the gut microbiota to produce SCFAs. Therefore, the effects of enzymolysis seaweed powder and *S. boulardii* on SCFAs were nonsignificant under the test conditions.

The gut mucosal barrier is pivotal in preventing the invasion of pathogenic antigens and maintaining a normal intestinal function. Increased intestinal permeability indicates damage to the gut epithelial barrier, causing the penetration of toxins and pathogens [67]. D-LA and LPS are intestinal bacterial metabolites that disrupt the intestinal mucosal barrier, invade the bloodstream, and promote the release of inflammatory factors [68,69]. DAO is a highly active enzyme in the epithelial villi of the mammalian intestinal mucosa, and iFABP mainly exists in the mature intestinal epithelial cells of the small intestine. Under normal conditions, they are present at very low levels in the blood. Damage to the gut mucosal barrier increases permeability, resulting in the release of large quantities of the above substances into the bloodstream [70,71]. This study found that the dietary addition of enzymolysis seaweed powder significantly reduced plasma levels of D-LA, LPS, DAO, and iFABP in kittens compared with the basal diet and the *S. boulardii* diet. Consistent with our results, sodium alginate improved the gut mucosal barrier function in cyclophosphamide-induced immunosuppressed mice, which reduced serum concentrations of D-LA and LPS [72]. Furthermore, dietary supplementation with seaweed-derived polysaccharides could promote the levels of claudin-1, occludin, and ZO-1 in the jejunal mucosa and significantly reduced DAO activity and serum D-LA concentrations in weaned piglets [41]. Seaweed extract may improve intestinal health by strengthening the gut barrier function.

The gut is the largest immune organ of the body. Intestinal physical, chemical, and microbial barriers influence and alter host inflammation by regulating homeostasis as well as tolerating and preventing pathological immune responses. Cytokines, a class of endogenous polypeptides generated mainly by immune system cells, have many biological effects and can mediate various immune responses. The dynamic balance between pro-inflammatory and anti-inflammatory cytokines is the key for maintaining the body's normal immune status and physiological activities [73]. IL-1β, IL-6, and TNF-α are typical pro-inflammatory cytokines. Among them, IL-1β and TNF-α are pleiotropic, and their local activation can lead to elevated levels of secondary inflammatory mediators including IL-6 [74]. Laminarin can significantly downregulate the expressions of colon mucosal inflammatory factors IL-6, IL-17, and IL-1β, thereby improving the intestinal health of piglets [75]. Additionally, the brown seaweed extract reduced IL-6, IL-8, and TNF-α expressions in piglets with LPS-induced colitis, thereby inhibiting the pro-inflammatory factor response [76]. Consistent with the findings of other studies, we found that a dietary addition of enzymolysis seaweed powder in kittens reduced TNF-α, IL-1β, and IL-6 levels, probably by inhibiting NF-κB activation to reduce the inflammatory response [77]. Seaweed polysaccharides were reported to inhibit NF-κB activation and downregulate the mRNA expressions of TLR-4, MyD88, and IκBα, thereby inhibiting IL-6 and TNF-α expressions [77]. IL-10 is a pleiotropic cytokine mainly secreted by antigen-presenting cells. It has powerful anti-inflammatory properties that inhibit the levels of pro-inflammatory cytokines by activating macrophages [78]. In our study, compared with CON diet- and *S. boulardii* diet-fed kittens, the serum IL-10 level in kittens fed with enzymatic seaweed powder significantly increased, indicating that enzymolysis seaweed powder improved the anti-inflammatory ability of kittens to some extent.

5. Conclusions

In this study, dietary supplementation with enzymolysis seaweed powder increased plasma immune and antioxidant capacities, improved intestinal barrier function, and reduced the inflammation level in kittens. In addition, the aforementioned diet enriched the abundance of Bacteroidetes, Lachnospiraceae, Prevotellaceae, Muribaculaceae, *Faecalibacterium*, *Bacteroides*, and *Eubacterium_hallii_group* in the guts of kittens. Our findings provide a basis for the application and value of enzymolysis seaweed powder in kitten food.

Author Contributions: Conceptualization, M.Z. and Y.W.; methodology, M.Z. and Y.W.; software, R.M. and H.W.; validation, M.Z., M.L. and P.L.; formal analysis, M.Z.; investigation, R.M. and T.L.; resources, M.L. and Y.Q.; data curation, M.Z.; writing—original draft preparation, M.Z., Y.W. and R.M.; writing—review and editing, R.M. and Y.Q.; visualization, H.W.; supervision, Y.W.; project administration, Y.W.; funding acquisition, Y.W. All authors have read and agreed to the published version of the manuscript.

Funding: This research received no external funding.

Institutional Review Board Statement: The animal study protocol was approved by the Care and Use of Experimental Animals Committee of the China Agricultural University (AW31402202-1-6).

Informed Consent Statement: Not applicable.

Data Availability Statement: For this study, data are available from the corresponding authors upon request. The data are not publicly available due to the related product is under development and patent application.

Acknowledgments: The authors would like to express their gratitude to Weihai Shidai Marine Biotechnology Co., Ltd., for this research. The authors thank the faculty and staff in the Ministry of Agriculture and Rural Affairs Feed Industry Centre (Beijing, China) for their support of this study.

Conflicts of Interest: The authors declare no conflict of interest. Mingtan Li and Yuankai Qu are employees of Shidai Marine Biotechnology Co., Ltd. (Weihai, China). The paper reflects the views of the scientists, and not the company.

Appendix A

Table A1. Ingredient composition and nutrient levels of the experimental diets (as-fed basis).

Items	Content
Ingredients, %	
Chicken meal	46.25
Potato starch	19.00
Poultry fat	9.50
Fish meal	5.50
Chicken livers	4.50
Rice	4.00
Potato	3.00
Fish oil	3.00
Sweet potato	3.50
Salt	0.45
Taurine	0.15
Choline chloride	0.40
Magnesium sulfate	0.25
Vitamin Premix [1]	0.30
Mineral premix [2]	0.20
Total	100.00
Analyzed nutrient levels [3]	
Dry matter, %	90.61
Total energy, MJ/kg	23.02
Crude protein, %	39.98
Ether extract, %	22.39
Ash, %	7.98

Note: [1] Vitamin premix provided the following per kilogram of feed: vitamin A (15,000 IU), vitamin B_1 (30 mg), vitamin B_2 (28 mg), vitamin B_3 (110 mg), vitamin B_5 (85 mg), vitamin B_6 (12 mg), vitamin B_{12} (0.19 mg), vitamin D_3 (15 IU), and vitamin E (75,300 IU); [2] Mineral premix provided the following per kilogram of feed: Ca (CaI_2) 20 mg, Co ($CoSO_4$) 0.10 mg, Cu ($CuSO_4$) 3 mg, Fe ($FeSO_4$) 50 mg, I (CaI_2) 40 mg, Mn ($MnSO_4$) 18 mg, Na (Na_2SeO_3) 0.05 mg, Zn ($ZnSO_4$) 38 mg, Se (Na_2SeO_3) 260 mg; [3] The nutrient levels of the diets were analyzed.

References

1. Alessandri, G.; Argentini, C.; Milani, C.; Turroni, F.; Cristina Ossiprandi, M.; van Sinderen, D.; Ventura, M. Catching a glimpse of the bacterial gut community of companion animals: A canine and feline perspective. *Microb. Biotechnol.* **2020**, *13*, 1708–1732. [CrossRef] [PubMed]
2. Zimmer, J.F.; Pollock, R.V. Esophageal, gastric, and intestinal disorders of young dogs and cats. *Vet. Clin. N. Am. Small Anim. Pract.* **1987**, *17*, 641–661. [CrossRef] [PubMed]
3. Kalliomäki, M.; Salminen, S.; Isolauri, E. Positive interactions with the microbiota: Probiotics. *Adv. Exp. Med. Biol.* **2008**, *635*, 57–66. [CrossRef] [PubMed]
4. Adlerberth, I. Factors influencing the establishment of the intestinal microbiota in infancy. *Nestle Nutr. Workshop Ser. Pediatr. Program* **2008**, *62*, 13–29. [CrossRef] [PubMed]
5. Pilla, R.; Suchodolski, J.S. The Gut Microbiome of Dogs and Cats, and the Influence of Diet. *Vet. Clin. N. Am. Small Anim. Pract.* **2021**, *51*, 605–621. [CrossRef]
6. David, L.A.; Maurice, C.F.; Carmody, R.N.; Gootenberg, D.B.; Button, J.E.; Wolfe, B.E.; Ling, A.V.; Devlin, A.S.; Varma, Y.; Fischbach, M.A.; et al. Diet rapidly and reproducibly alters the human gut microbiome. *Nature* **2014**, *505*, 559–563. [CrossRef]
7. Hooda, S.; Vester Boler, B.M.; Kerr, K.R.; Dowd, S.E.; Swanson, K.S. The gut microbiome of kittens is affected by dietary protein:carbohydrate ratio and associated with blood metabolite and hormone concentrations. *Br. J. Nutr.* **2013**, *109*, 1637–1646. [CrossRef]
8. Bermingham, E.N.; Young, W.; Kittelmann, S.; Kerr, K.R.; Swanson, K.S.; Roy, N.C.; Thomas, D.G. Dietary format alters fecal bacterial populations in the domestic cat (*Felis catus*). *Microbiologyopen* **2013**, *2*, 173–181. [CrossRef]
9. Deusch, O.; O'Flynn, C.; Colyer, A.; Swanson, K.S.; Allaway, D.; Morris, P. A Longitudinal Study of the Feline Faecal Microbiome Identifies Changes into Early Adulthood Irrespective of Sexual Development. *PLoS ONE* **2015**, *10*, e0144881. [CrossRef]
10. Gomez-Zavaglia, A.; Prieto Lage, M.A.; Jimenez-Lopez, C.; Mejuto, J.C.; Simal-Gandara, J. The Potential of Seaweeds as a Source of Functional Ingredients of Prebiotic and Antioxidant Value. *Antioxidants* **2019**, *8*, 406. [CrossRef]
11. Khan, M.I.; Shin, J.H.; Kim, J.D. The promising future of microalgae: Current status, challenges, and optimization of a sustainable and renewable industry for biofuels, feed, and other products. *Microb. Cell Fact.* **2018**, *17*, 36. [CrossRef] [PubMed]

12. Roohinejad, S.; Koubaa, M.; Barba, F.J.; Saljoughian, S.; Amid, M.; Greiner, R. Application of seaweeds to develop new food products with enhanced shelf-life, quality and health-related beneficial properties. *Food Res. Int.* **2017**, *99*, 1066–1083. [CrossRef] [PubMed]
13. O'Sullivan, L.; Murphy, B.; McLoughlin, P.; Duggan, P.; Lawlor, P.G.; Hughes, H.; Gardiner, G.E. Prebiotics from marine macroalgae for human and animal health applications. *Mar. Drugs* **2010**, *8*, 2038–2064. [CrossRef]
14. Corino, C.; Modina, S.C.; Di Giancamillo, A.; Chiapparini, S.; Rossi, R. Seaweeds in Pig Nutrition. *Animals* **2019**, *9*, 1126. [CrossRef]
15. Hong, Z.S.; Kim, E.J.; Jin, Y.C.; Lee, J.S.; Choi, Y.J.; Lee, H.G. Effects of Supplementing Brown Seaweed By-products in the Diet of Holstein Cows during Transition on Ruminal Fermentation, Growth Performance and Endocrine Responses. *Asian Australas J. Anim. Sci.* **2015**, *28*, 1296–1302. [CrossRef] [PubMed]
16. Akinyemi, F.; Adewole, D. Effects of brown seaweed products on growth performance, plasma biochemistry, immune response, and antioxidant capacity of broiler chickens challenged with heat stress. *Poult. Sci.* **2022**, *101*, 102215. [CrossRef] [PubMed]
17. Liu, Z.; Sun, X. A Critical Review of the Abilities, Determinants, and Possible Molecular Mechanisms of Seaweed Polysaccharides Antioxidants. *Int. J. Mol. Sci.* **2020**, *21*, 7774. [CrossRef]
18. Liu, T.; Li, Q.; Xu, X.; Li, G.; Tian, C.; Zhang, T. Molecular mechanisms of anti-cancer bioactivities of seaweed polysaccharides. *Chin. Herb. Med.* **2022**, *14*, 528–534. [CrossRef]
19. Pinna, C.; Vecchiato, C.G.; Grandi, M.; Stefanelli, C.; Zannoni, A.; Biagi, G. Seaweed Supplementation Failed to Affect Fecal Microbiota and Metabolome as Well as Fecal IgA and Apparent Nutrient Digestibility in Adult Dogs. *Animals* **2021**, *11*, 2234. [CrossRef]
20. Derrien, M.; van Hylckama Vlieg, J.E. Fate, activity, and impact of ingested bacteria within the human gut microbiota. *Trends Microbiol.* **2015**, *23*, 354–366. [CrossRef]
21. Zmora, N.; Zilberman-Schapira, G.; Suez, J.; Mor, U.; Dori-Bachash, M.; Bashiardes, S.; Kotler, E.; Zur, M.; Regev-Lehavi, D.; Brik, R.B.; et al. Personalized Gut Mucosal Colonization Resistance to Empiric Probiotics Is Associated with Unique Host and Microbiome Features. *Cell* **2018**, *174*, 1388–1405.e1321. [CrossRef] [PubMed]
22. Doron, S.; Snydman, D.R. Risk and safety of probiotics. *Clin. Infect. Dis.* **2015**, *60*, S129–S134. [CrossRef] [PubMed]
23. Santacroce, L.; Charitos, I.A.; Bottalico, L. A successful history: Probiotics and their potential as antimicrobials. *Expert Rev. Anti. Infect. Ther.* **2019**, *17*, 635–645. [CrossRef] [PubMed]
24. D'Angelo, S.; Fracassi, F.; Bresciani, F.; Galuppi, R.; Diana, A.; Linta, N.; Bettini, G.; Morini, M.; Pietra, M. Effect of *Saccharomyces boulardii* in dog with chronic enteropathies: Double-blinded, placebo-controlled study. *Vet. Rec.* **2018**, *182*, 258. [CrossRef] [PubMed]
25. Meineri, G.; Martello, E.; Atuahene, D.; Miretti, S.; Stefanon, B.; Sandri, M.; Biasato, I.; Corvaglia, M.R.; Ferrocino, I.; Cocolin, L.S. Effects of *Saccharomyces boulardii* Supplementation on Nutritional Status, Fecal Parameters, Microbiota, and Mycobiota in Breeding Adult Dogs. *Vet. Sci.* **2022**, *9*, 389. [CrossRef]
26. Laflamme, D. Nutritional management. *Vet. Clin. N. Am. Small Anim. Pract.* **1997**, *27*, 1561–1577. [CrossRef]
27. Westermarck, E.; Frias, R.; Skrzypczak, T. Effect of diet and tylosin on chronic diarrhea in beagles. *J. Vet. Intern. Med.* **2005**, *19*, 822–827. [CrossRef]
28. Makkar, H.P.S.; Tran, G.; Heuzé, V.; Giger-Reverdin, S.; Lessire, M.; Lebas, F.; Ankers, P. Seaweeds for livestock diets: A review. *Anim. Feed Sci. Technol.* **2016**, *212*, 1–17. [CrossRef]
29. Øverland, M.; Mydland, L.T.; Skrede, A. Marine macroalgae as sources of protein and bioactive compounds in feed for monogastric animals. *J. Sci. Food Agric.* **2019**, *99*, 13–24. [CrossRef]
30. Kraehenbuhl, J.P.; Neutra, M.R. Molecular and cellular basis of immune protection of mucosal surfaces. *Physiol. Rev.* **1992**, *72*, 853–879. [CrossRef]
31. De Haan, N.; Reiding, K.R.; Driessen, G.; van der Burg, M.; Wuhrer, M. Changes in healthy human IgG Fc-Glycosylation after birth and during early childhood. *J. Proteome Res.* **2016**, *15*, 1853–1861. [CrossRef] [PubMed]
32. Brown, E.M.; Sadarangani, M.; Finlay, B.B. The role of the immune system in governing host-microbe interactions in the intestine. *Nat. Immunol.* **2013**, *14*, 660–667. [CrossRef] [PubMed]
33. Choi, Y.J.; Lee, S.R.; Oh, J.W. Effects of dietary fermented seaweed and seaweed fusiforme on growth performance, carcass parameters and immunoglobulin concentration in broiler chicks. *Asian Australas J. Anim. Sci.* **2014**, *27*, 862–870. [CrossRef] [PubMed]
34. Kim, J.Y.; Kwon, Y.M.; Kim, I.S.; Kim, J.A.; Yu, D.Y.; Adhikari, B.; Lee, S.S.; Choi, I.S.; Cho, K.K. Effects of the Brown Seaweed Laminaria japonica Supplementation on Serum Concentrations of IgG, Triglycerides, and Cholesterol, and Intestinal Microbiota Composition in Rats. *Front. Nutr.* **2018**, *5*, 23. [CrossRef] [PubMed]
35. Leonard, S.G.; Sweeney, T.; Bahar, B.; Lynch, B.P.; O'Doherty, J.V. Effect of maternal fish oil and seaweed extract supplementation on colostrum and milk composition, humoral immune response, and performance of suckled piglets. *J. Anim. Sci.* **2010**, *88*, 2988–2997. [CrossRef]
36. Leonard, S.G.; Sweeney, T.; Bahar, B.; O'Doherty, J.V. Effect of maternal seaweed extract supplementation on suckling piglet growth, humoral immunity, selected microflora, and immune response after an ex vivo lipopolysaccharide challenge. *J. Anim. Sci.* **2012**, *90*, 505–514. [CrossRef]
37. Zhang, J.; Wang, X.; Vikash, V.; Ye, Q.; Wu, D.; Liu, Y.; Dong, W. ROS and ROS-mediated cellular signaling. *Oxid. Med. Cell. Longev.* **2016**, *2016*, 4350965. [CrossRef]

38. Apostolopoulos, A.; Nakamura, A.; Yokoyama, S.; Aoshima, M.; Fujimoto, R.; Nakamura, K.; Ito, R.; Goto, K. Nuclear Accumulation of HSP70 in Mouse Skeletal Muscles in Response to Heat Stress, Aging, and Unloading with or Without Reloading. *Front. Genet.* **2018**, *9*, 617. [CrossRef]
39. Liu, X.; Liu, H.; Zhai, Y.; Li, Y.; Zhu, X.; Zhang, W. Laminarin protects against hydrogen peroxide-induced oxidative damage in MRC-5 cells possibly via regulating NRF2. *PeerJ* **2017**, *5*, e3642. [CrossRef]
40. Hong, S.W.; Lee, H.S.; Jung, K.H.; Lee, H.; Hong, S.S. Protective effect of fucoidan against acetaminophen-induced liver injury. *Arch. Pharm. Res.* **2012**, *35*, 1099–1105. [CrossRef]
41. Zou, T.; Yang, J.; Guo, X.; He, Q.; Wang, Z.; You, J. Dietary seaweed-derived polysaccharides improve growth performance of weaned pigs through maintaining intestinal barrier function and modulating gut microbial populations. *J. Anim. Sci. Biotechnol.* **2021**, *12*, 28. [CrossRef]
42. Mondo, E.; Marliani, G.; Accorsi, P.A.; Cocchi, M.; Di Leone, A. Role of gut microbiota in dog and cat's health and diseases. *Open Vet. J.* **2019**, *9*, 253–258. [CrossRef] [PubMed]
43. Wernimont, S.M.; Radosevich, J.; Jackson, M.I.; Ephraim, E.; Badri, D.V.; MacLeay, J.M.; Jewell, D.E.; Suchodolski, J.S. The Effects of Nutrition on the Gastrointestinal Microbiome of Cats and Dogs: Impact on Health and Disease. *Front. Microbiol.* **2020**, *11*, 1266. [CrossRef] [PubMed]
44. Martinez, K.B.; Leone, V.; Chang, E.B. Microbial metabolites in health and disease: Navigating the unknown in search of function. *J. Biol. Chem.* **2017**, *292*, 8553–8559. [CrossRef] [PubMed]
45. Heintz-Buschart, A.; Wilmes, P. Human Gut Microbiome: Function Matters. *Trends Microbiol.* **2018**, *26*, 563–574. [CrossRef]
46. Deusch, O.; O'Flynn, C.; Colyer, A.; Morris, P.; Allaway, D.; Jones, P.G.; Swanson, K.S. Deep Illumina-based shotgun sequencing reveals dietary effects on the structure and function of the fecal microbiome of growing kittens. *PLoS ONE* **2014**, *9*, e101021. [CrossRef]
47. Ritchie, L.E.; Burke, K.F.; Garcia-Mazcorro, J.F.; Steiner, J.M.; Suchodolski, J.S. Characterization of fecal microbiota in cats using universal 16S rRNA gene and group-specific primers for *Lactobacillus* and *Bifidobacterium* spp. *Vet. Microbiol.* **2010**, *144*, 140–146. [CrossRef]
48. Kumar, M.; Babaei, P.; Ji, B.; Nielsen, J. Human gut microbiota and healthy aging: Recent developments and future prospective. *Nutr. Heal. Aging* **2016**, *4*, 3–16. [CrossRef]
49. Fischer, M.M.; Kessler, A.M.; Kieffer, D.A.; Knotts, T.A.; Kim, K.; Wei, A.; Ramsey, J.J.; Fascetti, A.J. Effects of obesity, energy restriction and neutering on the faecal microbiota of cats. *Br. J. Nutr.* **2017**, *118*, 513–524. [CrossRef]
50. Suchodolski, J.S.; Dowd, S.E.; Wilke, V.; Steiner, J.M.; Jergens, A.E. 16S rRNA gene pyrosequencing reveals bacterial dysbiosis in the duodenum of dogs with idiopathic inflammatory bowel disease. *PLoS ONE* **2012**, *7*, e39333. [CrossRef]
51. Suchodolski, J.S.; Markel, M.E.; Garcia-Mazcorro, J.F.; Unterer, S.; Heilmann, R.M.; Dowd, S.E.; Kachroo, P.; Ivanov, I.; Minamoto, Y.; Dillman, E.M.; et al. The fecal microbiome in dogs with acute diarrhea and idiopathic inflammatory bowel disease. *PLoS ONE* **2012**, *7*, e51907. [CrossRef] [PubMed]
52. Minamoto, Y.; Minamoto, T.; Isaiah, A.; Sattasathuchana, P.; Buono, A.; Rangachari, V.R.; McNeely, I.H.; Lidbury, J.; Steiner, J.M.; Suchodolski, J.S. Fecal short-chain fatty acid concentrations and dysbiosis in dogs with chronic enteropathy. *J. Vet. Intern. Med.* **2019**, *33*, 1608–1618. [CrossRef] [PubMed]
53. Xu, J.; Verbrugghe, A.; Lourenço, M.; Janssens, G.P.; Liu, D.J.; Van de Wiele, T.; Eeckhaut, V.; Van Immerseel, F.; Van de Maele, I.; Niu, Y.; et al. Does canine inflammatory bowel disease influence gut microbial profile and host metabolism? *BMC Vet. Res.* **2016**, *12*, 114. [CrossRef] [PubMed]
54. Ormerod, K.L.; Wood, D.L.; Lachner, N.; Gellatly, S.L.; Daly, J.N.; Parsons, J.D.; Dal'Molin, C.G.; Palfreyman, R.W.; Nielsen, L.K.; Cooper, M.A.; et al. Genomic characterization of the uncultured Bacteroidales family S24-7 inhabiting the guts of homeothermic animals. *Microbiome* **2016**, *4*, 36. [CrossRef]
55. Du Preez, R.; Magnusson, M.; Majzoub, M.E.; Thomas, T.; Praeger, C.; Glasson, C.R.K.; Panchal, S.K.; Brown, L. Brown seaweed sargassum siliquosum as an intervention for diet-induced obesity in male wistar rats. *Nutrients* **2021**, *13*, 1754. [CrossRef]
56. Scott, K.P.; Martin, J.C.; Duncan, S.H.; Flint, H.J. Prebiotic stimulation of human colonic butyrate-producing bacteria and bifidobacteria, in vitro. *FEMS Microbiol. Ecol.* **2014**, *87*, 30–40. [CrossRef]
57. Inness, V.L.; McCartney, A.L.; Khoo, C.; Gross, K.L.; Gibson, G.R. Molecular characterisation of the gut microflora of healthy and inflammatory bowel disease cats using fluorescence in situ hybridisation with special reference to *Desulfovibrio* spp. *J. Anim. Physiol. Anim. Nutr.* **2007**, *91*, 48–53. [CrossRef]
58. Cai, S.; Yang, Y.; Kong, Y.; Guo, Q.; Xu, Y.; Xing, P.; Sun, Y.; Qian, J.; Xu, R.; Xie, L.; et al. Gut Bacteria Erysipelatoclostridium and Its Related Metabolite Ptilosteroid A Could Predict Radiation-Induced Intestinal Injury. *Front. Public Health* **2022**, *10*, 862598. [CrossRef]
59. Blaak, E.E.; Canfora, E.E.; Theis, S.; Frost, G.; Groen, A.K.; Mithieux, G.; Nauta, A.; Scott, K.; Stahl, B.; van Harsselaar, J.; et al. Short chain fatty acids in human gut and metabolic health. *Benef. Microbes* **2020**, *11*, 411–455. [CrossRef]
60. Bajury, D.M.; Rawi, M.H.; Sazali, I.H.; Abdullah, A.; Sarbini, S.R. Prebiotic evaluation of red seaweed (*Kappaphycus alvarezii*) using in vitro colon model. *Int. J. Food Sci. Nutr.* **2017**, *68*, 821–828. [CrossRef]
61. Li, Y.; Ali, I.; Lei, Z.; Li, Y.; Yang, M.; Yang, C.; Li, L. Effect of a Multistrain Probiotic on Feline Gut Health through the Fecal Microbiota and Its Metabolite SCFAs. *Metabolites* **2023**, *13*, 228. [CrossRef] [PubMed]

62. Li, B.; Zhang, H.; Shi, L.; Li, R.; Luo, Y.; Deng, Y.; Li, S.; Li, R.; Liu, Z. Saccharomyces boulardii alleviates DSS-induced intestinal barrier dysfunction and inflammation in humanized mice. *Food Funct.* **2022**, *13*, 102–112. [CrossRef] [PubMed]
63. Zoran, D.L. The carnivore connection to nutrition in cats. *J. Am. Vet. Med. Assoc.* **2002**, *221*, 1559–1567. [CrossRef] [PubMed]
64. Laflamme, D.P. Understanding the Nutritional Needs of Healthy Cats and Those with Diet-Sensitive Conditions. *Vet. Clin. N. Am. Small Anim. Pract.* **2020**, *50*, 905–924. [CrossRef]
65. Sunvold, G.D.; Fahey, G.C., Jr.; Merchen, N.R.; Reinhart, G.A. In vitro fermentation of selected fibrous substrates by dog and cat fecal inoculum: Influence of diet composition on substrate organic matter disappearance and short-chain fatty acid production. *J. Anim. Sci.* **1995**, *73*, 1110–1122. [CrossRef]
66. Rochus, K.; Janssens, G.P.; Hesta, M. Dietary fibre and the importance of the gut microbiota in feline nutrition: A review. *Nutr. Res. Rev.* **2014**, *27*, 295–307. [CrossRef]
67. Suzuki, T. Regulation of intestinal epithelial permeability by tight junctions. *Cell. Mol. Life Sci.* **2013**, *70*, 631–659. [CrossRef]
68. Wang, W.; Zhao, J.; Gui, W.; Sun, D.; Dai, H.; Xiao, L.; Chu, H.; Du, F.; Zhu, Q.; Schnabl, B.; et al. Tauroursodeoxycholic acid inhibits intestinal inflammation and barrier disruption in mice with non-alcoholic fatty liver disease. *Br. J. Pharmacol.* **2018**, *175*, 469–484. [CrossRef]
69. Liu, B.; Zhang, J.; Sun, P.; Yi, R.; Han, X.; Zhao, X. Raw Bowl Tea (Tuocha) Polyphenol Prevention of Nonalcoholic Fatty Liver Disease by Regulating Intestinal Function in Mice. *Biomolecules* **2019**, *9*, 435. [CrossRef]
70. Kamei, H.; Hachisuka, T.; Nakao, M.; Takagi, K. Quick recovery of serum diamine oxidase activity in patients undergoing total gastrectomy by oral enteral nutrition. *Am. J. Surg.* **2005**, *189*, 38–43. [CrossRef]
71. Timmermans, K.; Sir, Ö.; Kox, M.; Vaneker, M.; de Jong, C.; Gerretsen, J.; Edwards, M.; Scheffer, G.J.; Pickkers, P. Circulating iFABP Levels as a marker of intestinal damage in trauma patients. *Shock* **2015**, *43*, 117–120. [CrossRef] [PubMed]
72. Huang, J.; Huang, J.; Li, Y.; Wang, Y.; Wang, F.; Qiu, X.; Liu, X.; Li, H. Sodium Alginate Modulates Immunity, Intestinal Mucosal Barrier Function, and Gut Microbiota in Cyclophosphamide-Induced Immunosuppressed BALB/c Mice. *J. Agric. Food Chem.* **2021**, *69*, 7064–7073. [CrossRef] [PubMed]
73. Liu, S.; Deng, Z.; Chen, K.; Jian, S.; Zhou, F.; Yang, Y.; Fu, Z.; Xie, H.; Xiong, J.; Zhu, W. Cartilage tissue engineering: From proinflammatory and anti-inflammatory cytokines to osteoarthritis treatments (Review). *Mol. Med. Rep.* **2022**, *25*, 99. [CrossRef] [PubMed]
74. Kwan Tat, S.; Padrines, M.; Théoleyre, S.; Heymann, D.; Fortun, Y. IL-6, RANKL, TNF-alpha/IL-1: Interrelations in bone resorption pathophysiology. *Cytokine Growth Factor Rev.* **2004**, *15*, 49–60. [CrossRef] [PubMed]
75. Walsh, A.M.; Sweeney, T.; O'Shea, C.J.; Doyle, D.N.; O'Doherty, J.V. Effects of supplementing dietary laminarin and fucoidan on intestinal morphology and the immune gene expression in the weaned pig. *J. Anim. Sci.* **2012**, *90* (Suppl. 4), 284–286. [CrossRef]
76. Bahar, B.; O'Doherty, J.V.; Hayes, M.; Sweeney, T. Extracts of brown seaweeds can attenuate the bacterial lipopolysaccharide-induced pro-inflammatory response in the porcine colon ex vivo. *J. Anim. Sci.* **2012**, *90*, 46–48. [CrossRef]
77. Guo, X.; Chen, J.; Yang, J.; He, Q.; Luo, B.; Lu, Y.; Zou, T.; Wang, Z.; You, J. Seaweed polysaccharide mitigates intestinal barrier dysfunction induced by enterotoxigenic Escherichia coli through NF-κB pathway suppression in porcine intestinal epithelial cells. *J. Anim. Physiol. Anim. Nutr.* **2021**, *105*, 1063–1074. [CrossRef]
78. Nagata, K.; Nishiyama, C. IL-10 in Mast Cell-Mediated Immune Responses: Anti-Inflammatory and Proinflammatory Roles. *Int. J. Mol. Sci.* **2021**, *22*, 4972. [CrossRef]

metabolites

MDPI

Article

Effects of Dietary Supplementation with Mulberry Leaf Powder on the Growth Performance, Lipid Metabolism Parameters, Immunity Indicators, and Gut Microbiota of Dogs

Aiying Yu [1,2], Cuiming Tang [1], Sutian Wang [3], Yuan Wang [1], Lian Chen [1], Zhiyi Li [1], Guoqing Luo [1], Jianwu Zhong [1], Zhengfeng Fang [2], Zhenjiang Wang [1,*] and Sen Lin [1,*]

1 Key Laboratory of Urban Agriculture in South China, Sericultural & Agri-Food Research Institute, Guangdong Academy of Agricultural Sciences, Guangzhou 510640, China; 2022214001@stu.sicau.edu.cn (A.Y.); tangcuiming@gdaas.cn (C.T.); wangyuan18789@163.com (Y.W.); luoguoqing@gdaas.cn (G.L.); zhongjianwu@gdaas.cn (J.Z.)
2 Key Laboratory for Animal Disease Resistance Nutrition of the Ministry of Education, Animal Nutrition Institute, Sichuan Agricultural University, Chengdu 611130, China; zfang@sicau.edu.cn
3 State Key Laboratory of Livestock and Poultry Breeding, Guangdong Key Laboratory of Animal Breeding and Nutrition, Institute of Animal Science, Guangdong Academy of Agricultural Sciences, Guangzhou 510640, China; wangsutian@gdaas.cn
* Correspondence: wzj@gdaas.cn (Z.W.); linsen@gdaas.cn (S.L.)

Abstract: Overfeeding and a lack of exercise are increasingly causing obesity in dogs, which has become a big problem threatening the health of dogs. Therefore, it is necessary to investigate how dietary regulations can help to improve dogs' body conditions and minimize obesity. This study was carried out to investigate the effects of dietary mulberry leaf powder (MLP) supplementation on the growth performance, lipid metabolism parameters, and gut microbiota of Chinese indigenous dogs. Fifteen Chinese indigenous dogs (6.34 ± 0.56 kg) were randomly assigned to three treatment groups and received either the control diet (CON), high-fat diet (HF), or high-fat diet containing 6% Mulberry leaf powder (MLP) for four weeks. The CON group received a basal diet, the HF group received a basal diet supplemented with 10% lard, and the MLP group received a basal diet supplemented with 10% lard and 6% MLP. The trial lasted for four weeks. The growth performance, lipid metabolism parameters, immune globulins, cytokines, and fecal microbiota were measured. Results showed that there was no significant difference in growth performance. The MLP group appeared to have decreased ($p < 0.05$) the serum level of low-density lipoprotein cholesterol (LDL-C) and apoliprotein-A1(APO-A1) in serum. The MLP group appeared to have higher ($p < 0.05$) serum immune globulin A (IgA) levels. UPGMA results showed that the MLP group was closer to the CON group than to the HF group. LEfSe analysis showed that dietary supplementation with MLP contributed to an alteration in the genus *Alloprevotella*, *Sarcina*, and species belonging to the *Bacteroides* and *Lactobacillus* genus. Overall, the dietary supplementation of 6% MLP can improve lipid metabolism conditions and immunity in high-fat-diet-fed dogs, and can alter the gut microbial composition of dogs.

Keywords: dogs; mulberry leaf powder; lipid metabolism; gut microbiota; immunity

Citation: Yu, A.; Tang, C.; Wang, S.; Wang, Y.; Chen, L.; Li, Z.; Luo, G.; Zhong, J.; Fang, Z.; Wang, Z.; et al. Effects of Dietary Supplementation with Mulberry Leaf Powder on the Growth Performance, Lipid Metabolism Parameters, Immunity Indicators, and Gut Microbiota of Dogs. *Metabolites* **2023**, *13*, 918. https://doi.org/10.3390/metabo13080918

Academic Editors: Lian Li, Baichuan Deng and Giulia Pignataro

Received: 30 June 2023
Revised: 29 July 2023
Accepted: 3 August 2023
Published: 4 August 2023

1. Introduction

In a society where aging and oligonucleolization are becoming more prevalent [1], the importance of dogs as companion animals cannot be overstated. Dogs play an important role in people's lives, and interaction with animals may positively impact their owners [2]. Previous research has shown that owning a dog can positively impact mental health by reducing feelings of loneliness and depression and increasing overall happiness and well-being [3]. Likewise, there have been a few studies highlighting the fact that having a pet has a certain positive effect on depression, anxiety, and behavioral disturbances [4–6]. Furthermore, for older adults who may be living alone, dogs can provide a sense of security and protection, and can also provide peace of mind [7]. Not only has it had an impact on psychology, but also on physical health. Dog owners exercise more and fall asleep more easily than non-dog owners [8]. Dogs may be beneficial in reducing cardiovascular risk in their owners by providing social support and motivation for physical activity [9,10].

Obesity is the most common nutritional disease which is characterized by the accumulation of adipose tissue to the point that health is adversely affected in dogs. When a dog is overweight or obese, it puts additional stress on its body and can lead to a variety of health problems, including joint problems, diabetes, respiratory problems, cardiovascular disease, and even shortened lifespan [11–13]. In people and animals, obesity is a multifactorial condition involving diet, a level of physical activity, behavioral factors, socioeconomic factors, environmental exposures, genetics, metabolism, and the microbiome [14]. Factors that cause obesity in dogs are related to the owners' awareness of obesity-related health risks [15]. The primary cause of obesity in dogs is a combination of overfeeding and a lack of exercise. To prevent obesity in dogs, it is important to feed them a balanced diet, restrict feeding, provide plenty of exercise, and monitor their weight regularly [16].

Given the importance of dietary regulation in maintaining dogs' health and preventing obesity, an increasing number of recent studies focused on investigating the effects of plant-derived bioactive components on dogs' health. In recent years, mulberry leaves have been used in agricultural animals and have been shown to play important roles in regulating lipid metabolism, antioxidative capacity, immune function, and muscle development [17–19]. The existing body of research on mulberry leaf demonstrates that mulberry leaf exhibits a variety of biological activities, including antioxidant [20–22], anticancer [23], anti-obesity [24–26], anti-inflammation [26–28], and antibacterial [29] effects. Based on their beneficial biological activities, mulberry leaves are also used in Chinese traditional herbal medicine to treat diseases [30,31]. Furthermore, previous research has established that mulberry leaf can be a potential protein source or beneficial additive for domestic animal feeds [18,32,33]. In recent years, there has been a dramatic increase in the lipid-lowering effect of mulberry leaf powder. The effective components of mulberry leaf polyphenols inhibited adipogenesis through elevating leptin-stimulated lipolysis [34,35], ameliorating obesity through enhancing brown adipose tissue activity, and modulating gut microbiota [26]. The dietary fiber in mulberry leaves may be another critical factor regulating lipid metabolism in animals. High dietary fiber intake was shown to decrease the plasma total cholesterol (T Chol), triglyceride (TG), and free fatty acid (FFA) levels in rats [36]. In addition, the dietary fiber may interact with gut microbes and produce metabolites, like short-chain fatty acids (SCFAs), which can be absorbed into the circulation and affect metabolic regulation [37]. As gut microbiota is strongly associated with obesity [38], it is worth studying the effects of mulberry leaves on gut microbiota. Therefore, we hypothesized that the addition of mulberry leaf powder to the diet can affect lipid metabolism and intestinal flora composition in dogs. In this study, Chinese indigenous dogs were fed with a control diet, a high-fat diet (HF), or a high-fat diet containing 6% Mulberry leaf powder (MLP) for four weeks to investigate the impact of mulberry leaf powder (MLP) on growth performance, lipid metabolism parameters, immunity indicators, and gut microbiota.

2. Materials and Methods

The Animal Care and Use Committee of Guangdong Academy of Agricultural Sciences granted approval to the protocols and procedures governing the care and handling of animals in this study.

2.1. MLP Preparation

The mulberry leaves utilized in this study were acquired from the Southern China distribution of the National Germplasm Garden of Mulberry at the Guangdong Academy of Agricultural Sciences. After collection, the leaves were dehydrated via a drying apparatus, pulverized with a feed grinder, and sifted through 90 mesh sieves. The nutrient composition in mulberry leaf powders are as follows: crude protein: 14.0%, crude fat: 4.3%, crude fiber: 6.0%, crude ash: 8.3%, total calcium: 1.3%, and total phosphorous: 0.3%. The bioactive components in mulberry leaf powders are as follows: 1-deoxynojirimycin: 7.06 μg/g, gallic acid: 2.51 μg/g, protocatechuic acid: 9.44 μg/g, gallocatechin gallate: 0.14 μg/g, catechin: 0.04 μg/g, rutin: 657.88 μg/g, quercetin: 3.48 μg/g, chlorogenic acid: 3449.12μg/g, and naringenin: 0.35 μg/g.

2.2. Animals, Experimental Design, and Diets

The study involved fifteen 2-month-old Chinese indigenous dogs with similar initial body weight, which came from 4 litters of the same breeds. The animal study was conducted in the Huadu test base of Guangdong Academy of Agricultural Sciences. The 15 dogs were randomly assigned to three treatment groups, each comprising five dogs. The treatment groups were designated as follows: the control (CON) group, the high-fat (HF) group, and the mulberry leaf powder (MLP) group. The CON group received a basal diet, the HF group received a basal diet supplemented with 10% lard, and the MLP group received a basal diet supplemented with 10% lard and 6% MLP. The trial lasted for four weeks, during which all dogs had unlimited access to water and feed. Throughout the trial period, the initial body weight (BW), the body weight (at experimental days 7, 14, and 21), and the final BW of each dog were recorded, and the average daily gain (ADG) was calculated for each dog.

2.3. Sample Collection

At the end of the study, dogs were fasted overnight before blood was collected from their lateral saphenous vein. The collected blood was allowed to sit for 30 min before serum samples were separated by centrifugation at 3000× g and 4 °C for 10 min. The serum samples were then stored in −20 °C until analysis. In order to determine the changes in gut microbiota following different dietary treatment, the fecal samples were collected form each dog on the last day of the trial. The fecal samples were stored in −80 °C after they were snap-frozen until analysis.

2.4. Measurement of Lipid Metabolism Parameters

Triglycerides (TGs), total cholesterol (TCHO), high-density lipoprotein cholesterol (HDL-C), low-density lipoprotein cholesterol (LDL-C), Apolipoprotein-B (APO-B), and apoliprotein-A1 (APO-A1) were determined in serum collected from dogs in different groups at the end of the study using ELISA kits (Jiangsu Meimian Industrial Co., Ltd., Nanjing, China) in accordance with the manufacturer's instructions.

2.5. Fecal Microbiota Analysis

The total DNA of each fecal sample was extracted using a DNA extraction kit (Omega Bio-Tek Inc., Norcross, GA, USA). After DNA extraction, specific primers with barcodes were synthesized based on the full-length primer sequences. The V3–V4 region of the bacterial 16S rRNA gene was amplified by PCR using primers 341F and 806R. When PCR amplification was performed, the products were purified, quantified, and normalized to form a sequencing library (SMRT Bell). The constructed library was first subjected to library quality control, and libraries that pass quality control were sequenced using PacBio Sequel. The circular consensus sequencing (CCS) sequences were obtained using the smrtlink tool provided by PacBio. Furthermore, lima v1.7.0 software was used to identify CCS sequences based on barcodes, generating raw CCS sequence data, and cutadapt 1.9.1 software was used to identify and remove adapter sequences and perform length filtering and generate clean CCS sequences without adapter sequences. Effective CCS sequences were generated using the UCHIME v4.2 software to identify and eliminate chimeric sequences. The effective CCS sequences were clustered/denoised to define OTUs/ASVs (referred to as "Features" hereafter), and their taxonomic classification was determined based on the sequence composition of features. At the 97% similarity level, the sequences were clustered into operational taxonomic units (OTUs) using USEARCH (version 10.0). Taxonomic analysis was performed on the sample at various classification levels. Alpha diversity analysis was conducted to investigate the species diversity within individual samples. Alpha diversity indices, including ACE, Shannon, Simpson, and Chao 1 indices, were calculated using QIIME2 software (version 2020.6). Beta diversity analysis was used to compare the differences in species diversity between different samples. Beta diversity was analyzed using Bray–Curtis distances and visualized using principal co-ordinates analysis (PCOA). Functional prediction analysis was performed to predict the gene functions or phenotypes of samples, as well as to calculate the abundance of functional genes or phenotypes.

2.6. *Analysis of Immunity Indicators*

Immune globulin A (IgA), immune globulin G (IgG), immune globulin M (IgM), interlukin-6 (IL-6), and tumor necrosis factor-α (TNF-α) in serum collected from dogs in different groups at the end of the study were determined using ELISA kits (Jiangsu Meimian Industrial Co., Ltd., Nanjing, China), in accordance with the manufacturer's instructions.

2.7. *Statistical Analysis*

All data were firstly examined for normal distribution purposes and then analyzed via one-way ANOVA analysis in GraphPad Prism 8.0 (GraphPad Prism Software Inc., San Diego, CA, USA). Tukey's test was performed to carry out post hoc testing. Differences between treatments were considered significant when $p < 0.05$. All data are expressed as means ± standard errors of the mean (SEM).

3. Results

3.1. *Growth Performance*

The growth performance results of the dogs throughout the entire experimental period are displayed in Figures 1 and 2. It was found that, on experimental days 7, 14, 21, and 28, the BW of dogs in the HF and MLP groups was similar to that of the control group. Although the dogs in the MLP group seemed to have gained less body weight in the first half, second half, and overall experimental period, no statistical differences were detected between the groups.

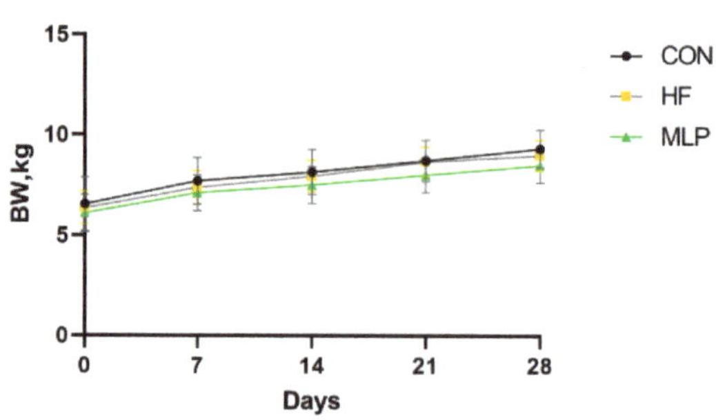

Figure 1. Effects of dietary supplementation with mulberry leaf powder (MLP) on the body weights of dogs. The body weights of all dogs were weighed and recorded every week. MLP, mulberry leaf powder; HF, high-fat dies; BW, body weight.

3.2. *Lipid Metabolism Parameters*

In order to assess the effects of the dietary inclusion of MLP on lipid metabolism status in Chinese indigenous dogs, parameters reflecting the blood lipids, cholesterol, and their transportation were determined. As shown in Figure 3, the triglyceride (TG) and total cholesterol (TCHO) levels in serum were not different among the three treatment groups. Moreover, the dogs fed three different diets appeared to have similar serum concentrations of high-density lipoprotein cholesterol (HDL-C) and Apolipoprotein-B

(APO-B). However, compared with dogs in the control group, dogs in the MLP group appeared to have decreased ($p < 0.05$) low-density lipoprotein cholesterol (LDL-C) levels in the serum. Additionally, the serum apoliprotein-A1(APO-A1) was significantly higher ($p < 0.05$) in HF dogs than in CON dogs, whereas MLP supplementation could significantly decrease ($p < 0.05$) the serum levels of APO-A1.

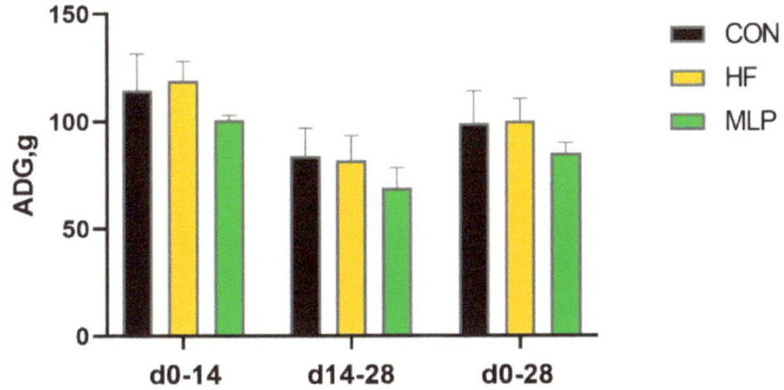

Figure 2. Effects of the dietary inclusion of MLP on average daily gain (ADG) of dogs during different experimental periods. MLP, mulberry leaf powder; HF, high-fat dies; ADG, average daily gain.

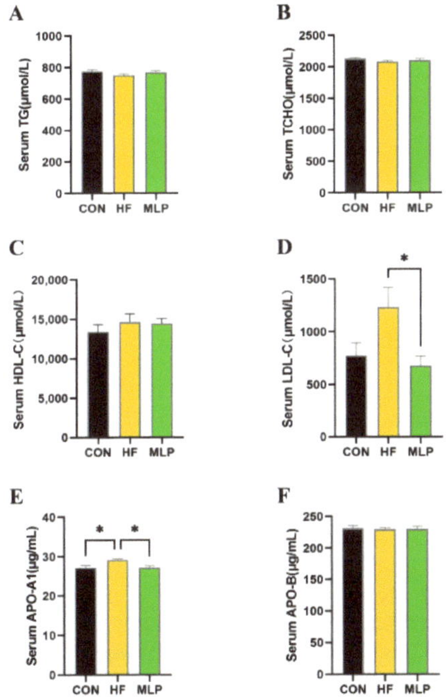

Figure 3. Effects of the dietary inclusion of MLP on lipid metabolism status in Chinese indigenous dogs. (**A**) Serum concentrations of triglycerides. (**B**) Serum concentrations of total cholesterol. (**C**) Serum concentrations of high-density lipoprotein cholesterol. (**D**) Serum concentrations of low-density lipoprotein cholesterol. (**E**) Serum concentrations of apoliprotein-A1. (**F**) Serum concentrations of Apolipoprotein-B. Data are expressed as means ± SEM. * indicates a significant difference ($p < 0.05$) between the two different groups.

3.3. Immunity Indicators

As a traditional Chinese medicine, mulberry leaf has been considered to be capable of improving immunity. In order to assess the effects of the dietary inclusion of MLP on immune status in Chinese indigenous dogs, immune globulins and cytokines in blood were determined. As shown in Figure 4, in comparison to the dogs in the HF group, dogs in the MLP group appeared to have higher ($p < 0.05$) serum immune globulin A (IgA) levels. However, the serum levels of IgG and IgM were not different among the three groups. The serum concentrations of interlukin-6 (IL-6) and tumor necrosis factor-α (TNF-α) in the three groups were also similar.

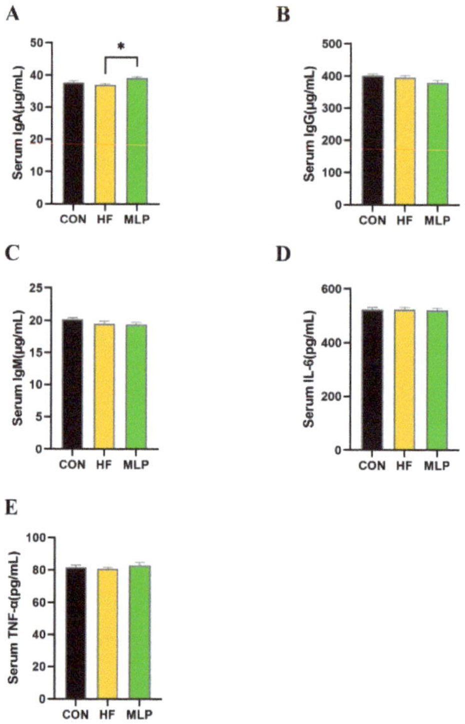

Figure 4. Effects of the dietary inclusion of MLP on immunity indicators in dogs. (**A**) Serum concentrations of immune globulin A. (**B**) Serum concentrations of immune globulin G. (**C**) Serum concentrations of immune globulin M. (**D**) Serum concentrations of interleukin-6. (**E**) Serum concentrations of tumor necrosis factor-α. Data are expressed as means ± SEM. * indicates a significant difference ($p < 0.05$) between two different groups.

3.4. Gut Microbiota Operational Taxonomic Unit (OTU) Analysis

As depicted in Figure 5, the control group had 196 OTUs, the HF group had 188 OTUs, and the MLP group had 175 OTUs. The number of shared OTUs between the CON and HF groups was 182, while there were 164 between the HF and MLP groups. Furthermore, the CON group and the MLP group had 169 common OTUs, and all three groups shared 161 common OTUs.

3.5. Species Annotation, Taxonomic Analysis, and Alpha Diversity Analysis

A naïve Bayes classifier combined with alignment was used to classify feature sequences based on the Silva reference database and obtain the taxonomic annotation for each feature. Then, the community composition of each sample at various taxonomic levels was assessed. At the phylum level, taxonomic information on the dominant species is shown in Figure 6. The dominant phylum species were *Firmicute*, *Bacteroidete*, *Proteobacteria*, *Fusobacteriota*, and *Fusobacteria*. Taxonomic information on the dominant species at the genus level comprised *Megamonas*, *Prevotella*, *Fusobacterium*, *Bacteroides*, and *Alloprevotella*. Alpha diversity is a major indicator illustrating the diversity of the gut microbiota. In this study, the ACE indices, Chao1 indices, Simpson indices, and Shannon indices were all calculated (Table 1), whereas none of these indices showed any difference among the groups.

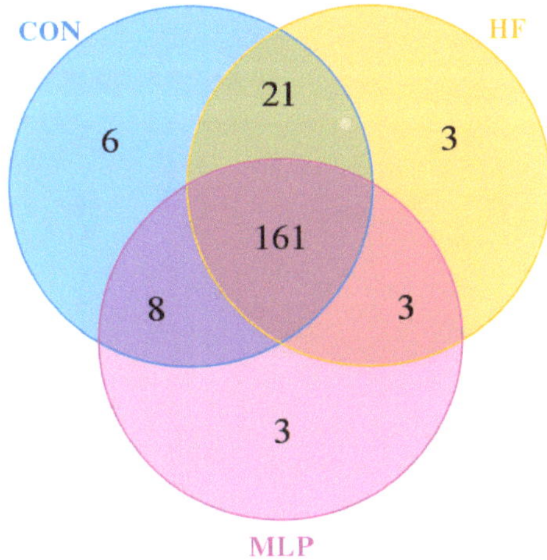

Figure 5. OTU Venn diagram.

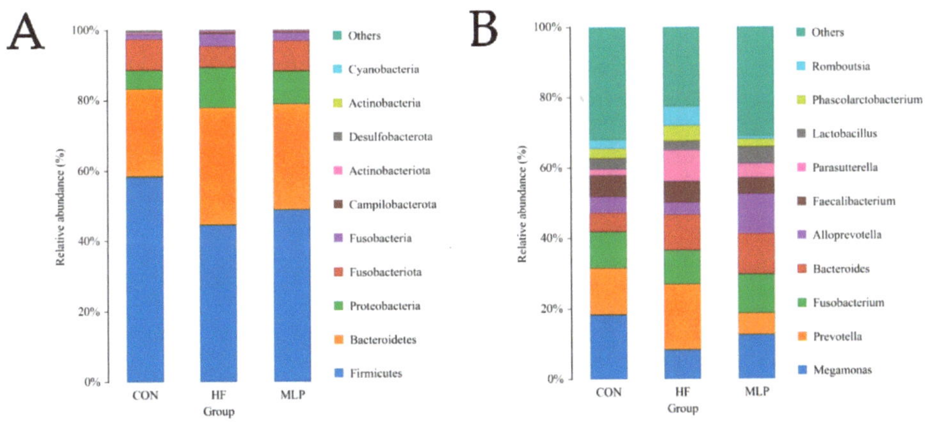

Figure 6. Phylum-level structural analysis. (**A**) Relative abundance of gut microbiota at the phylum level. (**B**) Relative abundance of the gut microbiota at the genus level.

Table 1. Statistical analysis of microbial alpha diversity.

Group	ACE	Chao1	Simpson	Shannon
CON	149.92 ± 15.36	149.00 ± 15.72	0.85 ± 0.05	4.30 ± 0.45
HF	139.80 ± 7.94	136.10 ± 10.04	0.84 ± 0.03	3.87 ± 0.30
MLP	129.5 ± 7.38	129.2 ± 7.79	0.90 ± 0.01	4.33 ± 0.22

Data are presented as mean ± SEM. There was no significant difference among the treatment groups ($p > 0.05$).

3.6. Beta Diversity Analysis

To measure the degree of similarity between microbial communities, β-diversity was further evaluated using Bray–Curtis PCoA. In this study, PCoA was used to analyze the fecal samples of the dogs from different groups. The results showed that there was no significant separation among the three groups (Figure 7A). However, the UPGMA results showed that the MLP group was closer to the CON group than to the HF group (Figure 7B).

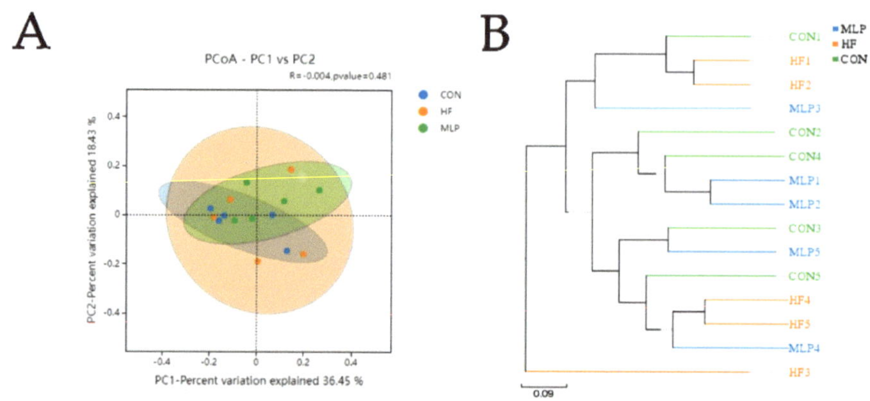

Figure 7. Dimension reduction analysis. Principal component analysis (PCA) for: (**A**) principal coordinate analysis (PCoA) of weighted UniFrac distances of 16S rRNA genes and (**B**) unweighted pair group method with arithmetic mean (UPGMA) analysis.

3.7. LEFSe Difference Analysis

To identify the differences in abundance between the groups, linear discriminant analysis (LDA) and LDA effect size (LEfSe) analysis were performed, and thee results are shown in Figure 8 with the histogram length indicating the contribution of various species. Notably, significantly higher numbers of 12 species belonging to the Firmicutes phylum, mainly Clostridium, were observed in the control group compared to the MLP group. Conversely, the MLP group exhibited a significant abundance of *Alloprevotella, Sarcina,* and *Lactobacillus,* which belong to the phylum *Bacteroidetes* and *Firmicutes.* Figure 8B presents the clade of different species, with nodes representing taxonomic levels ranging from the phylum to the genus or species. Red nodes signify the microbial taxa that played a crucial role in the control group, while green nodes represent those in the MLP group. In the control group, *Ruminiclostridium, Clostridium, Faecalitalea, Terrisporobacter,* and *Candidatus* were identified as crucial genera. In contrast, genera including *Alloprevotella, Bacteroides, Sarcina,* and *Lactobacillus* played a crucial role in the MLP group. However, no bacteria in the HF group were observed to play crucial roles compared to the other two groups.

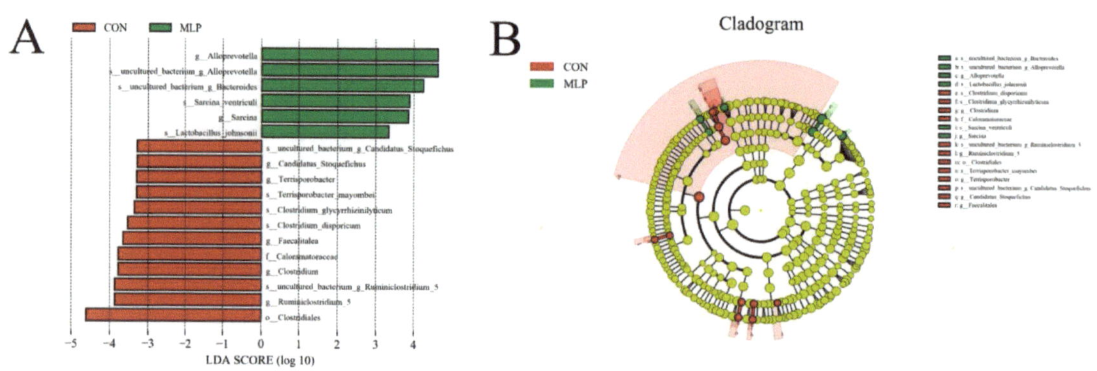

Figure 8. LEFSe difference analysis. (**A**) LDA scores for microbial taxa with significant roles in both groups. The LDA score represents the effect size. The histogram of the LDA scores presents species (biomarker) whose abundance showed significant differences between different groups. The LDA score represents the effect size. (**B**) LDA scores for microbial taxa with significant roles in both groups. In the cladogram, circles radiating from the inner side to the outer side represent taxonomic levels from the phylum to the genus (species). Each circle's diameter is proportional to the taxon's relative abundance.

4. Discussion

In today's society, on the one hand, aging and oligonucleolization are becoming increasingly prevalent. On the other hand, the trend of young people choosing not to marry or have children has become mainstream. Therefore, the companionship of pets shows increased importance than in previous eras. More and more pet owners are regarding their pets as family members and thus are starting to pay more attention to the health of pets. Similar to humans, pets are also facing increasingly prominent "three highs" issues. A recent study showed that the overall prevalence rates of overweight and obesity in a dog group were 21.1% and 20.2%, which would contribute to a variety of disease processes and negatively affect the quality of life in dogs [39]. MLP has been recognized as a traditional Chinese medicine that could decrease fat deposition and improve health status. Previously, we have shown that dietary MLP could decrease backfat thickness and improve the lean percentage of pigs [17]. The case of whether dietary MLP could improve lipid metabolism and bring other beneficial outcomes in dogs is worthy of further research.

In this study, high-fat diets were given to dogs to create a hyperlipidemia model. The growth performance results indicated that, although dogs in the MLP group seemed to have slightly lower ADG in terms of numbers, no significant difference was observed among the three groups, which might be attributed to the fact that the feeding trial was not long enough to highlight the difference in growth rates. In comparison, the lipid metabolism indicators did present differences among the groups. The HF group tended to have higher LDL-C than the CON group. Compared with the high-fat-diet-fed dogs, the dogs in the MLP group appeared to have lower serum LDL-C concentrations. LDL-C is often called the bad cholesterol because it can build up on the arterial walls to initiate the formation of atherosclerotic plaques [40], raising the chances of health problems like a heart attack. Therefore, lowered LDL-C levels have beneficial effects on preventing cardiovascular diseases in MLP-fed dogs. Apo-A1, the major component of high-density lipoprotein, plays a crucial role in promoting reverse cholesterol transport by effectively incorporating phospholipids and cholesterol [41]. In this study, high-fat-diet-fed dogs appeared to have higher serum APO-A1 levels, while MLP consumption could decrease the level of APO-A1. A high-fat-diet-induced increase in APO-A1 has been proven in previous studies [42], and it may indicate that the body itself attempts to remove excess cholesterol. In the context of the dogs' lipid profile, the difference in serum levels of LDL-C and APO-A1 in these groups reflects that MLP can improve lipid metabolism conditions in high-fat-diet-fed dogs.

The effects of mulberry leaf powder on regulating immunity in animal studies have also been well illustrated [43–45]. Herein, immune indicators were determined, and mulberry leaf consumption appeared to improve the serum levels of IgA. Immunoglobulin A (IgA), as a crucial component among the five major immunoglobulins, exerts a central role in maintaining mucosal homeostasis in the gastrointestinal, respiratory, and genitourinary systems [46]. It serves as the primary antibody in these contexts, governing an immune response. Therefore, MLP can improve dogs' immunity through enhancing IgA secretion, whereas the underlying mechanisms need further research. Inflammatory cytokines directly reflect inflammatory status. In this study, the inflammatory cytokine levels in serum seemed to be the same among the three groups, indicating that the inflammatory status in different groups is similar.

In dogs, gastrointestinal health is also a very important issue that pet owners should pay attention to. Of note, the gut microbial environment is strongly associated with dogs' body conditions [47]. The gut microbiota plays a pivotal role in extraintestinal disorders, including obesity [48]. Therefore, gut microbial composition was determined by 16S rRNA gene sequencing. In accordance with previous studies [49,50], Firmicutes and Bacteroidetes were identified as the predominant bacterial phyla in experimental dogs.

The UPGMA algorithm calculates the pairwise genetic distances between samples using a simple arithmetic mean and then groups the most similar samples together. According to the UPGMA results in this study, the MLP group was closer to the CON group than to the HF group, indicating a higher microbial community similarity between the MLP group and the CON group. LEfSe analysis showed that dietary supplementation with MLP contributed to an alteration in the genus *Alloprevotella*, *Sarcina*, and the species belonging to the Bacteroides and Lactobacillus genus. *Alloprevotella* is recognized as a beneficial bacterium and can produce SCFA and promote an anti-inflammatory environment. Moreover, *Alloprevotella* was reported to be negatively correlated with the abnormal lipid metabolism parameters, including serum LDL-C levels [51]. An alteration in the genus *Alloprevotella* might also partly be attributed to the effects of fibers on MLP. As shown in a previous study, mixed dietary fiber intervention for 4 days significantly promoted the growth of *Alloprevotella* in young healthy people [52]. Bacteroides has been recognized as a bacterial genus related to lipid metabolism. It was shown in a previous study that the abundance of Bacteroides was greatly reduced in patients with nonalcoholic fatty liver disease [53]. Some species of gut Bacteroides have been demonstrated to downregulate the expression of genes involved in lipogenesis and ameliorate lipid metabolic disorders [54,55]. Lactobacillus is a genus of bacteria that is considered to be beneficial and is commonly referred to as probiotics.

In a mice study, lactobacilli supplementation could significantly improve blood lipid levels and liver function and alleviate liver oxidative stress [56]. Moreover, lactobacilli supplementation significantly inhibited lipid accumulation in the liver and regulate lipid metabolism in epididymal fat pads. Lactobacillus species have also been shown to be associated with immunity regulation. In a pig study, Lactobacillus-treated pigs were found to have increased immunoglobin secretion and regulate immune indices [57]. Therefore, the altered gut microbiota in MLP-fed dogs may play important roles in the regulation of lipid metabolism and immunity.

5. Conclusions

In summary, dietary supplementation with 6% MLP decreased serum LDL-C and APO-A1 levels, increased the secretion of serum immune globulin A, and altered the gut microbial composition of dogs, whereas it had no effects on the growth rates. The improved lipid metabolism conditions and immune indices may be associated with the altered abundances of the bacteria genus. This study shows that dietary supplementation with 6% MLP can improve lipid metabolism conditions and immunity in high-fat-diet-fed dogs, and can also alter the gut microbial composition of dogs.

Author Contributions: Conceptualization, S.L., Z.W. and Z.F.; methodology, C.T. and Y.W.; software, A.Y. and Z.L.; validation, C.T., L.C. and Z.L.; formal analysis, A.Y.; investigation, A.Y., S.W. and Y.W.; resources, C.T.; data curation, Z.L.; writing—original draft preparation, A.Y.; writing—review and editing, S.L. and S.W.; visualization, A.Y.; supervision, S.L., G.L. and J.Z.; project administration, A.Y. and Y.W.; funding acquisition, S.L. and Z.W. All authors have read and agreed to the published version of the manuscript.

Funding: This research was funded by Rural revitalization strategy special fund seed industry revitalization project of Guangdong Province (2022-NBH-00-009), Talent Training Project of Guangdong Academy of Agricultural Sciences (R2019YJ-YB3002), Agricultural competitive industry discipline team building project of Guangdong Academy of Agricultural Sciences (202119TD), and China Agriculture Research System of MOF and MARA.

Institutional Review Board Statement: The study was conducted in accordance with the Declaration of Helsinki, and approved by the Institutional Review Board of Animal Care and Use Committee of Guangdong Academy of Agricultural Sciences (protocol code GDAASSA-2021023).

Informed Consent Statement: Not applicable.

Data Availability Statement: The data presented in this study are available on request from the corresponding author. The data are not publicly available because the authors are applying for patent.

Conflicts of Interest: The authors declare no conflict of interest.

References

1. Deering, S. Clinical public health, climate change, and aging. *Can. Fam. Physician* **2023**, *69*, 233–235. [CrossRef] [PubMed]
2. Serpell, J.A. *In the Company of Animals: A Study of Human-Animal Relationships*; Cambridge University Press: Cambridge, UK, 1996.
3. Islam, A.; Towell, T. Cat and Dog Companionship and Well-being: A Systematic Review. *Int. J. Appl. Psychol.* **2013**, *2013*, 149–155. [CrossRef]
4. Virués-Ortega, J.; Pastor-Barriuso, R.; Castellote, J.M.; Población, A.; de Pedro-Cuesta, J. Effect of animal-assisted therapy on the psychological and functional status of elderly populations and patients with psychiatric disorders: A meta-analysis. *Health Psychol. Rev.* **2012**, *6*, 197–221. [CrossRef]
5. Hussein, S.M.; Soliman, W.S.; Khalifa, A.A. Benefits of pets' ownership, a review based on health perspectives. *J. Intern. Med. Emerg.Res.* **2021**, *2*, 1–9. [CrossRef]
6. McConnell, A.R.; Brown, C.M.; Shoda, T.M.; Stayton, L.E.; Martin, C.E. Friends with benefits: On the positive consequences of pet ownership. *J. Pers. Soc. Psychol.* **2011**, *101*, 1239–1252. [CrossRef]
7. Hughes, M.J.; Verreynne, M.L.; Harpur, P.; Pachana, N.A. Companion Animals and Health in Older Populations: A Systematic Review. *Clin. Gerontol.* **2020**, *43*, 365–377. [CrossRef]
8. Mein, G.; Grant, R. A cross-sectional exploratory analysis between pet ownership, sleep, exercise, health and neighbourhood perceptions: The Whitehall II cohort study. *BMC Geriatr.* **2018**, *18*, 176. [CrossRef]
9. Kramer, C.K.; Mehmood, S.; Suen, R.S. Dog Ownership and Survival: A Systematic Review and Meta-Analysis. *Circ. Cardiovasc. Qual. Outcomes* **2019**, *12*, e005554. [CrossRef]
10. Mubanga, M.; Byberg, L.; Nowak, C.; Egenvall, A.; Magnusson, P.K.; Ingelsson, E.; Fall, T. Dog ownership and the risk of cardiovascular disease and death—A nationwide cohort study. *Sci. Rep.* **2017**, *7*, 15821. [CrossRef]
11. German, A.J. The growing problem of obesity in dogs and cats. *J. Nutr.* **2006**, *136*, 1940s–1946s. [CrossRef]
12. Marshall, W.G.; Hazewinkel, H.A.; Mullen, D.; de Meyer, G.; Baert, K.; Carmichael, S. The effect of weight loss on lameness in obese dogs with osteoarthritis. *Vet. Res. Commun.* **2010**, *34*, 241–253. [CrossRef]
13. Montoya, M.; Morrison, J.A.; Arrignon, F.; Spofford, N.; Charles, H.; Hours, M.-A.; Biourge, V. Life expectancy tables for dogs and cats derived from clinical data. *Front Vet. Sci.* **2023**, *10*, 1082102. [CrossRef]
14. Chandler, M.; Cunningham, S.; Lund, E.M.; Khanna, C.; Naramore, R.; Patel, A.; Day, M.J. Obesity and Associated Comorbidities in People and Companion Animals: A One Health Perspective. *J. Comp. Pathol.* **2017**, *156*, 296–309. [CrossRef]
15. Courcier, E.A.; Thomson, R.M.; Mellor, D.J.; Yam, P.S. An epidemiological study of environmental factors associated with canine obesity. *J. Small Anim. Pract.* **2010**, *51*, 362–367. [CrossRef]
16. Oba, P.M.; Kelly, J.; Kostiuk, D.; Swanson, K.S. Effects of weight loss and feeding specially formulated diets on the body composition, blood metabolite profiles, voluntary physical activity, and fecal metabolites and microbiota of obese dogs. *J. Anim. Sci.* **2023**, *101*, skad073. [CrossRef]
17. Wang, S.T.; Tang, C.M.; Li, J.H.; Wang, Z.J.; Meng, F.M.; Luo, G.Q.; Xin, H.Y.; Zhong, J.W.; Wang, Y.; Li, B.H.; et al. The Effects of Dietary Inclusion of Mulberry Leaf Powder on Growth Performance, Carcass Traits and Meat Quality of Tibetan Pigs. *Animals* **2022**, *12*, 13. [CrossRef]
18. Liu, Y.; Li, Y.; Xiao, Y.; Peng, Y.; He, J.; Chen, C.; Xiao, D.; Yin, Y.; Li, F. Mulberry leaf powder regulates antioxidative capacity and lipid metabolism in finishing pigs. *Anim. Nutr.* **2021**, *7*, 421–429. [CrossRef]
19. Zhang, B.; Wang, Z.; Huang, C.; Wang, D.; Chang, D.; Shi, X.; Chen, Y.; Chen, H. Positive effects of Mulberry leaf extract on egg quality, lipid metabolism, serum biochemistry, and antioxidant indices of laying hens. *Front. Vet. Sci.* **2022**, *9*, 1005643. [CrossRef]
20. Ji, T.; Su, S.L.; Zhu, Y.; Guo, J.M.; Qian, D.W.; Tang, Y.P.; Duan, J.A. The mechanism of mulberry leaves against renal tubular interstitial fibrosis through ERK1/2 signaling pathway was predicted by network pharmacology and validated in human tubular epithelial cells. *Phytother. Res.* **2019**, *33*, 2044–2055. [CrossRef]
21. Hu, L.; Wang, C.; Guo, X.; Chen, D.; Zhou, W.; Chen, X.; Zhang, Q. Flavonoid Levels and Antioxidant Capacity of Mulberry Leaves: Effects of Growth Period and Drying Methods. *Front. Plant Sci.* **2021**, *12*, 684974. [CrossRef]
22. Wang, B.; Luo, H. Effects of mulberry leaf silage on antioxidant and immunomodulatory activity and rumen bacterial community of lambs. *BMC Microbiol.* **2021**, *21*, 250. [CrossRef] [PubMed]
23. Kumkoon, T.; Srisaisap, M.; Boonserm, P. Biosynthesized Silver Nanoparticles Using *Morus alba* (White Mulberry) Leaf Extract as Potential Antibacterial and Anticancer Agents. *Molecules* **2023**, *28*, 1213. [CrossRef] [PubMed]
24. Li, Q.; Liu, F.; Liu, J.; Liao, S.; Zou, Y. Mulberry Leaf Polyphenols and Fiber Induce Synergistic Antiobesity and Display a Modulation Effect on Gut Microbiota and Metabolites. *Nutrients* **2019**, *11*, 1017. [CrossRef] [PubMed]
25. Mahboubi, M. *Morus alba* (mulberry), a natural potent compound in management of obesity. *Pharmacol. Res.* **2019**, *146*, 104341. [CrossRef]
26. Park, M.; Jaiswal, V.; Kim, K.; Chun, J.; Lee, M.J.; Shin, J.H.; Lee, H.J. Mulberry Leaf Supplements Effecting Anti-Inflammatory Genes and Improving Obesity in Elderly Overweight Dogs. *Int. J. Mol. Sci.* **2022**, *23*, 15215. [CrossRef]
27. Lin, Z.; Gan, T.; Huang, Y.; Bao, L.; Liu, S.; Cui, X.; Wang, H.; Jiao, F.; Zhang, M.; Su, C.; et al. Anti-Inflammatory Activity of Mulberry Leaf Flavonoids In Vitro and In Vivo. *Int. J. Mol. Sci.* **2022**, *23*, 7694. [CrossRef]
28. Liang, H.W.; Yang, T.Y.; Teng, C.S.; Lee, Y.J.; Yu, M.H.; Lee, H.J.; Hsu, L.S.; Wang, C.J. Mulberry leaves extract ameliorates alcohol-induced liver damages through reduction of acetaldehyde toxicity and inhibition of apoptosis caused by oxidative stress signals. *Int. J. Med. Sci.* **2021**, *18*, 53–64. [CrossRef]

29. Kim, D.; Kang, K.H. Anti-Inflammatory and Anti-Bacterial Potential of Mulberry Leaf Extract on Oral Microorganisms. *Int. J. Environ. Res. Public Health* **2022**, *19*, 4984. [CrossRef]
30. Jung, S.H.; Han, J.H.; Park, H.S.; Lee, D.H.; Kim, S.J.; Cho, H.S.; Kang, J.S.; Myung, C.S. Effects of unaltered and bioconverted mulberry leaf extracts on cellular glucose uptake and antidiabetic action in animals. *BMC Complement. Altern. Med.* **2019**, *19*, 55. [CrossRef]
31. Tang, C.; Bao, T.; Zhang, Q.; Qi, H.; Huang, Y.; Zhang, B.; Zhao, L.; Tong, X. Clinical potential and mechanistic insights of mulberry (*Morus alba* L.) leaves in managing type 2 diabetes mellitus: Focusing on gut microbiota, inflammation, and metabolism. *J. Ethnopharmacol.* **2023**, *306*, 116143. [CrossRef]
32. Huang, Z.; Dai, H.; Jiang, J.; Ye, N.; Zhu, S.; Wei, Q.; Lv, Z.; Shi, F. Dietary mulberry-leaf flavonoids improve the eggshell quality of aged breeder hens. *Theriogenology* **2022**, *179*, 177–186. [CrossRef]
33. Huang, Z.; Lv, Z.; Dai, H.; Li, S.; Jiang, J.; Ye, N.; Zhu, S.; Wei, Q.; Shi, F. Dietary mulberry-leaf flavonoids supplementation improves liver lipid metabolism and ovarian function of aged breeder hens. *J. Anim. Physiol. Anim. Nutr.* **2022**, *106*, 1321–1332. [CrossRef]
34. Li, Q.; Liao, S.; Pang, D.; Li, E.; Liu, T.; Liu, F.; Zou, Y. The transported active mulberry leaf phenolics inhibited adipogenesis through PPAR-γ and Leptin signaling pathway. *J. Food Biochem.* **2022**, *46*, e14270. [CrossRef]
35. Fan, L.; Peng, Y.; Wu, D.; Hu, J.; Shi, X.; Yang, G.; Li, X. Dietary supplementation of *Morus nigra* L. leaves decrease fat mass partially through elevating leptin-stimulated lipolysis in pig model. *J. Ethnopharmacol.* **2020**, *249*, 112416. [CrossRef]
36. Li, J.; Kaneko, T.; Qin, L.-Q.; Wang, J.; Wang, Y.; Sato, A. Long-term effects of high dietary fiber intake on glucose tolerance and lipid metabolism in GK rats: Comparison among barley, rice, and cornstarch. *Metabolism* **2003**, *52*, 1206–1210. [CrossRef]
37. Myhrstad, M.C.W.; Tunsjø, H.; Charnock, C.; Telle-Hansen, V.H. Dietary Fiber, Gut Microbiota, and Metabolic Regulation—Current Status in Human Randomized Trials. *Nutrients* **2020**, *12*, 859. [CrossRef]
38. Liu, B.N.; Liu, X.T.; Liang, Z.H.; Wang, J.H. Gut microbiota in obesity. *World J. Gastroenterol.* **2021**, *27*, 3837–3850. [CrossRef]
39. Chiang, C.-F.; Villaverde, C.; Chang, W.-C.; Fascetti, A.J.; Larsen, J.A. Prevalence, Risk Factors, and Disease Associations of Overweight and Obesity in Dogs that Visited the Veterinary Medical Teaching Hospital at the University of California, Davis from January 2006 to December 2015. *Top. Companion Anim. Med.* **2022**, *48*, 100640. [CrossRef]
40. Elshourbagy, N.A.; Meyers, H.V.; Abdel-Meguid, S.S. Cholesterol: The Good, the Bad, and the Ugly—Therapeutic Targets for the Treatment of Dyslipidemia. *Med. Princ. Pract.* **2013**, *23*, 99–111. [CrossRef]
41. Yiu, J.H.C.; Chan, K.-S.; Cheung, J.; Li, J.; Liu, Y.; Wang, Y.; Fung, W.W.L.; Cai, J.; Cheung, S.W.M.; Dorweiler, B.; et al. Gut Microbiota-Associated Activation of TLR5 Induces Apolipoprotein A1 Production in the Liver. *Circ. Res.* **2020**, *127*, 1236–1252. [CrossRef]
42. Meksawan, K.; Pendergast, D.R.; Leddy, J.J.; Mason, M.; Horvath, P.J.; Awad, A.B. Effect of Low and High Fat Diets on Nutrient Intakes and Selected Cardiovascular Risk Factors in Sedentary Men and Women. *J. Am. Coll. Nutr.* **2004**, *23*, 131–140. [CrossRef] [PubMed]
43. Aziz-Aliabadi, F.; Noruzi, H.; Hassanabadi, A. Effect of different levels of green tea (*Camellia sinensis*) and mulberry (*Morus alba*) leaves powder on performance, carcass characteristics, immune response and intestinal morphology of broiler chickens. *Vet. Med. Sci.* **2023**, *9*, 1281–1291. [CrossRef] [PubMed]
44. Deng, L.; Zhong, G.; Peng, Y.; Zhu, Z. Mulberry leaf and konjac compound powder improves the metabolic capacity of old mice on a high-protein diet by regulating the structure of the intestinal microbiota. *J. Sci. Food Agric* **2023**, *103*, 5949–5957. [CrossRef] [PubMed]
45. Cui, X.; Yang, Y.; Zhang, M.; Liu, S.; Wang, H.; Jiao, F.; Bao, L.; Lin, Z.; Wei, X.; Qian, W.; et al. Transcriptomics and metabolomics analysis reveal the anti-oxidation and immune boosting effects of mulberry leaves in growing mutton sheep. *Front. Immunol.* **2022**, *13*, 1088850. [CrossRef]
46. Breedveld, A.; van Egmond, M. IgA and FcαRI: Pathological Roles and Therapeutic Opportunities. *Front. Immunol.* **2019**, *10*, 553. [CrossRef]
47. You, I.; Kim, M.J. Comparison of Gut Microbiota of 96 Healthy Dogs by Individual Traits: Breed, Age, and Body Condition Score. *Animals* **2021**, *11*, 2432. [CrossRef]
48. Alessandri, G.; Argentini, C.; Milani, C.; Turroni, F.; Cristina Ossiprandi, M.; van Sinderen, D.; Ventura, M. Catching a glimpse of the bacterial gut community of companion animals: A canine and feline perspective. *Microb. Biotechnol.* **2020**, *13*, 1708–1732. [CrossRef]
49. Handl, S.; Dowd, S.E.; Garcia-Mazcorro, J.F.; Steiner, J.M.; Suchodolski, J.S. Massive parallel 16S rRNA gene pyrosequencing reveals highly diverse fecal bacterial and fungal communities in healthy dogs and cats. *FEMS Microbiol. Ecol.* **2011**, *76*, 301–310. [CrossRef]
50. Alessandri, G.; Milani, C.; Mancabelli, L.; Mangifesta, M.; Lugli, G.A.; Viappiani, A.; Duranti, S.; Turroni, F.; Ossiprandi, M.C.; van Sinderen, D.; et al. The impact of human-facilitated selection on the gut microbiota of domesticated mammals. *FEMS Microbiol. Ecol.* **2019**, *95*, fiz121. [CrossRef]
51. Guo, W.-L.; Pan, Y.-Y.; Li, L.; Li, T.-T.; Liu, B.; Lv, X.-C. Ethanol extract of Ganoderma lucidum ameliorates lipid metabolic disorders and modulates the gut microbiota composition in high-fat diet fed rats. *Food Funct.* **2018**, *9*, 3419–3431. [CrossRef]
52. Tian, T.; Zhang, X.; Luo, T.; Wang, D.; Sun, Y.; Dai, J. Effects of Short-Term Dietary Fiber Intervention on Gut Microbiota in Young Healthy People. *Diabetes Metab. Syndr. Obes.* **2021**, *14*, 3507–3516. [CrossRef]
53. Mouzaki, M.; Comelli, E.M.; Arendt, B.M.; Bonengel, J.; Fung, S.K.; Fischer, S.E.; McGilvray, I.D.; Allard, J.P. Intestinal microbiota in patients with nonalcoholic fatty liver disease. *Hepatology* **2013**, *58*, 120–127. [CrossRef]
54. Qiao, S.; Bao, L.; Wang, K.; Sun, S.; Liao, M.; Liu, C.; Zhou, N.; Ma, K.; Zhang, Y.; Chen, Y.; et al. Activation of a Specific Gut Bacteroides-Folate-Liver Axis Benefits for the Alleviation of Nonalcoholic Hepatic Steatosis. *Cell Rep.* **2020**, *32*, 108005. [CrossRef]
55. Liu, R.; Hong, J.; Xu, X.; Feng, Q.; Zhang, D.; Gu, Y.; Shi, J.; Zhao, S.; Liu, W.; Wang, X.; et al. Gut microbiome and serum metabolome alterations in obesity and after weight-loss intervention. *Nat. Med.* **2017**, *23*, 859–868. [CrossRef]

56. Li, H.; Liu, F.; Lu, J.; Shi, J.; Guan, J.; Yan, F.; Li, B.; Huo, G. Probiotic Mixture of Lactobacillus plantarum Strains Improves Lipid Metabolism and Gut Microbiota Structure in High Fat Diet-Fed Mice. *Front. Microbiol.* **2020**, *11*, 512. [CrossRef]
57. Zhang, D.; Wang, S.; Li, G.; Chen, M.; Liu, H.; Ji, H. Two doses of Lactobacillus induced different microbiota profiles and serum immune indices in pigs. *J. Funct. Foods* **2023**, *102*, 105405. [CrossRef]

metabolites

Article

Baicalin Exhibits a Protective Effect against Cisplatin-Induced Cytotoxic Damage in Canine Renal Tubular Epithelial Cells

Yao Wang [1,†], Xiao Li [2,†], Chuanguo Yan [1], Liuwei Xie [3,*] and Yang Yang [3,*]

1 College of Police Dog Technology, Criminal Investigation Police University of China, Shenyang 110854, China; 2018990094@cipuc.edu.cn (Y.W.); 2018990026@cipuc.edu.cn (C.Y.)
2 Institute of Special Wild Economic Animals and Plants, Chinese Academy of Agricultural Sciences, Changchun 130112, China; 82101212443@caas.cn
3 The Second Affiliated Hospital of Shenyang Medical College, Shenyang 110031, China
* Correspondence: 2021990057@cipuc.edu.cn (L.X.); yang_yang913@126.com (Y.Y.)
† These authors contributed equally to this work.

Abstract: Renal failure is a common chronic disease in dogs that substantially affects both their quality of life and longevity. The objective of this study was to assess the protective mechanisms of baicalin in cisplatin-induced Madin–Darby canine kidney (MDCK) epithelial cells' apoptosis model and explore the impacts of baicalin at varying doses on various indexes, such as cisplatin-induced MDCK cell apoptosis, oxidation and antioxidation, and inflammatory factors. (Methods) MDCK cells in the logarithmic growth phase were randomly divided into a control group, a model group (20 μmol/L cisplatin), and a baicalin-protection group (20 μmol/L cisplatin + 50, 25 μmol/L baicalin) and received the corresponding treatments for 24 h. The effects of cisplatin on MDCK cell apoptosis, oxidation and antioxidation, inflammatory factors, and other indicators were studied, and the relieving effect of baicalin on cisplatin-induced MDCK cell damage was explored. Calcein/PI staining and Annexin V-FITC/PI staining showed that cisplatin induced the apoptosis of MDCK cells, while baicalin effectively reduced the damage caused by cisplatin. The ELISA results demonstrated a significant elevation in the nitric oxide (NO) and malondialdehyde (MDA) levels within the MDCK cells following treatment with cisplatin ($p < 0.01$). In addition, superoxide dismutase (SOD), glutathione peroxidase (GSH), and catalase (CAT) activities remarkably declined ($p < 0.01$), while tumor necrosis factor α (TNF-α), interleukin-1β (IL-1β), and interleukin-6 (IL-6) expression within the MDCK cells were apparently elevated ($p < 0.01$). However, baicalin treatment resulted in opposite changes in these factors. The findings suggested that baicalin exhibits potential in mitigating cisplatin-induced oxidative stress and inflammation in MDCK cells. As revealed with the Western blot results, cisplatin promoted P62, P53, and BAX protein levels, increased mTOR phosphorylation, inhibited AMPK phosphorylation, and reduced Beclin1 and BCL-2 protein levels. However, a contrasting trend was observed following baicalin treatment. Cisplatin can inhibit the activity of MDCK cells, lead to abnormalities in oxidation and antioxidation functions and cell inflammatory factors, and accelerate cell apoptosis. Moreover, baicalin can significantly alleviate the damage of cisplatin to MDCK cells.

Keywords: baicalin; cisplatin; canine renal tubular epithelial cells; toxic injury

Citation: Wang, Y.; Li, X.; Yan, C.; Xie, L.; Yang, Y. Baicalin Exhibits a Protective Effect against Cisplatin-Induced Cytotoxic Damage in Canine Renal Tubular Epithelial Cells. *Metabolites* **2023**, *13*, 1173. https://doi.org/10.3390/metabo13121173

Academic Editor: Akiyoshi Hirayama

Received: 19 October 2023
Revised: 14 November 2023
Accepted: 20 November 2023
Published: 24 November 2023

1. Introduction

In recent years, the growing demand for companion animals, influenced by changes in population structure and lifestyle [1], has propelled the rapid development of the pet industry [2]. Pets, particularly dogs, have assumed an important role in families due to the physiological and psychological companionship they offer. However, pet owners frequently exhibit excessive indulgence of their pets, such as providing unrestricted diets, which often elevates the risk of various chronic diseases in dogs, including kidney failure [3]. Consequently, there is an increasing focus on exploring effective strategies for preventing and curing kidney failure, improving the overall health status of dogs, and ultimately aiming to extend their lifespan.

Renal failure can lead to the rapid accumulation of metabolic waste products within the body, electrolyte imbalances, acid–base disturbances, and, eventually, a high mortality [4]. At present, kidney failure can hardly be cured and is only treated symptomatically to prolong the lifespan of patients, and such symptomatic treatment is associated with great side effects [5]. Kidney failure is accompanied by immune dysregulation, inflammatory response, and renal dysfunction in dogs, which have greatly affected the life and living conditions of dogs in some severe cases. Under the background of canine fine feeding, improving the body's antioxidative capacity and immune function in dogs through natural drug components has been an effective way of reducing the occurrence of disease in the current canine feeding industry. However, feed supplements or veterinary medicinal products that are extracted from natural medicinal plants and can effectively improve canine immune system's development are lacking in the industry for the time being. Therefore, it is of great significance to search for natural drug components with significant preventive or therapeutic effects and no side effects to improve the overall performance of dogs.

Cisplatin, a broad-spectrum anticancer drug, has been well-recognized to induce hepatorenal toxicity and serves as a model drug for investigating the protection on the liver and kidneys [6,7]. The cytotoxic effect of cisplatin is mainly induced by the cytotoxic effect resulting from its accumulation in the kidney. Pharmacokinetics revealed that the accumulation amount of cisplatin in the kidney is higher than that in the plasma or other tissues. When cisplatin enters the body, 90% of it can bind to plasma proteins, and it is mostly excreted directly via the kidney. Moreover, it displays the high-aggregation, high-excretion, and high-metabolism characteristics of the kidney. A large dose or continuous administration of cisplatin may induce irreversible, acute necrosis of proximal renal tubular epithelial cells, renal structural and functional disorders, and renal failure or even death in some severe cases [8].

Baicalin is a flavonoid compound with a relatively high content in the *Scutellaria baicalensis* Georgi, a perennial herb of the Labiaceae family. Baicalin possesses various properties such as antioxidant, anti-inflammatory, and autophagy-regulating effects [9]. It can exert protective effects on cellular toxicity. Baicalin can mitigate cellular damage, maintain cellular function and stability, and contribute to the normal functioning of tissues and organs [10]. Modern research has suggested that baicalin exhibits diverse effects, such as clearing away heat and toxic materials, anti-inflammation, antioxidation, and immunoregulation [11,12]; therefore, it has been used to treat a variety of inflammatory diseases, including periodontitis, hepatitis, and ulcerative colitis [13,14]. Moreover, baicalin also exerts a certain renal-protective effect. Zhang et al. [15] discovered that baicalin improved lead poisoning-induced oxidative injury in the kidneys of mice. Lin et al. [16] found that baicalin alleviated oxidative stress and renal ischemia–reperfusion injury to improve renal function. However, research on the role of baicalin in canine renal failure-related inflammation is rare. The role and mechanism of baicalin in ameliorating the inflammation associated with canine renal failure remain unclear. Therefore, the aim of this study was to evaluate the protective mechanisms of baicalin in cisplatin-induced Madin–Darby canine kidney (MDCK) epithelial cells' apoptosis model. The study explored the impacts of baicalin at varying doses on various indexes, such as the cisplatin-induced MDCK cellular structure, cell apoptosis, oxidation and antioxidation, and inflammatory factors, and it assessed its effect on alleviating MDCK cell injury. The study will provide an important theoretical foundation and reference for the application of baicalin in feed as a feed additive or veterinary medicinal product.

2. Materials and Methods

2.1. Cells and Materials

This work acquired MDCK cells from the Cell Bank of the Shanghai Institutes for Biological Sciences, in the Chinese Academy of Sciences. Baicalin, a standard substance, was procured from Solarbio Biotechnology Co., Ltd. (CAS: 21967-41-9, Molecular Formula: $C_{21}H_{18}O_{11}$; Beijing, China).

2.2. Cell Culture

The MDCK cells were cultivated within DMEM and maintained in a 5% CO_2 incubator under 37 °C, with regular medium change. Cells in the logarithmic growth stage and reaching 90% of confluence were collected for 0.25% trypsin digestion. The cells (approximately 9×10^3/well) were inoculated into a 96-well plate for a 24 h period to allow cell adherence, followed by a 24 h treatment using cisplatin at varying doses (10–160 μmol/L). The DMEM (Cat# KGM12800-500) was purchased from KeyGEN Biotech Co., Ltd. (Nanjing, China). The trypsin (Cat# BL512B) was purchased from Biosharp Co., Ltd. (Hefei, China). The cisplatin (Cat# 61825-94-3) was purchased from Sigma-Aldrich Co. (Shanghai, China).

2.3. Drug Treatment

To establish the cell model, cisplatin at 20 μmol/L was added for cell treatment. Before cell adherence, baicalin at 50, 25, 20, and 15 μmol/L was added for a 12 h cell pretreatment to determine the optimal protective concentration. Each treatment was repeated six times, and the formula below was employed for calculating cell viability: cell viability = (As − Ab)/(Ac − Ab) × 100%. In this formula, As stands for the experimental well's absorbance (including cells, culture medium, and CCK-8 reagent); Ab indicates the blank well's absorbance (including culture medium and CCK-8 reagent), whereas Ac is the control well's absorbance (including cells and culture medium, together with CCK-8 reagent). The CCK-8 reagent (Cat# C0038) was purchased from Shanghai Beyotime Biotechnology Co., Ltd. (Shanghai, China).

2.4. Apoptosis Detection

After a sequential drug treatment in the culture plate, PBS was added to rinse the cells thrice. Then, to investigate the effect of baicalin on apoptotic cell death, cells were stained with Calcein/PI staining (LIVE/DEAD® Viability/Cytotoxicity Kit for mammalian cells, Molecular Probes, Eugene, OR, USA; Cat# L-3224) and fluorescence microscopy (Olympus, Japan). Moreover, the following were carried out: calxanthin AM staining of living cells with green fluorescence and propyl iodide (PI) staining of dead cells with red fluorescence. The cells at the logarithmic growth phase were digested and inoculated into a 6-well cell culture plate at 9×10^3 cells/mL, followed by a 24 h treatment using 20 μmol/L of cisplatin (continued culture), with the addition of baicalin for the protective group. After collection, cells from each group were subjected to flow cytometric analysis with an Annexin V-FITC/PI staining kit (Wanlei Biotechnology Co., Ltd., Shenyang, China; Cat# WLA001b) in line with specific protocols. This examination process was completed within 1 h [17].

2.5. Oxidative and Antioxidant Function Assessment

The MDCK cells at the logarithmic growth phase were treated with cisplatin prior to their transfer to a CO_2 incubator for an additional 24 h incubation. Baicalin (25 and 50 μmol/L) was then added to the protection group. The cells were collected after trypsin digestion, and the levels of oxidative indicators (MDA) and antioxidant indicators (SOD, GSH, and CAT) were analyzed with enzyme-linked immunosorbent assay (ELISA) kits. The absorbance values at OD_{450} were measured, and the MDA levels together with the SOD, GSH, and CAT activities were calculated [18]. The MDA (Cat# A003-1-2), SOD (Cat# A001-1-2), GSH (Cat# A006-2-1), and CAT (Cat# A007-1-1) assay kits were purchased from the Nanjing Jiancheng Bioengineering Institute (Nanjing, China).

2.6. Detection of Cellular Inflammatory Cytokine Levels

Cells from each group were collected using a cell scraper. Later, the TNF-α, IL-1β, IL-6, and NO levels were analyzed using the corresponding kits in line with their specific protocols, and the levels of inflammatory cytokines in each group were calculated [19]. The TNF-α (Cat# JL22455), IL-1β (Cat# JL22367), and IL-6 (Cat# JL22371) ELISA kits were purchased from Shanghai Jianglai Biotechnology Co., Ltd. (Shanghai, China). NO (Cat# A012-1-2) assay kit was purchased from Nanjing Jiancheng Bioengineering Institute (Nanjing, China).

2.7. Western Blot Analysis

The cellular proteins were extracted using a protein extraction kit (Beyotime Biotechnology Co., Ltd., Nanjing, China; Cat# P0033) in line with the specific protocols. The protein contents were analyzed using the BCA approach (Beyotime Biotechnology Co., Ltd., Nanjing, China; Cat# P0010). The protein samples were denatured at 100 °C for 5 min, and, later, a portion of proteins were separated through sodium dodecyl sulfate-polyacrylamide gel electrophoresis (SDS-PAGE) and transferred onto a PVDF membrane (Beyotime Biotechnology Co., Ltd., Nanjing, China; Cat# FFP71). After blocking with 5% of defatted milk, the membrane was subjected to overnight antibody incubation (1:1000, Cell Signaling Technology, Inc., Danvers, MA, USA), including anti-P62 (Cat# 5114), anti-Beclin1 (Cat# 3738), anti-AMPK (Cat# 2532), anti-P-AMPK (Cat# 2531), anti-mTOR (Cat# 2983), anti-P-mTOR (Cat# 2971), anti-BAX (Cat# 2772), anti-BCL-2 (Cat# 15071), anti-P-53 (Cat# 9282), and anti-GAPDH (1:2000, Cell Signaling Technology, Inc.; Cat# 2118), under 4 °C. On the following day, the membrane was further incubated with secondary antibodies for a 1.5 h period under ambient temperature. After washing with TBST thrice, the membrane was further incubated with secondary antibodies for 2 h. After washing, protein expression was detected using an ECL substrate (Shanghai TianNeng Technology Co., Ltd., Shanghai, China; Cat# 180-501), and the protein expression's intensity was analyzed using the Quantity One software (version 4.6.8) [20].

2.8. Data Analysis

The results were represented with mean ± standard deviation. The GraphPad Prism 6.0.4 software was employed for statistical analysis [21]. The Shapiro–Wilk normality test was used for the normality test prior to the analyses. The intergroup differences were evaluated using a one-way analysis of variance (ANOVA), where $p < 0.05$ indicated significant differences among diverse treatments, while $p < 0.01$ suggested extremely significant differences.

3. Results

3.1. Modeling Concentration and Safe Concentration of Baicalin

To determine the safe and protective concentrations of baicalin, the MDCK cells were treated with baicalin at concentrations of 15, 20, 25, 50, and 100 μmol/L, respectively, according to the canine metabolic body weights. Compared to the CON group, at the concentration of 50 μmol/L, the cells' proliferative ability was not affected, and the cells' survival rate was high. At the concentration of 100 μmol/L, the cells' proliferation was significantly affected ($p < 0.01$) (Figure 1A); therefore, 50 μmol/L was selected as the safety concentration of baicalin. In comparison, the cells treated with 20 μmol/L of cisplatin exhibited a significantly decreased cell viability relative to the CON group ($p < 0.01$), and there was little difference among the high-concentration groups as the cisplatin concentration increased, with a high cell lethality rate. Such result indicated that 20 μmol/L of cisplatin already caused a significant toxicity to cells' growth, so 20 μmol/L of cisplatin was chosen as the modeling concentration. On this basis, baicalin at 15, 20, 25, and 50 μmol/L was added for a 12 h cell pretreatment, followed by the induction of injury with 20 μmol/L of cisplatin. Among these concentrations, the cells treated with baicalin at 25 and 50 μmol/L exhibited a significantly increased cell survival rate relative to the MOD group, and the survival rates of the cells treated with baicalin at 15 and 20 μmol/L were not significantly different from that of the MOD group, indicating the protective effect of baicalin on the cisplatin-induced MDCK cells, with the best effects being observed at 25 and 50 μmol/L ($p < 0.01$).

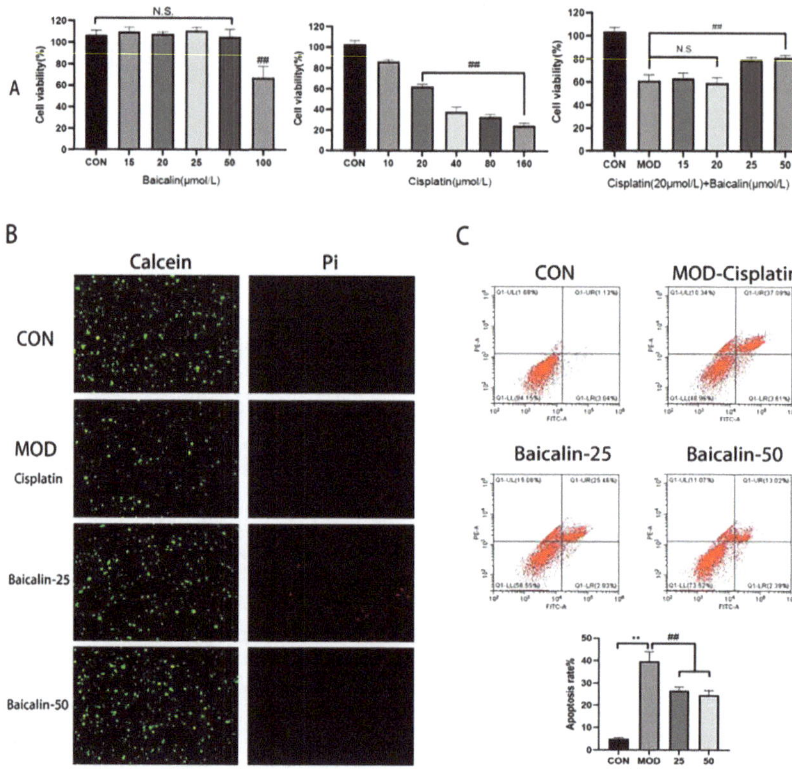

Figure 1. (**A**): Safe concentration of baicalin, modeling concentration of cisplatin, effective protective concentration of baicalin; (**B**): Calcein/PI staining; (**C**): Flow cytometry detection of cell apoptosis. CON: MDCK cells without treatment; MOD: cisplatin 20 μmol/L; Baicalin-25: cisplatin 20 μmol/L + baicalin 25 μmol/L; Baicalin-50: cisplatin 20 μmol/L + baicalin 50 μmol/L. N.S., no significance; ** and ##, $p < 0.01$.

3.2. Baicalin Improved Cisplatin-Mediated MDCK Cell Apoptosis

Calcein-AM is a virtually non-fluorescent cell-permeant neutral dye that is con-verted through cell esterase into its negative, impermeant, green-fluorescent analogue [22]. Living cells are distinguished by the presence of extensive esterase activity and an intact cellular membrane. However, PI enters cells characterized by membrane degradation, a feature of late apoptosis, and cells without any membrane, a feature of necrosis (reviewed in Atale et al. [23]). The results indicated that, following cisplatin treatment, there was an increase in cells with PI staining, while cells with Calcein staining showed a decrease, suggesting that the cells underwent late apoptosis or necrosis. However, after treatment with baicalin, the number of cells with PI staining decreased, with a more significant effect observed especially at a concentration of 50 μmol/L (Figure 1B).

The cells with Annexin V-FITC/PI staining were analyzed using flow cytometry. From the flow cytometry results, compared to the CON group, the apoptosis rate of the MOD group increased by seven times, while, after baicalin intervention, the proportion of apoptotic cells remarkably decreased. Typically, the best apoptosis-inhibition effect was achieved at the concentration of 50 μmol/L, and the number of cells in the late apoptosis stage or necrotic cells apparently decreased. This outcome demonstrated a conformity with the findings obtained from the Calcein/PI staining technique. Such results suggested that cisplatin induction increased the apoptosis rate, whereas baicalin at different concentrations provided protection through reducing the apoptosis ratio of the MDCK cells, and the ability to reduce cell apoptosis was in direct proportion to the baicalin's concentration (Figure 1C).

3.3. Baicalin Alleviated Cisplatin-Mediated Oxidative Stress and Inflammation of MDCK Cells

In a comparison with the CON group, the SOD, CAT, and GSH activities of the MOD group remarkably declined ($p < 0.01$), whereas the MDA level was evidently elevated ($p < 0.01$) (Figure 2A–D). However, relative to the MOD group, baicalin at 50 μmol/L significantly enhanced the SOD, CAT, and GSH activities ($p < 0.01$), but remarkably decreased the MDA level ($p < 0.01$). In a comparison with the CON group, the TNF-α, IL-1β, IL-6, and NO levels of the MOD group significantly increased ($p < 0.01$). However, relative to the MOD group, baicalin at 50 μmol/L significantly reduced the TNF-α, IL-1β, IL-6, and NO levels ($p < 0.01$) (Figure 2E–H). The above research results indicated that, after cisplatin induction, the MDCK cells were damaged, accompanied by an oxidative stress reaction and an inflammatory response, and that the two might be closely associated with each other. After the combined use of baicalin, the content of the oxidizing agent decreased, and the antioxidant activity was enhanced, suggesting that the oxidative stress reaction in the cells was weakened and that the oxidative toxicity was reduced. The inflammatory factor levels also dramatically decreased after baicalin application, indicating that baicalin alleviated the cisplatin-induced oxidative stress and inflammatory response in the MDCK cells.

3.4. Baicalin Effectively Counteracted Cisplatin-Mediated MDCK Cells Apoptosis

In comparison with the CON group, autophagy-related protein Beclin1 expression significantly decreased, while p62 expression apparently increased in the MOD group (Figure 3), indicating a decrease in the autophagic activity of the MDCK cells after cisplatin induction. In comparison to the MOD group, Beclin1 expression dramatically increased in the baicalin intervention group, while p62 expression apparently decreased, indicating that baicalin improved the autophagic function in the MDCK cells after cisplatin-induced injury.

The AMPK–mTOR axis plays an important role in energy metabolism and cell autophagy, and p53 can regulate the mTOR upstream regulatory factors to promote autophagy. Compared to the CON group, the *p*-AMPK/AMPK protein's expression ratio decreased, whereas the *p*-mTOR/mTOR protein's expression ratio increased in the protein samples from the MOD group, indicating a decrease in the autophagic activity. However, when baicalin was administered, the *p*-AMPK/AMPK ratio increased, while the *p*-mTOR/mTOR ratio declined, suggesting that baicalin promoted the autophagy of the MDCK cells by up-regulating *p*-AMPK or down-regulating *p*-mTOR. Additionally, compared to the CON group, the p53 and pro-apoptotic protein BAX expression apparently increased, whereas the anti-apoptotic protein BCL-2 expression remarkably decreased in the MOD group. After baicalin intervention, the p53 and BAX levels apparently decreased, while BCL-2 expression markedly increased, indicating that baicalin markedly reduced the cisplatin-induced apoptosis of the MDCK cells; the best effect was achieved at the baicalin concentration of 50 μmol/L.

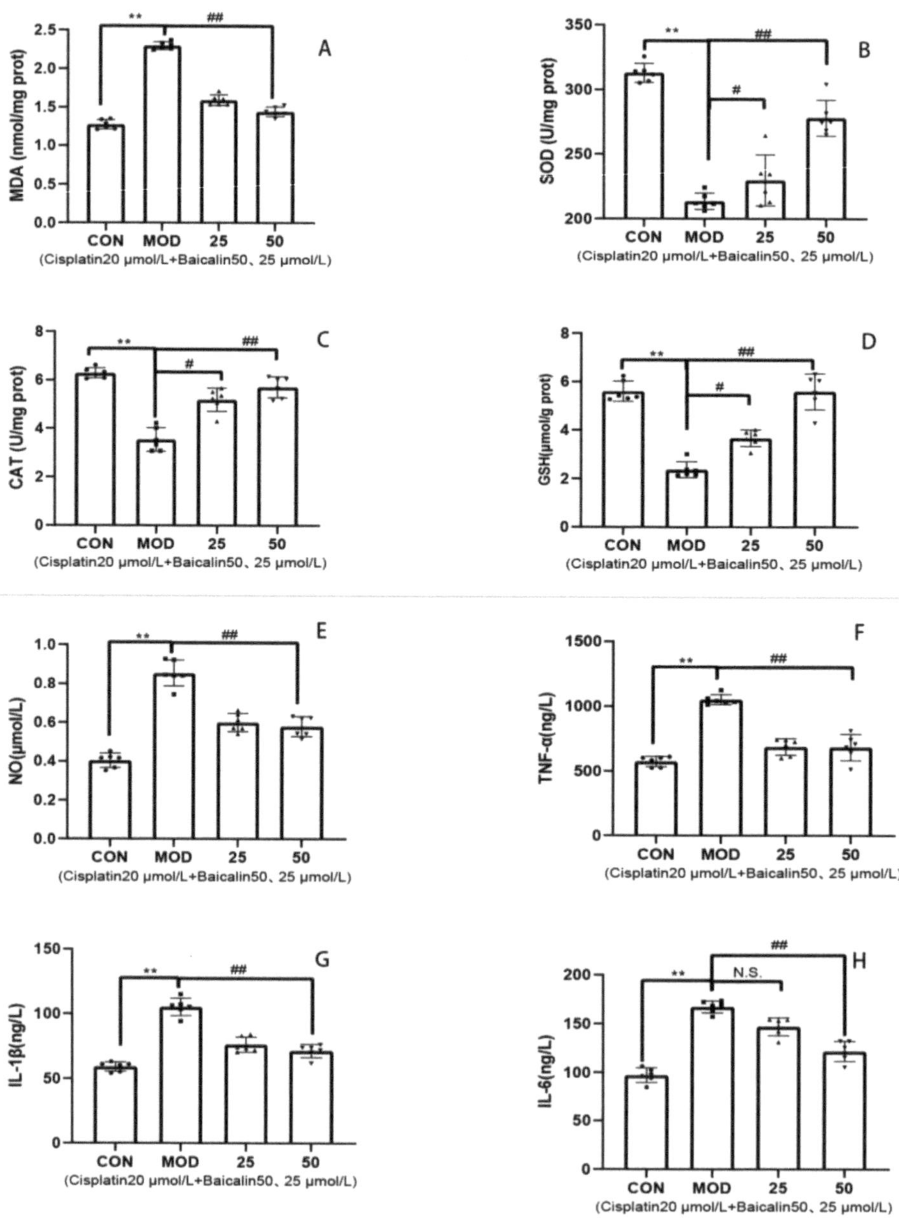

Figure 2. (**A**): MDA content in cells; (**B**): SOD content in cells; (**C**): CAT content in cells; (**D**): GSH content in cells; (**E**): NO content in cells; (**F**): TNF-α content in cells; (**G**): IL-1β content in cells; and (**H**): IL-6 content in cells. CON: MDCK cells without treatment. MOD: cisplatin 20 μmol/L; baicalin-25—Cisplatin 20 μmol/L + Baicalin 25 μmol/L; and baicalin-50—Cisplatin 20 μmol/L + Baicalin 50 μmol/L. N.S., no significance; #, $p < 0.05$; ** and ##, $p < 0.01$.

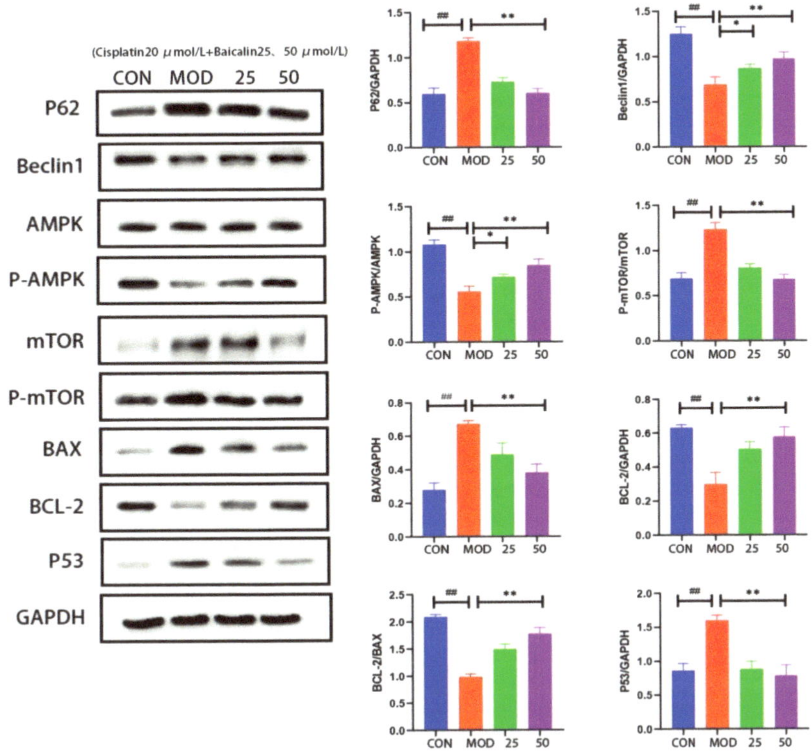

Figure 3. Expression levels of P62, Beclin1, AMPK, P-AMPK, mTOR, P-mTOR, BAX, BCL-2, and P53 in the cells. CON: MDCK cells without treatment. MOD: cisplatin 20 μmol/L; baicalin-25—cisplatin 20 μmol/L + baicalin 25 μmol/L; and baicalin-50—cisplatin 20 μmol/L + baicalin 50 μmol/L. *, $p < 0.05$; ** and ##, $p < 0.01$. Blue: CON group; Red: MOD group; Green: cisplatin 20 μmol/L + baicalin 25 μmol/L; Purple: isplatin 20 μmol/L + baicalin 50 μmol/L.

4. Discussion

Renal failure is a common clinical disease in dogs, ranking among the top three causes of canine mortality, with a prevalence of 2–5% [24]. Although the further deterioration of the disease can be controlled in the late stage through symptomatic treatment, it significantly affects the quality of life and lifespan of the dogs [25,26]. Renal failure is often accompanied by immune dysregulation and inflammatory responses in the body [27]. Moreover, long-term antibiotic application can have adverse effects on the health of dogs. Baicalin has a low price and possesses favorable pharmacological activities related to anti-inflammation, antioxidation, and immunoregulation, with a low toxicity and minimal side effects; thus, it is a high-quality natural drug component [11]. However, research on the role of baicalin in canine renal failure-related diseases is lacking. Consequently, baicalin was selected as the object of study in this research to explore its role in mitigating MDCK cell injury.

Cisplatin is one of the first-generation platinum-based cyclin-specific antitumor drugs, exhibiting a favorable anticancer activity [28]. Despite its broad-spectrum and efficient anti-cancer properties, cisplatin is also associated with severe renal toxicity, making it a model drug for researching kidney protection drugs [29]. In this study, we utilized cisplatin to induce MDCK cell injury, and the results indicated that cisplatin at 20 μmol/L significantly inhibited the growth of MDCK cells, increased the apoptosis rate, induced oxidative stress and inflammation, and reduced autophagy activity in MDCK cells. It has been reported that the renal toxicity of cisplatin is primarily induced by its accumulation in the kidney, resulting in cytotoxicity [30]. Modern pharmacological research suggests that the major mechanisms of cisplatin in inducing cell death may include apoptosis, autophagy, and necrosis [31,32], which aligns with our findings. Therefore, the present results demonstrate the successful construction of a cisplatin-induced MDCK cell injury model.

This study explored the experimental concentrations of baicalin and the modeling concentrations of cisplatin based on canine metabolic body weights. Our findings revealed that cisplatin concentrations exceeding 20 μmol/L significantly inhibited MDCK cell proliferation and reduced cell survival rates. Consequently, we selected 20 μmol/L as the modeling concentration for cisplatin. Our results showed that baicalin concentrations below 50 μmol/L had no impact on the proliferation and survival rates of MDCK cells. Additionally, baicalin at 25 μmol/L and 50 μmol/L significantly promoted cisplatin-induced MDCK cell proliferation and increased survival rates. Thus, we established 50 μmol/L as the safe concentration for baicalin, with the experimental concentrations chosen as 25 μmol/L and 50 μmol/L.

This experiment utilized Calcein/PI staining and Annexin V-FITC/PI staining to assess cell apoptosis or necrosis. Our results indicated that cisplatin induction significantly promoted late apoptosis or necrosis in MDCK cells, while baicalin markedly reduced the number of late apoptotic or necrotic cells in a dose-dependent manner. The mitochondrial pathway is the classical cell apoptotic pathway, where the anti-apoptotic protein Bcl-2 and the pro-apoptotic protein BAX regulate mitochondrial structural and functional stability [33]. The results showed that, after cisplatin intervention, the BCL-2 expression level in the MDCK cells decreased, and the Bax expression level increased, indicating that cisplatin increased the apoptosis of the MDCK cells. In contrast, baicalin treatment led to opposite results in a dose-dependent manner. Thus, baicalin can mitigate cisplatin-induced MDCK cell apoptosis by regulating the mitochondrial pathway. An animal experiment revealed that *Scutellaria baicalensis* Georgi extract can ameliorate mice renal tubular cell necrosis and sloughing induced by cisplatin, with baicalin being identified as the main active component [34]. Baicalin has also been demonstrated to exert a protective effect against apoptosis in renal cells induced by various chemicals such as lead, H_2O_2, and methylglyoxal, and this protective effect is associated with baicalin's reduction of the BAX/BCL2 ratio, activation of the Nrf2 pathway, and prevention of the mitochondrial membrane's potential loss [15,35,36]. These findings suggest that baicalin can improve renal cell apoptosis induced by different chemicals through various molecular pathways.

The organism has developed a comprehensive antioxidant system. Excessive production of MDA in cells is a sign of oxidative damage, leading to an imbalance in cellular oxidation [37]. GSH consumption indicates the presence of excessive free radicals or other oxidative toxic factors in the body [38]. The levels of SOD, CAT, and GSH also reflect the metabolism of oxygen free radicals and the body's antioxidant ability [39]. Studies conducted by Hu et al. [11] and Jia et al. [40] revealed that supplementing baicalin in a diet enhances the activities of CAT and SOD while reducing MDA content in pre-weaned calves and tilapia. Baicalin also plays an important regulatory role in ameliorating cells' oxidative stress induced by various factors [41–43]. Moreover, the antioxidant activity of baicalin has been demonstrated to be mediated through the Nrf2 signaling pathway [44,45]. Our study obtained similar results; the results revealed that cisplatin treatment for 24 h significantly increased the production of MDA in the MDCK cells and decreased the activities of SOD, CAT, and GSH, while baicalin treatment significantly reversed these adverse effects in a dose-dependent manner. This further indicates that baicalin displays apparent pharmacological characteristics and animal universality in enhancing the body's antioxidant capacity. Inflammatory responses are triggered when cells and tissues are damaged. Inflammatory cytokines have essential effects on immunoregulation and other processes [46]. The changes in inflammatory factors reflect the severity of inflammation in the body and the immune response to inflammation [47]. In this study, TNF-α, IL-6, IL-1β, and NO were selected as the indexes for evaluating inflammation. According to our findings, cisplatin treatment significantly increased the TNF-α, IL-6, IL-1β, and NO contents in the MDCK cells, indicating the occurrence of cellular inflammatory injury. Baicalin effectively repaired the inflammatory damage caused through cisplatin by reducing inflammatory factor levels, and these results were similar to those of previous research [48,49]. This suggests that baicalin, in ameliorating cisplatin-induced MDCK cell injury, not only improves oxidative stress but also exerts immunomodulatory effects.

In mammalian cells, ATP homeostasis is closely related to cell growth, proliferation, autophagy, and apoptosis [50]. Adenosine monophosphoprotein kinase (AMPK) is a key regulatory factor for ATP homeostasis, extensively distributed in metabolic organs throughout the body, and plans an important role in maintaining energy, metabolism, and cell apoptosis [51,52]. Cell autophagy is a tightly regulated process where the cancer suppressor gene p53 can up-regulate the mammalian target of rapamycin (mTOR)'s upstream regulatory factors to promote autophagy through the transcription-dependent pathway [53]. mTOR regulates cell autophagy, while p62 is the autophagic substrate, and the disruption of cell apoptosis leads to the accumulation of p62. Consequently, p62, p53, and Beclin1 protein levels are indicative of autophagic activity [54]. Our research found that cisplatin treatment reduced Beclin1 expression but elevated p62 and p53 expression in the MDCK cells, suggesting a decrease in autophagic function. However, baicalin intervention showed the opposite effect, indicating that baicalin improved autophagic function in the MDCK cells. Additionally, the AMPK/mTOR pathway has been extensively found to be involved in autophagy regulation [55,56]. Cisplatin intervention enhanced mTOR phosphorylation and reduced AMPK phosphorylation, indicating a decreased autophagic activity. Baicalin reversed these changes, suggesting its ability to attenuate autophagy in cisplatin-induced MDCK cells by regulating the AMPK/mTOR pathway.

Based on the current results, cisplatin-induced MDCK cell injury involves oxidative stress, inflammatory response, autophagy, and apoptosis. Baicalin, in turn, regulates and mitigates cisplatin-induced MDCK cell injury through multiple pathways and cellular physiological processes. This cellular level study will provide a reference for future research on baicalin as a feed additive in the daily diet of dogs. Overall, baicalin shows promise as a therapeutic agent for alleviating the toxic side effects associated with cisplatin treatment. However, further research is required to elucidate the underlying molecular mechanisms and assess the efficacy of baicalin in vivo. This research holds significant importance in the development of a natural drug to prevent and improve canine renal failure.

Author Contributions: Conceptualization, Y.Y. and Y.W.; methodology, X.L. and Y.W.; formal analysis, C.Y. and L.X.; writing—review and editing, X.L., L.X. and Y.Y.; project management, Y.W. All authors have read and agreed to the published version of the manuscript.

Funding: 1. Research on the Application of Police Dog Autonomous Search Technology in High Pole Crops, Caves, and Pit Environments, a Key Project of the Ministry of Public Security (2022JSZ10), and 2. Industry Standard Search Dog Field Search Use Specification (2022BZ126).

Institutional Review Board Statement: Not applicable.

Informed Consent Statement: Not applicable.

Data Availability Statement: The data presented in this study are available on request from the corresponding author. The data are not publicly available due to privacy.

Acknowledgments: We thank Hengxing Ba for his critique of the paper and valuable suggestions.

Conflicts of Interest: The authors declare no conflict of interest.

References

1. Friedmann, E.; Gee, N.R.; Simonsick, E.M.; Studenski, S.; Resnick, B.; Barr, E.; Kitner-Triolo, M.; Hackney, A. Pet Ownership Patterns and Successful Aging Outcomes in Community Dwelling Older Adults. *Front. Vet. Sci.* **2020**, *7*, 293. [CrossRef]
2. Moon, D.C.; Choi, J.H.; Boby, N.; Kim, S.J.; Song, H.J.; Park, H.S.; Gil, M.C.; Yoon, S.S.; Lim, S.K. Prevalence of Bacterial Species in Skin, Urine, Diarrheal Stool, and Respiratory Samples in Cats. *Pathogens* **2022**, *11*, 324. [CrossRef]
3. Stokes, J.E.; Forrester, S.D. New and unusual causes of acute renal failure in dogs and cats. *Vet. Clin. N. Am.-Small* **2004**, *34*, 909–922. [CrossRef] [PubMed]

4. Collaboration, G.B.D.C.K.D. Global, regional, and national burden of chronic kidney disease, 1990-2017: A systematic analysis for the Global Burden of Disease Study 2017. *Lancet* **2020**, *395*, 709–733. [CrossRef]
5. Banerji, B.; Pramanik, S.K. Binding studies of creatinine and urea on iron-nanoparticle. *Springerplus* **2015**, *4*, 708. [CrossRef] [PubMed]
6. Waseem, M.; Tabassum, H.; Bhardwaj, M.; Parvez, S. Ameliorative efficacy of quercetin against cisplatin-induced mitochondrial dysfunction: Study on isolated rat liver mitochondria. *Mol. Med. Rep.* **2017**, *16*, 2939–2945. [CrossRef] [PubMed]
7. Xie, Y.H.; Wang, L.; Li, M.L.; Gong, Z.C.; Du, J. Role of myo-inositol in acute kidney injury induced by cisplatin. *Toxicology* **2023**, *499*, 153653. [CrossRef]
8. Pabla, N.; Dong, Z. Cisplatin nephrotoxicity: Mechanisms and renoprotective strategies. *Kidney Int.* **2008**, *73*, 994–1007. [CrossRef]
9. Bao, M.; Ma, Y.; Liang, M.; Sun, X.; Ju, X.; Yong, Y.; Liu, X. Research progress on pharmacological effects and new dosage forms of baicalin. *Vet. Med. Sci.* **2022**, *8*, 2773–2784. [CrossRef]
10. Yingrui, W.; Zheng, L.; Guoyan, L.; Hongjie, W. Research progress of active ingredients of Scutellaria baicalensis in the treatment of type 2 diabetes and its complications. *Biomed. Pharmacother.* **2022**, *148*, 112690. [CrossRef]
11. Hu, F.M.; Bi, Y.L.; Zheng, X.L.; Lu, M.; Diao, Q.Y.; Tu, Y. Effect of baicalin supplementation on the growth, health, antioxidant and anti-inflammatory capacity, and immune function of preweaned calves. *Anim. Feed. Sci. Tech.* **2023**, *298*, 115598. [CrossRef]
12. Ahmadi, A.; Mortazavi, Z.; Mehri, S.; Hosseinzadeh, H. Scutellaria baicalensis and its constituents baicalin and baicalein as antidotes or protective agents against chemical toxicities: A comprehensive review. *Naunyn Schmiedebergs Arch. Pharmacol.* **2022**, *395*, 1297–1329. [CrossRef]
13. Jang, J.Y.; Im, E.; Kim, N.D. Therapeutic Potential of Bioactive Components from Scutellaria baicalensis Georgi in Inflammatory Bowel Disease and Colorectal Cancer: A Review. *Int. J. Mol. Sci.* **2023**, *24*, 1954. [CrossRef] [PubMed]
14. Luo, W.; Wang, C.Y.; Jin, L. Baicalin downregulates Porphyromonas gingivalis lipopolysaccharide-upregulated IL-6 and IL-8 expression in human oral keratinocytes by negative regulation of TLR signaling. *PLoS ONE* **2012**, *7*, e51008. [CrossRef]
15. Zhang, Z.; Gao, X.; Guo, M.; Jiang, H.; Cao, Y.; Zhang, N. The Protective Effect of Baicalin Against Lead-Induced Renal Oxidative Damage in Mice. *Biol. Trace Elem. Res.* **2017**, *175*, 129–135. [CrossRef] [PubMed]
16. Lin, M.; Li, L.; Li, L.; Pokhrel, G.; Qi, G.; Rong, R.; Zhu, T. The protective effect of baicalin against renal ischemia-reperfusion injury through inhibition of inflammation and apoptosis. *BMC Complement. Altern. Med.* **2014**, *14*, 19. [CrossRef] [PubMed]
17. Shen, H.; He, Z.; Pei, H.; Zhai, L.; Guan, Q.; Wang, G. Aurantiamide promotes M2 polarization of microglial cells to improve the cognitive ability of mice with Alzheimer's disease. *Phytother. Res.* **2023**, *37*, 101–110. [CrossRef] [PubMed]
18. Liu, W.; Zhang, H.; Hou, Y.Y.; Hu, R.Y.; Zhang, J.J.; Chen, X.; Wang, S.; Hu, J.N.; Wang, Z.; Li, W. Arginyl-fructosyl-glucose, a major Maillard reaction product of red ginseng mitigates cisplatin-evoked intestinal toxicity in vivo and in vitro. *Food Funct.* **2022**, *13*, 11283–11297. [CrossRef]
19. Zhang, M.; Chen, W.; Zong, Y.; Shi, K.; Li, J.; Zeng, F.; He, Z.; Du, R. Cognitive-enhancing effects of fibrauretine on Aβ(1-42)-induced Alzheimer's disease by compatibilization with ginsenosides. *Neuropeptides* **2020**, *82*, 102020. [CrossRef] [PubMed]
20. Hao, M.; Ding, C.; Peng, X.; Chen, H.; Dong, L.; Zhang, Y.; Chen, X.; Liu, W.; Luo, Y. Ginseng under forest exerts stronger anti-aging effects compared to garden ginseng probably via regulating PI3K/AKT/mTOR pathway, SIRT1/NF-κB pathway and intestinal flora. *Phytomedicine* **2022**, *105*, 154365. [CrossRef] [PubMed]
21. Zhang, J.J.; Wang, J.Q.; Xu, X.Y.; Yang, J.Y.; Wang, Z.; Jiang, S.; Wang, Y.P.; Zhang, J.; Zhang, R.; Li, W. Red ginseng protects against cisplatin-induced intestinal toxicity by inhibiting apoptosis and autophagy via the PI3K/AKT and MAPK signaling pathways. *Food Funct.* **2020**, *11*, 4236–4248. [CrossRef] [PubMed]
22. Neri, S.; Mariani, E.; Meneghetti, A.; Cattini, L.; Facchini, A. Calcein-acetyoxymethyl cytotoxicity assay: Standardization of a method allowing additional analyses on recovered effector cells and supernatants. *Clin. Diagn. Lab. Immunol.* **2001**, *8*, 1131–1135. [CrossRef] [PubMed]
23. Atale, N.; Gupta, S.; Yadav, U.C.; Rani, V. Cell-death assessment by fluorescent and nonfluorescent cytosolic and nuclear staining techniques. *J. Microsc.* **2014**, *255*, 7–19. [CrossRef] [PubMed]
24. Lund, E.M.; Armstrong, P.J.; Kirk, C.A.; Kolar, L.M.; Klausner, J.S. Health status and population characteristics of dogs and cats examined at private veterinary practices in the United States. *J. Am. Vet. Med. Assoc.* **1999**, *214*, 1336–1341.
25. Stevens, K.B.; O'Neill, D.; Jepson, R.; Holm, L.P.; Walker, D.J.; Cardwell, J.M. Signalment risk factors for cutaneous and renal glomerular vasculopathy (Alabama rot) in dogs in the UK. *Vet. Rec.* **2018**, *183*, 448. [CrossRef]
26. O'Neill, D.G.; Elliott, J.; Church, D.B.; McGreevy, P.D.; Thomson, P.C.; Brodbelt, D.C. Chronic kidney disease in dogs in UK veterinary practices: Prevalence, risk factors, and survival. *J. Vet. Intern. Med.* **2013**, *27*, 814–821. [CrossRef]
27. Martínez-Hernández, S.L.; Muñoz-Ortega, M.H.; Ávila-Blanco, M.E.; Medina-Pizaño, M.Y.; Ventura-Juárez, J. Novel Approaches in Chronic Renal Failure without Renal Replacement Therapy: A Review. *Biomedicines* **2023**, *11*, 2828. [CrossRef] [PubMed]
28. Diéras, V.; Guastalla, J.P.; Ferrero, J.M.; Curé, H.; Weber, B.; Winckel, P.; Lortholary, A.; Mayer, F.; Paraiso, D.; Magherini, E.; et al. A multicenter phase II study of cisplatin and docetaxel (Taxotere) in the first-line treatment of advanced ovarian cancer: A GINECO study. *Cancer Chemother. Pharmacol.* **2004**, *53*, 489–495. [CrossRef]
29. de Godoy Torso, N.; Visacri, M.B.; Quintanilha, J.C.F.; Cursino, M.A.; Pincinato, E.C.; Moriel, P. Assessment of Renal Function in Head and Neck Cancer Patients Treated with Cisplatin: Different Biomarkers and Acute Kidney Injury Classifications. *Int. J. Mol. Sci.* **2022**, *24*, 141. [CrossRef]

30. Ueno, M.; Morizane, C.; Okusaka, T.; Mizusawa, J.; Kataoka, T.; Ikeda, M.; Ozaka, M.; Okano, N.; Sugimori, K.; Todaka, A.; et al. Comparison of gemcitabine-based chemotherapies for advanced biliary tract cancers by renal function: An exploratory analysis of JCOG1113. *Sci. Rep.* **2021**, *11*, 12885. [CrossRef]
31. Cavallo, F.; Feldman, D.R.; Barchi, M. Revisiting DNA damage repair, p53-mediated apoptosis and cisplatin sensitivity in germ cell tumors. *Int. J. Dev. Biol.* **2013**, *57*, 273–280. [CrossRef] [PubMed]
32. Zhang, H.Q.; He, B.; Fang, N.; Lu, S.; Liao, Y.Q.; Wan, Y.Y. Autophagy inhibition sensitizes cisplatin cytotoxicity in human gastric cancer cell line SGC7901. *Asian Pac. J. Cancer Prev.* **2013**, *14*, 4685–4688. [CrossRef] [PubMed]
33. Llambi, F.; Wang, Y.M.; Victor, B.; Yang, M.; Schneider, D.M.; Gingras, S.; Parsons, M.J.; Zheng, J.H.; Brown, S.A.; Pelletier, S.; et al. BOK Is a Non-canonical BCL-2 Family Effector of Apoptosis Regulated by ER-Associated Degradation. *Cell* **2016**, *165*, 421–433. [CrossRef]
34. Huang, T.H.; Wu, T.H.; Guo, Y.H.; Li, T.L.; Chan, Y.L.; Wu, C.J. The concurrent treatment of Scutellaria baicalensis Georgi enhances the therapeutic efficacy of cisplatin but also attenuates chemotherapy-induced cachexia and acute kidney injury. *J. Ethnopharmacol.* **2019**, *243*, 112075. [CrossRef] [PubMed]
35. Lin, M.; Li, L.; Zhang, Y.; Zheng, L.; Xu, M.; Rong, R.; Zhu, T. Baicalin ameliorates H2O2 induced cytotoxicity in HK-2 cells through the inhibition of ER stress and the activation of Nrf2 signaling. *Int. J. Mol. Sci.* **2014**, *15*, 12507–12522. [CrossRef]
36. Singh, J.; Chaudhari, B.P.; Kakkar, P. Baicalin and chrysin mixture imparts cyto-protection against methylglyoxal induced cytotoxicity and diabetic tubular injury by modulating RAGE, oxidative stress and inflammation. *Environ. Toxicol. Pharmacol.* **2017**, *50*, 67–75. [CrossRef]
37. Li, X.; Chen, S.; Li, J.E.; Wang, N.; Liu, X.; An, Q.; Ye, X.M.; Zhao, Z.T.; Zhao, M.; Han, Y.; et al. Chemical Composition and Antioxidant Activities of Polysaccharides from Yingshan Cloud Mist Tea. *Oxid. Med. Cell Longev.* **2019**, *2019*, 1915967. [CrossRef]
38. Yang, C.; Li, L.; Ma, Z.; Zhong, Y.; Pang, W.; Xiong, M.; Fang, S.; Li, Y. Hepatoprotective effect of methyl ferulic acid against carbon tetrachloride-induced acute liver injury in rats. *Exp. Ther. Med.* **2018**, *15*, 2228–2238. [CrossRef]
39. Shao, D.; Wu, Z.; Bai, S.; Fu, G.; Zou, Z. The function of miRNA-153 against isoflurane-induced neurotoxicity via Nrf2/ARE cytoprotection. *Mol. Med. Rep.* **2019**, *19*, 4001–4010. [CrossRef]
40. Jia, R.; Du, J.; Cao, L.; Feng, W.; Xu, P.; Yin, G. Effects of dietary baicalin supplementation on growth performance, antioxidative status and protection against oxidative stress-induced liver injury in GIFT tilapia (*Oreochromis niloticus*). *Comp. Biochem. Physiol. C Toxicol. Pharmacol.* **2021**, *240*, 108914. [CrossRef]
41. Wang, X.; Yu, J.Y.; Sun, Y.; Wang, H.; Shan, H.; Wang, S. Baicalin protects LPS-induced blood-brain barrier damage and activates Nrf2-mediated antioxidant stress pathway. *Int. Immunopharmacol.* **2021**, *96*, 107725. [CrossRef] [PubMed]
42. Gao, W.; Xu, B.; Zhang, Y.; Liu, S.; Duan, Z.; Chen, Y.; Zhang, X. Baicalin Attenuates Oxidative Stress in a Tissue-Engineered Liver Model of NAFLD by Scavenging Reactive Oxygen Species. *Nutrients* **2022**, *14*, 541. [CrossRef]
43. Zhang, L.; Yu, G.; Yu, Q.; Wang, L.; Wu, L.; Tao, Z.; Ding, J.; Lin, D. Baicalin promotes random-pattern skin flap survival by inducing autophagy via AMPK-regulated TFEB nuclear transcription. *Phytother. Res.* **2023**, *37*, 3926–3938. [CrossRef] [PubMed]
44. Bao, M.; Liang, M.; Sun, X.; Mohyuddin, S.G.; Chen, S.; Wen, J.; Yong, Y.; Ma, X.; Yu, Z.; Ju, X.; et al. Baicalin Alleviates LPS-Induced Oxidative Stress via NF-κB and Nrf2-HO1 Signaling Pathways in IPEC-J2 Cells. *Front. Vet. Sci.* **2021**, *8*, 808233. [CrossRef] [PubMed]
45. Li, H.; Cong, X.; Yu, W.; Jiang, Z.; Fu, K.; Cao, R.; Tian, W.; Feng, Y. Baicalin inhibits oxidative injures of mouse uterine tissue induced by acute heat stress through activating the Keap1/Nrf2 signaling pathway. *Res. Vet. Sci.* **2022**, *152*, 717–725. [CrossRef] [PubMed]
46. Hu, J.; Wu, H.; Wang, D.; Yang, Z.; Zhuang, L.; Yang, N.; Dong, J. Weicao capsule ameliorates renal injury through increasing autophagy and NLRP3 degradation in UAN rats. *Int. J. Biochem. Cell Biol.* **2018**, *96*, 1–8. [CrossRef]
47. Gu, X.; Ma, Z.; Fang, J.; Cai, D.; Zuo, Z.; Liang, S.; Cui, H.; Deng, J.; Ma, X.; Ren, Z.; et al. Obesity Enhances Antioxidant Capacity and Reduces Cytokine Levels of the Spleen in Mice to Resist Splenic Injury Challenged by Escherichia coli. *J. Immunol. Res.* **2020**, *2020*, 5948256. [CrossRef]
48. Li, L.; Dong, J.M.; Ye, H.H.; Jiang, M.J.; Yang, H.H.; Liang, L.P.; Ning, L.J.; Wu, Y. Baicalin promotes antiviral IFNs production and alleviates type I IFN-induced neutrophil inflammation. *J. Nat. Med.* **2023**, *77*, 677–687. [CrossRef]
49. Wang, X.; Xie, L.; Long, J.; Liu, K.; Lu, J.; Liang, Y.; Cao, Y.; Dai, X.; Li, X. Therapeutic effect of baicalin on inflammatory bowel disease: A review. *J. Ethnopharmacol.* **2022**, *283*, 114749. [CrossRef]
50. Dlamini, N.Z.; Somboro, A.M.; Amoako, D.G.; Arhin, I.; Khumalo, H.M.; Khan, R.B. Toxicogenicity and mechanistic pathways of aflatoxin B1 induced renal injury. *Environ. Toxicol.* **2021**, *36*, 1857–1872. [CrossRef]
51. Wei, X.; Mao, T.; Li, S.; He, J.; Hou, X.; Li, H.; Zhan, M.; Yang, X.; Li, R.; Xiao, J.; et al. DT-13 inhibited the proliferation of colorectal cancer via glycolytic metabolism and AMPK/mTOR signaling pathway. *Phytomedicine* **2019**, *54*, 120–131. [CrossRef]
52. Luo, Z.; Zhu, T.; Luo, W.; Lv, Y.; Zhang, L.; Wang, C.; Li, M.; Wu, W.; Shi, S. Metformin induces apoptotic cytotoxicity depending on AMPK/PKA/GSK-3β-mediated c-FLIP(L) degradation in non-small cell lung cancer. *Cancer Manag. Res.* **2019**, *11*, 681–689. [CrossRef]
53. Wang, C.; Mei, X.; Wu, Y.; Yang, Y.; Zeng, Z. Cinobufagin alleviates lipopolysaccharide-induced acute lung injury by regulating autophagy through activation of the p53/mTOR pathway. *Front. Pharmacol.* **2022**, *13*, 994625. [CrossRef] [PubMed]

54. Kong, Q.; Yan, X.; Cheng, M.; Jiang, X.; Xu, L.; Shen, L.; Yu, H.; Sun, L. p62 Promotes the Mitochondrial Localization of p53 through Its UBA Domain and Participates in Regulating the Sensitivity of Ovarian Cancer Cells to Cisplatin. *Int. J. Mol. Sci.* **2022**, *23*, 3290. [CrossRef] [PubMed]
55. Kim, J.; Kundu, M.; Viollet, B.; Guan, K.L. AMPK and mTOR regulate autophagy through direct phosphorylation of Ulk1. *Nat. Cell Biol.* **2011**, *13*, 132–141. [CrossRef] [PubMed]
56. Tasdemir, E.; Maiuri, M.C.; Galluzzi, L.; Vitale, I.; Djavaheri-Mergny, M.; D'Amelio, M.; Criollo, A.; Morselli, E.; Zhu, C.; Harper, F.; et al. Regulation of autophagy by cytoplasmic p53. *Nat. Cell Biol.* **2008**, *10*, 676–687. [CrossRef] [PubMed]

metabolites

MDPI

Review

Determination of Bile Acids in Canine Biological Samples: Diagnostic Significance

Krisztián Németh [1], Ágnes Sterczer [2,*], Dávid Sándor Kiss [1], Réka Katalin Lányi [3], Vivien Hemző [1], Kriszta Vámos [2], Tibor Bartha [1], Anna Buzás [4] and Katalin Lányi [4]

1 Department of Physiology and Biochemistry, University of Veterinary Medicine, István u. 2, H-1078 Budapest, Hungary; nemeth.krisztian@univet.hu (K.N.); kiss.david@univet.hu (D.S.K.); hemzo.vivien@student.univet.hu (V.H.); bartha.tibor@univet.hu (T.B.)
2 Department of Internal Medicine, University of Veterinary Medicine, István u. 2, H-1078 Budapest, Hungary; vamos.kriszta@student.univet.hu
3 Faculty of Pharmacy, University of Szeged, Zrínyi u. 9, H-6720 Szeged, Hungary; lanyi.reka.katalin@o365.u-szeged.hu
4 Institute of Food Chain Science, University of Veterinary Medicine, István u. 2, H-1078 Budapest, Hungary; buzas.anna@univet.hu (A.B.); lanyi.katalin@univet.hu (K.L.)
* Correspondence: sterczer.agnes@univet.hu

Abstract: The comprehensive examination of bile acids is of paramount importance across various fields of health sciences, influencing physiology, microbiology, internal medicine, and pharmacology. While enzymatic reaction-based photometric methods remain fundamental for total BA measurements, there is a burgeoning demand for more sophisticated techniques such as liquid chromatography–tandem mass spectrometry (LC-MS/MS) for comprehensive BA profiling. This evolution reflects a need for nuanced diagnostic assessments in clinical practice. In canines, a BA assessment involves considering factors, such as food composition, transit times, and breed-specific variations. Multiple matrices, including blood, feces, urine, liver tissue, and gallbladder bile, offer insights into BA profiles, yet interpretations remain complex, particularly in fecal analysis due to sampling challenges and breed-specific differences. Despite ongoing efforts, a consensus regarding optimal matrices and diagnostic thresholds remains elusive, highlighting the need for further research. Emphasizing the scarcity of systematic animal studies and underscoring the importance of ap-propriate sampling methodologies, our review advocates for targeted investigations into BA alterations in canine pathology, promising insights into pathomechanisms, early disease detection, and therapeutic avenues.

Keywords: bile acids; enterohepatic circulation; canine; LC-MS/MS

Citation: Németh, K.; Sterczer, Á.; Kiss, D.S.; Lányi, R.K.; Hemző, V.; Vámos, K.; Bartha, T.; Buzás, A.; Lányi, K. Determination of Bile Acids in Canine Biological Samples: Diagnostic Significance. *Metabolites* **2024**, *14*, 178. https://doi.org/10.3390/metabo14040178

Academic Editors: Lian Li, Baichuan Deng and Giulia Pignataro

Received: 24 February 2024
Revised: 19 March 2024
Accepted: 20 March 2024
Published: 22 March 2024

1. Introduction

Domesticated dogs (*Canis lupus familiaris*) were the first animals tamed by human beings around 20–40,000 years ago, according to the recent anthropological consensus. Canines were strict carnivores originally as their relatives living in the wilderness, wolves and dingoes, are still carnivores. Living with humans changed their access possibilities to feedstuffs; therefore, dogs turned into facultative carnivores slowly. For a long time in history, dogs were kept for their usefulness, such as protecting herds, assisting in hunting, and for other functions. Nowadays, dogs are companion animals possessing full family member status in many cases. Their health is brought to the attention of their owners—or human companions as many of the owners considers themselves—resulting in an increased interest in healthcare activities and prevention measures for protecting the well-being of their animal companions [1].

Bile acids (BAs) are steroids of amphipathic characteristics containing a 24-carbon atom core with a pentanoic acid substituent and playing important role in the metabolic processes of a series of living beings. They are significant factors in cholesterol homeostasis, lipid absorption, and maintaining the bile flow. Bile acids can be used as biomarkers and signaling compounds for diagnosing diseases, such as hepatic and intestinal diseases, malignant tumors within and outside the digestive system, and obesity. Consequently, changes in the concentrations of bile acids in blood plasma or serum, urine, or feces can provide important medical information for the prognosis, diagnosis, and follow-up of several diseases that involve bile acid metabolism [2].

While bile acid formation, circulation, metabolism, their roles, and their prospective diagnostic values have been studied intensively in human medicine, animal healthcare issues are underrepresented. In bile acid studies, animals are mainly used as "further species" to widen the biological scale of the studies or as model animals for studying closer human healthcare questions that would be difficult to study on human beings. Targeted studies on animals with the aim of clarifying animal healthcare issues are much less frequent and systematic.

The present review aims to explore healthy and clinical conditions, in which BA profile tests are applied to various species. Our objectives include proposing appropriate methodologies and matrices for the diagnosis of different diseases in canines while also confronting different methods and matrices utilized for the determination of BA profiles. We intend to assess the applicability of data obtained from various species including humans, mice, and rats for use in dogs as well as to explore how different dog breeds and body sizes may impact the composition of BA pools within dogs. We also seek to analyze the trustworthiness of available knowledge on BAs and BA profiles in dogs, as well as to investigate whether there is accurate information regarding the reported changes in healthy and ill animals concerning these characteristics. Finally, our aim is to summarize crucial elements from our review for the sake of emphasizing the significance of analyzing BAs with regard to the early detection of relevant disorders as well as to therapeutic purposes in companion animals.

2. Bile Acids in the Body: Physiological and Biochemical Aspects

2.1. Enterohepatic Circulation

In the last few decades, bile acids have been the focus of several studies due to their wide-ranging relations with numerous frequently studied fields of physiology, microbiology, internal medicine, and pharmacology. The extensive metabolic effects of these molecules with diverse characteristics are based on their continuous enterohepatic circulation (their specific route in the body) and also on their molecular structure.

Primary BAs are endogenously synthesized from cholesterol in the liver of the majority of mammals. Cholesterol is conjugated with glycine or taurine as the final step of a 17-step reaction sequence catalyzed by 16 enzymes. An ester, ether, or amide linkage on a side chain carboxyl group or to one of the ring hydroxyl groups increases the solubility of BAs. In humans, dogs, and cats, cholic acid (CA) and chenodeoxycholic acid (CDCA) are the major primary BAs. Immediately after synthetization, BAs are released into the bile in conjugated form, which is a secretion with a relatively complex composition and is concentrated and stored in the gallbladder. Not negligibly, bile salts make up the majority of the dry matter of the bile [3–10].

In the gallbladder, the bile consists of up to 15 distinct BAs, but the three main ones—72.8% taurocholic acid (TCA), 20.3% taurodeoxycholic acid (TDCA), and 6.2% taurochenodeoxycholic acid (TCDCA)—make up 99% of the overall pool [1].

When a food is consumed, the gallbladder is prompted by the entero-hormone cholecystokinin (CCK) to discharge bile into the duodenum via the bile duct. In this process, BAs assist in breaking down and absorbing lipids and fat-soluble vitamins [11–14].

Within the gut, these compounds, leastwise the fraction escaping absorption, undergo a multi-step molecular structural further transformation by the so-called Hylemon–Björkhem pathway of the resident gut microbiota (see Figure 1). Thus, BAs can be considered both substrates for microbial reactions and microbial nutritive factors, in this way having an impact on the survival possibilities of microorganisms. Since the description of this relationship, BAs have been recognized as being vital for maintaining an appropriate intestinal microbiota [7,15–17].

Microbial transformations involve four main pathways, i.e., deconjugation, dehydroxylation, oxidation, and epimerization. Furthermore, the microbiota has the ability for sulfate esterification and to reconjugate BAs, thereby augmenting the variety within BA pools [7,9].

Deconjugation (7 alpha-dehydroxylation) is the cornerstone of the aforementioned key processes (see Figure 1). 7α-dehydrogenase (7α-HSDH) is the primary enzyme responsible for the synthesis of secondary BAs. It is encoded by bile acid inducible (bai)-genes found in certain microorganisms present in the distal small and large intestines, particularly *Clostridium* species (e.g., C. or *Peptacetobacter hiranonis*, *C. scindens*, and *C. hylemonae*). When the biochemical reactions of BAs in the digestive tract are disturbed at certain points, the ratio of primary and secondary BAs shifts, causing changes in their toxicity index and hydrophobicity, as well as in the host metabolic function [4,7–9,15,17–31].

BAs are a sizable class of steroids containing a carboxyl group in the side chain. The structure of BAs is modified through the molecular relocation, removal, or entry of various functional groups during reactions; however, the number and position of hydroxyl groups on the sterol ring are the primary distinguishing characteristics [32].

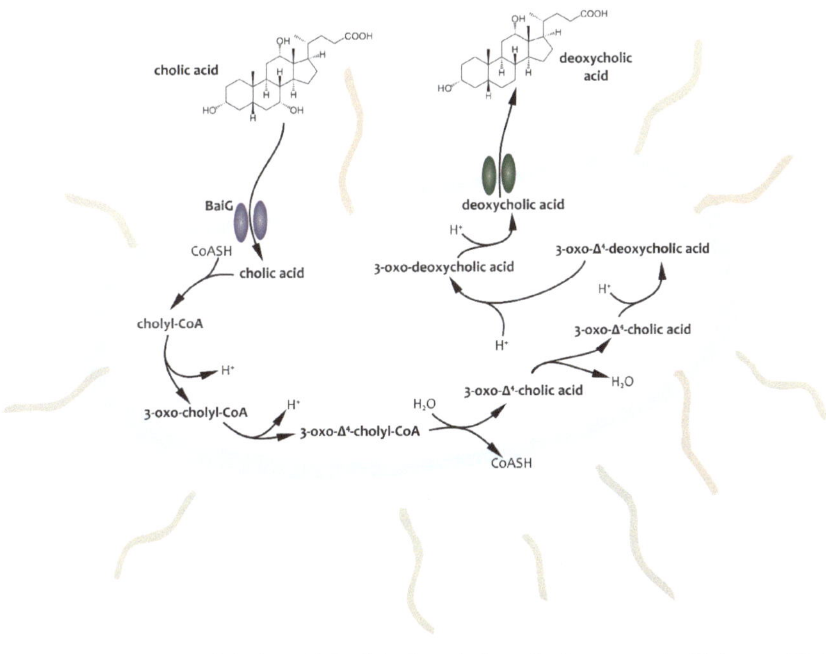

Figure 1. The process of cholic acid 7α-dehydroxylation, also known as the Hylemon–Björkhem pathway, occurs in intestinal anerobic microbes. Intermediate compounds formed in the synthesis of secondary bile acid deoxycholic acid are indicated [15,18,33,34].

During the enterohepatic circulation, the reabsorption of BAs is the step after microbial reactions. Some of their forms (deconjugated BAs) are absorbed passively from the jejunum to a small extent and by active transport from the terminal ileum and the colon in the main. The latter occurs via the apical bile salt transporter, or the apical sodium-dependent BA transporter (ASBT) expressed in the apical membrane of the enterocytes in the distal intestine epithelium, controlled by the so-called farnesoid X receptor (FXR) pathway, a nuclear receptor-binding mechanism of BAs. It may result in a potential defensive mechanism: the bile flow into the large intestine might be increased in response to an infection with invasive pathogens, which could modify FXR signaling and confer a degree of resistance to colonization. In the cytoplasm of enterocytes, BAs are capable of binding to the ileal BA-binding protein (IBABP) or the ileal lipid-binding protein (ILBP, fatty acid-binding protein, FABP) maintaining the transport and binding of BAs to FXR. Then, at the basolateral membrane, they are secreted into the blood by the heterodimeric organic solute transporter (OSTα/β) and return to the liver across the portal circulation. At this location, the cycle closes by getting BAs transported from the portal blood into the hepatocytes by the sodium taurocholate co-transporting polypeptide (NTCP) predominantly and by the organic anion-transporting polypeptides (OATPs) to a lesser extent. These polypeptides are located at the basolateral (sinusoidal) membrane, where they herewith activate the hepatic FXR system. FXR exerts an indirect negative feedback on de novo BA synthesis by regulating the expression of multiple genes implicated in primary BA synthesis. Then, initiating a new cycle, hepatocytes express an ATP-dependent efflux pump onto their membrane. This protein mediates the export of newly synthesized and recovered BAs to bile across the canalicular membrane (canalicular bile salt export pump, BSEP, ABCB11) [9,16,25,35–41].

A considerable proportion, precisely 95%, of the total BAs undergo reabsorption in the ileum before being returned to the liver for recycling and to modulate their de novo synthesis. The remaining 5% (the "loss") is leaving the cycle eliminated via feces after being metabolized by the gastrointestinal microbiota in the colon. This ratio corresponds to the quantity of hepatic synthesis [3,9,11,21,39,42–45].

By impeding passive intestinal absorption, the conjugation facilitates the involvement of BAs in the intestinal digestion process of dietary lipids. Moreover, 10 to 50% of reabsorbed BAs enter the systemic circulation via peripheral spillover [11,36,44,46–48].

The enterohepatic circulation of BAs is shown in Figure 2, including cellular processes.

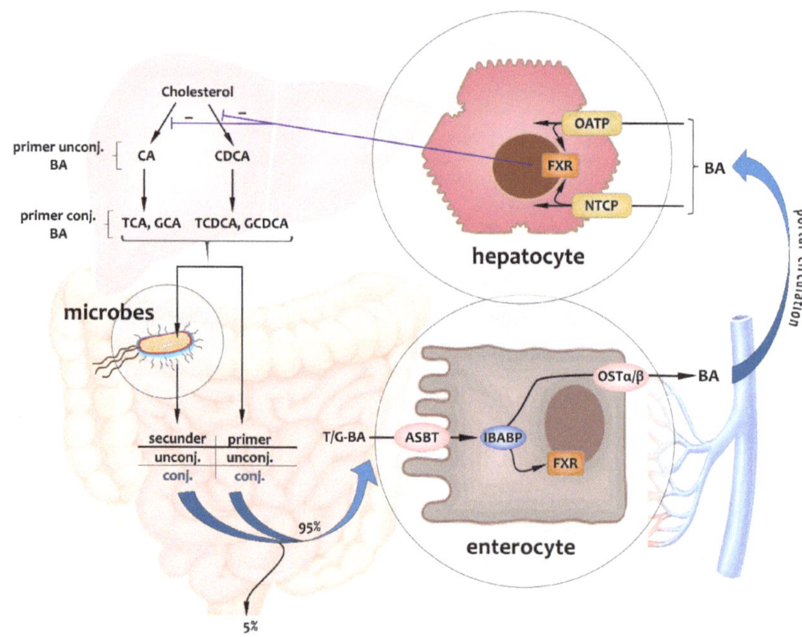

Figure 2. Enterohepatic circulation of bile acids (BAs). BAs are synthesized in the hepatocytes and then, after tauro- or glycoconjugation, secreted into the lumen of the small intestine via bile following their production by the liver. In the distal segment of the small intestine, enterocytes recover the majority of conjugated BAs through an active mechanism involving sodium-dependent BA transporter (ASBT) embedded in the apical membrane. Bile acids that are not absorbed (5% of the

total amount) then travel to the colon, where they are modified further by the gut microbes, while 95% of the BAs stay in the cycle. In ileal enterocytes, ileal BA-binding protein (IBABP) serves as the primary intracellular transporter for reabsorbed conjugated BAs. The heterodimeric basolateral membrane organic solute transporter OSTα/OSTβ facilitates the elimination of BAs from ileal enterocytes, allowing them to be transported back to the liver through the portal circulation. Bile acids are reabsorbed into the hepatocytes, assisted by sodium taurocholate co-transporting polypeptide (NTCP) and organic anion-transporting polypeptides (OATPs), which are located in the basolateral (sinusoidal) membrane of the liver in direct contact with the portal blood. The occurrence of this uptake mechanism has negative feedback on the process of endogenous synthesis from cholesterol. Since the cycle is continuous, the liver also excretes these reabsorbed BAs into the bile [20,36,37,49].

2.2. Cytotoxicity and Properties

As a substantial category of steroids, BAs have a carboxyl group attached to their side chains. During the preceding reactions, various functional groups are introduced, removed, or relocated molecularly to alter the structure of BAs. Nevertheless, the number and position of hydroxyl groups on the sterol ring serve as the principal differentiating features [50]. There are subsequent changes in their biochemical properties and effects on the host. However, it is relevant to mention critically that a contradiction can be found at this point from the side of different research groups, as well as in connection with the question of which fraction of BAs is toxic and which one is healthy.

In general, according to the most studies, secondary BAs are more hydrophilic than primary ones, and this fraction is meant to be the healthy one. Deoxycholic acid (DCA) and CDCA, which contain only two hydroxyl groups and are therefore less polar, are more hydrophobic, according to Zwicker and Agellon [20]. Biomembranes are typically more susceptible to cytotoxic effects as their hydrophobicity increases. Thus, as their cytotoxicity decreases, BAs are categorized as monohydroxy, dihydroxy, and trihydroxy, subsequently unconjugated, taurine-conjugated, and glycine-conjugated BAs [20,51]. Herstad et al. established a BA hydrophobicity scale that can be said basically as follows: ursodeoxycholic acid (UDCA) < cholic acid (CA) < CDCA < DCA < lithocolic acid (LCA) [52]. The significance of the hydroxyl group placement is further underscored by the protective nature of 7β-dihydroxy BA UDCA that is generated via bacterial transformation from the primary BA CDCA [20]. The secondary bile acids DCA and LCA are derived from CA and CDCA, respectively. UDCA is also produced when the primary bile acid CDCA undergoes bacterial transformation [1,52].

Nonetheless, Sitkin et al. conducted a study revealing that DCA could have either proinflammatory (bad) or anti-inflammatory (good) effects. Furthermore, their research team identified a Western diet as a risk factor for the onset of inflammatory bowel disease (IBD). In addition, the presence of elevated quantities of fecal DCA can have a crucial effect [53,54].

Concluding this chapter, one can establish that the classification of BAs into primary and secondary compounds pertains solely to their synthesis and does not necessarily indicate their positive or negative physiological effects. Questioning the classification of secondary BAs as either good or bad—as Lenci et al. stated in their review—is valid. Furthermore, the elimination of a hydroxyl group at the C-7 position leads generally to decreased water solubility and greater detergency compared to primary precursors. The correlation between the hydrophilic or hydrophobic nature of a particular BA and its potential for cytotoxicity and harmful consequences was revealed. From this standpoint, secondary BAs are commonly considered to be potentially harmful substances when they reach sufficient levels due to their detergent or destabilizing impact on cell membranes [55].

The most relevant canine BAs are shown in Table 1.

Table 1. Molecular structure, common name, chemical name, and abbreviated names of most relevant primary and secondary conjugated and unconjugated bile acids in dogs [15,18,33,34].

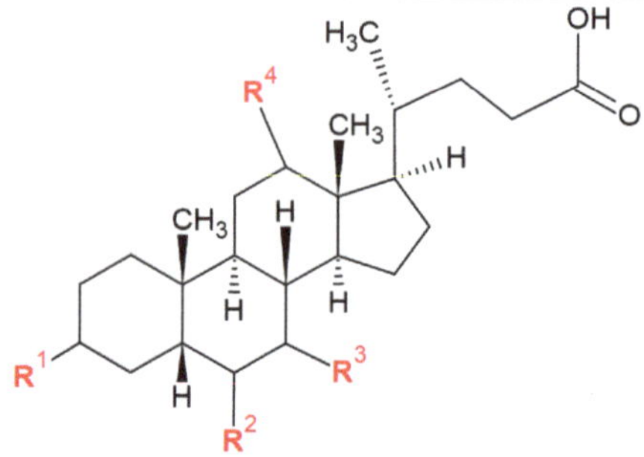

Common Name	Chemical Name	Abbreviation	
cholic acid	5b-cholanic acid-3a,7a,12a-triol	CA	PRIMARY BILE ACIDS
taurocholic acid	5b-cholanic acid-3a,7a,12a-triol-N-(2-sulpho-ethyl)-amide	TCA	
glycocholic acid	5b-cholanic acid-3a,7a,12a-triol-N-(carboxymethyl)-amide	GCA	
chenodeoxycholic acid	5b-cholanicacid-3a,7a-diol	CDCA	
taurochenodeoxycholic acid	5b-cholanic acid-3a,7a-diol-N-(2-sulpho-ethyl)-amide	TCDCA	
glycochenodeoxycholic acid	5b-cholanic acid-3a,7a-diol-N-(carboxymethyl)-amide	GCDCA	
deoxycholic acid	5b-cholanic acid-3a,12a-diol	DCA	SECONDARY BILE ACIDS
taurodeoxycholic acid	5b-cholanic acid-3a,12a-diol-N-(2-sulphoethyl)-amide	TDCA	
glycodeoxycholic acid	5b-cholanic acid-3a,12a-diol-N-(carboxymethyl)-amide	GDCA	
lithocholic acid	5b-cholanic acid-3a-ol	LCA	
taurolithocholic acid	5b-cholanicacid-3a-ol-N-(2-sulphoethyl)-amide	TLCA	
glycolithocholic acid	5b-cholanic acid-3a-ol-N-(carboxymethyl)-amide	GLCA	
ursodeoxycholic acid	5b-cholanic acid-3a,7b-diol	UDCA	
tauroursodeoxycholic acid	5b-cholanic acid-3a,7b-diol-N-(2-sulphoethyl)-amide	TUDCA	
glycoursodeoxycholic acid	5b-cholanic acid-3a,7b-diol-N-(carboxymethyl)-amide	GUDCA	

2.3. Roles

Apart from the fact that they are involved in digestive processes, called enteral lipid emulsification [46], depending on the state of the bile acid profile (BAP), BAs also have other local and systemic functions, or they can exert or amplify pathological effects.

- On the one hand, these molecules exert influence within the enterohepatic circulation. While certain bile salts increase the effectiveness of antibiotics produced by bacterial species, others are capable of stabilizing the composition of the intestinal microbiome because they can be used as substrates for microbes [56].

In addition to the above, bactericidal activities of BAs on their own were recorded. In support of a previous hypothesis that the hydrophobicity of a BA molecule is associated with its antibacterial activity, Watanabe et al. documented their findings using the tenfold disparity in bactericidal activities between CA and DCA as an illustration. The amphiphilic nature of the molecule—the steroid skeleton of a BA molecule creates a hydrophilic side (α-face) and a hydrophobic side (β-face)—provides the context for this mechanism. It can lead to the disturbance of the integrity of bacterial cell membranes by causing a leakage of proton and potassium ions on them and consequently leading to cell death. [20,57–59]. Indeed, BAs are involved in the pathomechanism of some different liver diseases and inflammatory diseases of the gastrointestinal tract as well [3,9,10,30,44,54,56,57,60–72].

- On the other hand, BAs were found to impact systems beyond the traditional scope of enterohepatic circulation. From some points of view, both the ab ovo synthesized and bacterially transformed BAs can be considered active metabolites with hormonal effects, the pathological patterns of which play a role in the manifestation of neuroinflammatory or neurodegenerative diseases mediated by the gut–brain axis [16,38,43,73–75]. These molecules can also lead to the improvement of barrier function and the mediation of anti-inflammatory mechanisms and can play a regulatory role at specific points of intermediate metabolism, e.g., in the metabolic regulatory system of carbohydrates or synthesis and oxidation of fatty acids [38,40,71,76–78].

Last but not least, the immunomodulating effect of BAs was also established [10,25,40,54,79]. As a brief insight into the mutual influence of BAs, the microbial environment of the gut, and the circumstances of the immune system, Ju et al. aimed to investigate the impact of gut BA levels on the gut microbiota and intestinal inflammation in subjects with IBD. The research team presented that the gut microbiota plays a crucial role in producing and modifying BAs, which can be disrupted by imbalances of, diversity in, and the quantity of gut microbes. BA metabolism issues affect the immune system and the function of intestinal barrier. The study found that alterations in gut microbiomes led to reduced levels of secondary BAs in serum and fecal samples, suggesting that the microbiome, BAs (and their metabolites), and IBD are correlated. DCA may cause alterations in the microbial community, which may have a significant role in the development of intestinal inflammation and is linked to disturbances in BA metabolism and the suppression of the FXR pathway. DCA also increased interleukin-1β concentration, decreased tuft cells, and enhanced cluster of differentiation (CD) 3+ and 4+ T cell expression in mice with drug-induced colitis, exacerbating intestinal inflammation. The regulation of intestinal BA levels is crucial for the clinical diagnosis and treatment of IBD [54].

Nevertheless, it is not to be forgotten either that their roles include not only those recorded recently but also trivial ones such as the regulation of cholesterol homeostasis. The process by which cholesterol is metabolized into BAs and bile alcohols is critical for its maintenance [32].

Over the last decade, the literature has been actively expanding with knowledge of the intestinal microbiota. Research data focus on the protective and pathological effects of the components of the microbiome on the gut–brain axis, as well as the influence of the bacterial composition with prebiotics and various exogenous substrates [9,26,29,63,73,75,80,81]. A complex parameter, the so-called dysbiosis index (DI), was developed to facilitate the progression in this research direction. While measuring the entire bacterial population quantitatively, this pioneering diagnostic parameter carries out also the quantitative characterization of seven specific bacterial groups separately (*Fecalibacterium* spp., *Turicibacter* spp., *Streptococcus* spp., *Escherichia coli*, *Blautia* spp., *Fusobacterium* spp., *Clostridium* or *Peptacetobacter hiranonis*) and provides information about the balance of the microbiome or its shift. The sensitivity of the procedure is low, and the specificity is high in terms of separating healthy dogs from patients [65,74,82,83].

3. Matrices to Be Analyzed for Bile Acid Composition

Since bile acids are synthesized in a well-defined area of the body and fulfil their role in another well-defined area, after which they are absorbed back into their synthesis place, the range of possible matrices for analysis is not too wide. Blood (serum or plasma), feces, urine, liver tissue, or gallbladder bile liquid are the most frequently studied matrices from this point of view. On the other hand, diagnostic purposes may require analyzing more than one matrix in order to gain a more complete view on the bile acid pool. Out of the almost a hundred bibliographic sources processed, only about half of them dealt with only one matrix, and approx. six percent had no matrix at all; they were articles focusing on basic method development working with pure solutions of bile acids. In all the remaining papers, more than one matrix was analyzed (see Figure 3).

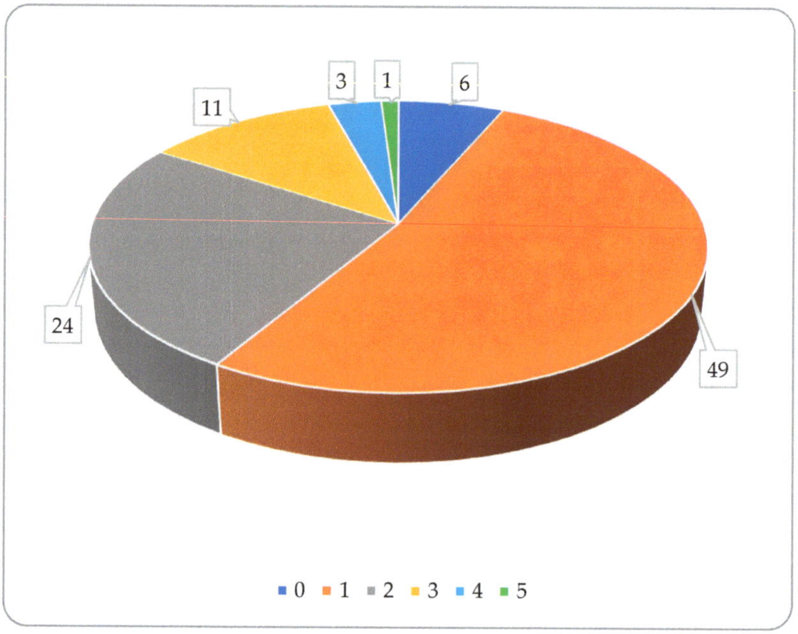

Figure 3. Number of matrices tested for BAs within a trial (n = 94) (number of relevant papers).

Blood sampling could be an obvious choice, even in spite of the fact that the majority of bile acids are removed from the blood during enterohepatic circulation [84,85]. Considering blood samples, it is always a question whether blood plasma or blood serum is more suitable for analysis before deciding about more detailed issues. In the bibliographic sources studied, no significant difference could be found between bile acid concentrations in plasma and serum, with only a few exceptions; for example, Sarafian et al. reported lower bile acid concentrations in plasma [86]. According to the previous reviews in this field [87,88], there is a disagreement among authors concerning the suitability of blood serum or plasma as a matrix for bile acid analysis. Serum can be an argumentative decision since recoveries were found to be higher here than from plasma samples. On the other hand, the characteristics of plasma are more similar to those of whole blood due to the quenched coagulation [87]. This uncertainty could be tracked also in our study: 39.9% and 22.8% of the papers processed referred to serum and plasma, respectively, as the matrix studied; however, several times (9.6% of papers), serum and plasma were analyzed together. For 27.7% of the papers studied, blood was not used as a matrix for analysis. See Figure 4 and Table 2. The decision whether to use blood serum or plasma may depend on several factors, ranging from the technical background available on the sampling site to the aim of analysis in the laboratory.

Bile acids in blood serum or plasma samples can stay stable for 24 h at room temperature and for 2 months at −80 °C. Dried blood spots (DBSs), consisting of a small amount of capillary blood plasma or serum collected on a sampling card made of filter paper and dried afterward, can be stored even from 1 year to more than 18 years [87].

Total bile acid concentrations in human blood serum samples were found to be 2.1–8.0 µmol/L. CDCA, DCA, LCA, CA, and UDCA make up 92.1% of total BAs in the blood. CDCA, either conjugated or unconjugated, was found to be the most abundant BA. Blood plasma is similar to serum; they differ from each other mainly in containing the coagulation factor. Total bile acid concentrations in human blood plasma were found to be 1.4 to 6.5 μmol/L, so approximately the same order of magnitude as in the serum. The composition of bile acid species in plasma is similar to the serum, too: CA, UDCA, CDCA, DCA, LCA, and their glycine and taurine conjugates are the most abundant ones [87].

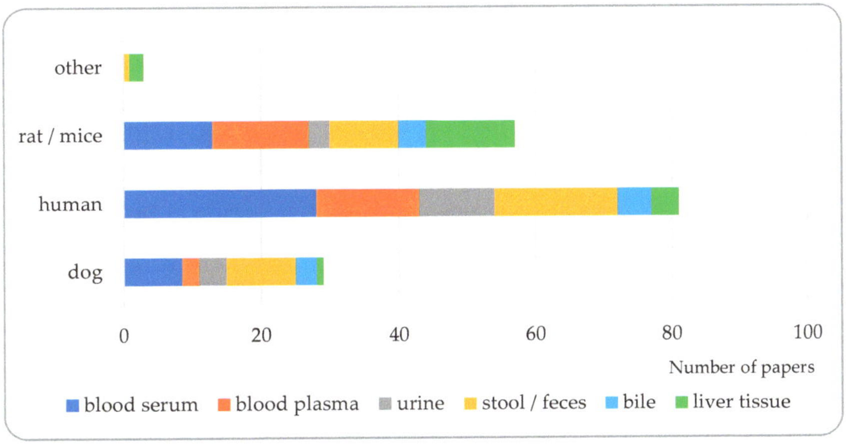

Figure 4. Matrices studied the most frequently for BA composition by species (in number of papers; $n = 94$).

In human medicine, urine is a quite popular and obvious biological sample because it allows non-invasive sample collections which are easily available in most cases. On the other hand, urine can be a good choice as the matrix for mainly polar (hydrophilic) compounds that are excreted in themselves or in metabolized form through the kidneys. The analysis of highly hydrophobic compounds from urine could require special sample treatment and measurement methods [88]. In veterinary practice, urine is also an evident choice as a matrix for bioanalysis. Equipment for collecting urine samples on the spot or at a veterinary clinic are easily available and usable [87].

Bile acids may be less stable in urine samples than in blood serum or plasma. Specifically, longer standing at room temperature or repeated freeze–thaw procedures may cause substantial losses in BAs. Total bile acid concentrations in human urine samples were found to be approximately 9 μmol/L in a similar order of magnitude as in serum or plasma. DCA and CDCA were to be found the most abundant urinary BAs, comprising approximately half of the urinary bile pool [87].

Feces are solid- or semi-solid-state excretory products of digestion containing only a small portion of non-absorbed BAs. They also may be an attractive matrix choice due to similar reasons as urine (the possibility of non-invasive sampling). On the other hand, feces excretion is irregular; therefore, bile acids are not evenly distributed in any particular sample [87].

In the feces of adults, the BA pool consists of mainly unconjugated secondary BAs [87]. In contrary to a recent review in this field [88], we found that feces was the matrix analyzed in the second highest number in the bibliographic sources. Total bile acid concentrations in feces were found to be 1 to 25 μmol/g of dry matter. The majority of BAs in feces is present in unconjugated form and as secondary bile acids due to the bacterial activity in the guts. DCA and LCA were found to be the most abundant BAs in feces [87].

As we discussed before, bile is a biofluid produced by the liver, which is composed of mainly primary BAs and conjugated with glycine or taurine. Sampling from bile, duodenal or gastric fluid, or other liquids from the digestive tract requires an invasive intervention; therefore, it is applied only with specific precautions in both human and veterinary medicine and only when detailed information about the alterations of the BA pool is required [87,88].

Although bile is the primary medium for bile acids, their concentration in it is far from being even. The total BA concentration in bile may range between 0 and 180 mM, thus showing off a much higher variability than in blood, urine, or feces. The total BA concentration in gastric juice may range between 0 and 86 mM with usual average concentrations between 5 and 12 mM. The BAs of bile are mostly conjugated; however, the conjugated moiety depends on the subject's age. While in fetuses, the majority of gallbladder BAs is conjugated with taurine, in the adults' bile, most of the BAs are glycine-conjugated. The two primary bile acids produced by the liver, CA and CDCA, are found generally in bile samples [87].

In addition to the biological samples used most frequently for BA analysis, discussed above, other media can also be tested for them. Liver tissue was analyzed in 15 papers [28,63,89–101], although only 1 [99] included also dogs in the studied species. Sampling liver is especially difficult and requires highly invasive interaction with the subject sampled. Generally, it is performed by sacrificing the experimental animal, and therefore, mainly rodents are the test subjects.

Sometimes, ileal, caecal, duodenal, and other intestinal contents were analyzed for BAs. These studies involved mainly rats and mice, removing the intestinal content after their sacrifice [28,75,90,91,94], or humans were studied using a specific device to remove samples from the duodenal liquid [102]. Furthermore, as well, meconium [103], ovaries [101], saliva, exhausted breath [104], eWAT (epididymal white adipose tissue), muscle, and hypothalamus [95] were special matrices studied for BA content on a case-by-case basis.

Table 2. Matrices studied the most frequently for BA composition.

Matrix		Bibliographic Sources
blood serum and plasma		[105–108]
	+ urine, feces, and bile	[109]
	+ urine	[86,110]
	+ feces	[111]
	+ liver tissue	[89]
blood serum and urine		[112–114]
	+ feces	[115]
	+ feces and bile	[116]
	+ feces and liver tissue	[90]
blood serum and feces		[54,64,67,117,118]
	+ bile	[75]
	+ liver tissue	[28,91]
blood serum		[119–132]
	+ bile	[133–135]
	+ liver tissue	[92,93]
blood plasma		[3,102,136–141]
	+ urine, feces and bile	[142]
	+ urine	[143]
	+ feces	[10]
	+ feces and liver tissue	[63,94]
	+ liver tissue	[95–98]
blood, without detailing whether it was serum or plasma		[144]
urine		[145–148]
feces		[19,52,65,83,149–162]
	+ bile	[103]
bile		[163,164]
liver tissue		[99–101]
no specific matrix (or pure solutions)		[18,57,104,165–167]

The sample preparation for BA measurements depends strongly on the choice of matrix and the range of compounds to be analyzed. The selection of analytical method may also influence the sample preparation procedure. Since all the biological samples used for BA analysis contain a rather significant amount of protein that may bind the bile acids and therefore modify the results of the analysis, deproteination is generally an important starting step usually made by adding organic solvents (methanol, acetonitrile), organic acids, or sometimes both of them together. Ultrasound homogenization or heating can also be applied together, possibly adding an alkaline agent to decrease BAs' protein binding. However, this latter one should be carried out with care since bases may catalyze the hydrolysis of conjugated BAs [87,88]. The use of solid-phase extraction (SPE) is frequent to further clean and concentrate the samples; however, it is far from being a fully general method. From the papers we studied, only 24% used SPE for cleaning the biological samples. Evaporating the extract to dryness and reconstituting it into an organic solvent suitable for the chosen analytical method is the final step generally.

Recently, the extraction of bile acids from biological samples using various methods has been described several times. However, the efficiency of these extractions is not always comparable due to the amphipathic characteristics of bile acid molecules. This central question was addressed in a study aimed at comparing the critical efficiency for very low BA concentrations [116].

The irregularity of bile acid excretion makes the analysis of fecal bile acids more complicated and increases uncertainty of the measurements. Some authors suggested pooled feces aliquots of 4–5 days to be analyzed [87]. On the other hand, the liver and the intestines are very dynamic biological systems, and their status will fundamentally influence the bile pool, the composition of which changes dynamically with the organs producing and using it. The motility of the subject and the general healthcare status will also be an influencing factor. For all these reasons, pooling feces from 4–5 days may lead to losing valuable information about the details of the status and processes of the digestive system or the whole organism in general.

Because of its phase, composition, and origin, feces proves to be a more difficult sample than the liquid-phase biological samples. The preparation of feces samples generally involves the steps of lyophilization, homogenization, saponification, and extraction–preconcentration.

BAs are frequently measured from lyophilized feces instead of fresh ones [10,28,52,65,67,83,89,90,94,101,103,109,111,115,116,149,151,153–158], thus expressing the BA concentrations based on the dry weight of the lyophilizate. At the same time, it is important to mention that lyophilization requires special equipment and elongates the sample processing time. The interchangeability of fresh and lyophilized feces samples was studied [153], demonstrating that the measured BA concentrations were comparable for original and lyophilized samples. Original, fresh feces can be a suitable matrix for BAs analysis except for in detailed investigations of trace bile acids. On the other hand, the BA composition of feces does not refer to a given time point but instead a longer time period. Therefore, the comparability of feces BA concentrations to those measured in biological fluids is still to be clarified.

4. Measurement Methods and Instruments Used

Measuring the BA concentration can be conducted via two basically different approaches: quantifying the total bile acid amount present in the given sample or measuring the individual BAs one by one. Total bile acid measurement is an important part of present clinical laboratory practices; however, the determination of the individual BA concentrations has gained increased importance recently. Several reviews have dealt with the issue of bile acid analysis in the past few decades [2,32,84,85,87,88,168]. The common finding of these reviews is the statement that analyzing more bile acids together is a challenging task in the laboratory because of their low concentrations in biological samples (except bile fluid), similarity in chemical structure, and physico-chemical characteristics. The development of separation techniques and combining them with mass spectrometry has opened new possibilities in the analysis of BAs.

In addition to the two approaches described above, methods currently used for BA analysis can be divided further into two main groups. One of them includes those ones where the analysis is made without separating the bile acids from each other, and in the other one, the separation is carried out before the actual analysis. Analytical methods belonging to the first group are enzymatic assays, direct infusion mass spectrometry, or nuclear magnetic resonance (NMR) spectroscopy, among others [32,87].

Enzymatic methods are rather popular in clinical studies, owing to their relative cost-effectiveness and simplicity. Since many other enzymatic methods are also used in clinical laboratory practice, the general tools and expertise needed to carry out these tests are available in most cases. On the other hand, enzymatic tests may have specificity problems, and components of the matrix may cause cross-effects, spoiling the results. These methods have an even more significant drawback, i.e., analyzing several compounds needs one test per compound, or one can measure only the sum of the concentration of the compounds detected via the test. Therefore, enzymatic assays are expected to keep their significance in determining total bile acids (TBA) concentrations; however, they may lose their role in determining individual bile acids in the future.

Direct infusion mass spectrometry can be used for the preliminary screening of biological samples to settle the mass/charge ratios of the sought-after BAs. It is useful for obtaining quick qualitative information about the sample without requiring complex sample preparation; by contrast, it lacks the ability to allow a more detailed analysis. NMR spectroscopy is a powerful tool in the structure elucidation of organic compounds, which can give valuable information on the composition of biological samples. However, quantifications may be difficult, and the method requires expensive equipment and specialized expertise; therefore, its use for BAs analysis is not widely spread.

In the second group of bile acid analysis methods are those involving the separation of the compounds before or during the analysis. In the past few decades, the necessity of analyzing individual bile acids has become stronger, highlighting the importance of applying separation techniques (chromatography methods) in this field. From the many available separation procedures, several ones were tested for BA analysis. Thin-layer chromatography (TLC), supercritical fluid chromatography (SFC), and electrophoresis were all used, to some extent, for this purpose [2,168]; however, these ones gained no significant role due to their limitations in analytical performance, availability, or cost-effectiveness.

Gas chromatography (GC) was used to analyze bile acids in a few cases in the past [87]. Since BAs are not volatile, these procedures required derivatization that made sample preparation difficult and time-consuming. Traditional GC instruments did not possess any significant strength in bile acid analysis, even with the most carefully selected derivatization, therefore remaining in the background when analyzing BAs. This situation may change when the gas chromatograph is combined with a mass spectrometer (GC-MS). There is still need for derivatization to make BAs volatile; however, the powerful structure analyzing capabilities of electron ionization (EI) mass spectrometry used in GC-MS instruments may compensate for this weakness. It is still important to note that the derivatization process includes the cleavage of conjugates, therefore causing a significant information loss on the detailed composition of the bile pool studied [2]. Until the high-resolution mass spectrometry (HRMS) became wider spread in the LCMS methods, GC-MS was indeed a well-utilizable tool in BA analysis. Nowadays, its role is changing as GC-MS is losing its role in favor of more sophisticated LCMS methods.

High-performance liquid chromatography (HPLC) with optical detection methods can be a useful tool in analyzing bile acids since it copes well with non-volatile compounds. On the other hand, most BAs have weak absorption in the UV range (although there are exceptions); therefore, derivatization may be needed to improve the sensitivity [87,169]. Moreover, the separation of the structurally very similar bile acids, such as CDCA, DCA, and UDCA, can be challenging on traditional HPLC systems and columns. The introduction of ultra-high-performance liquid chromatography (UHPLC) with much narrower internal sizes in the equipment tubing, as well as smaller particle sizes and internal diameters in the LC columns, enabled the effective separation of the structurally similar BAs.

Combining UPLC systems with mass spectrometry (LC-MS), especially in tandem mass spectrometry systems, opened new horizons in bile acid analysis, and this is expected to continue [2,32,84,85,87,88,168,169]. Electrospray ionization is used in most cases in negative polarity, detecting [M-H]- precursors. The so-called multiple reaction monitoring (MRM) or selected reaction monitoring (SRM) are used most frequently for quantifying bile acids. The approach uses one or more mass spectrometric transitions from the same precursor, thus enabling better sensitivity and selectivity of the measurement method. Conjugated BAs can be conveniently sensed in that way since glycine and taurine units give stable and significant fragments of m/z 74 and 80, respectively. Unconjugated bile acids lack the stable [M-H]- precursors; therefore, these ones are frequently measured using the selected ion monitoring (SIM) approach when only the precursor ion is detected, or using the so-called pseudo-MRM when the [molar mass-1 $\rightarrow$ molar mass-1] theoretical transition is followed. However, for some unconjugated bile acids, more complex adducts can be detected in positive polarity ESI [166]. Still, it can be said that the fragmentation of bile acids is rather weak among ESI circumstances; moreover, the same fragments may appear for several compounds containing the same specific chemical moieties. Therefore, effective chromatographic separation remains of high importance when analyzing bile acids. In the case of the chromatographic separation of BAs for LC-MS measurements, chiefly, C18 columns are used. Solvents used as eluents include acetonitrile and, to a lesser extent methanol, with or without adding formic acid or ammonium formate. Aqueous solutions of formic acid, ammonium formate, or other organic buffers are most frequently used as the aqueous phase.

Mass analyzers used in the instruments performing BA analysis include triple quadruple (QqQ) MS-MS, time-of-flight (TOF), and ion trap (IT) units, the former one frequently connected to a quadruple analyzer (Q-TOF). From the bibliographic sources processed, approximately three-quarters of the papers used LCMS for measuring BAs, QqQ and TOF or IT MS sharing an approximately 2:1 ratio within it (see Figure 5). Single quadruple analyzers are seldom used since the identical molecular masses, the very similar structures, and the co-elution originating from these, in the case of some bile acid compounds, render single quad MS instruments less useful in BA analysis.

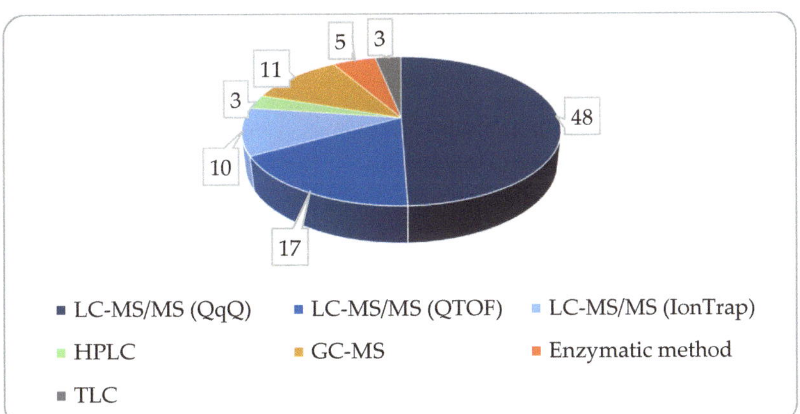

Figure 5. Ratio and number of measurement methods in the bibliographic sources processed (n = 94).

In addition to the targeted analysis of chemical compounds expected to appear in the sample, the semi-targeted or completely non-targeted analysis of additional related compounds, metabolites, or other transformation products has gained increased importance recently when analyzing bile acids in biological systems. While targeted analysis can be considered a routine process, the untargeted approach requires specific pre-processing methods and sophisticated statistical calculations to investigate the metabolite profiles of BAs [78].

5. Factors Influencing the Enterohepatic Circulation of BA

5.1. Effects of Feeding State on the Bile Acid Levels

The circulating total bile acid level is measured typically 8 to 12 h after feed deprivation and 2 h after consuming a small portion of canned feedstuff to stimulate gallbladder contraction. Fasting serum TBA in dogs is within the range 0 to 15 µmol/L, and the postprandial (PP) values range between 0 and 25 µmol/L, with a slight difference in cut-off values per laboratory [46,170]. During food intake, the degree of luminal tension to gastric wall elevates, and cholecystokinin (CCK) mRNS is upregulated in duodenal mucosal enterocytes, which is even more enhanced by the high lipid content of the bolus. Consequently, hormone synthesis increases, thus elevating the CCK level in the bloodstream. As a result of this process, the so-called choleretic effect succeeds to be manifested through an endocrine way, via contracting the gallbladder and relaxing the Oddi sphincter, which leads to the emptying of the gallbladder. The occurrence of this phenomenon can be prolonged, resulting in the formation of the highest concentration of PP serum BAs after more than 2 h following the release of CCK [46,171].

However, in healthy dogs, the gallbladder is shown to become empty in an even shorter period of time via the application of a CCK analogue rather than a simple test meal. Bridger et al. compared the effect of feeding-induced CCK provocation with CCK-analogue ceruletide stimulation on TBA levels. They found that the ceruletide-induced BA stimulation had been more sensitive than feeding-induced CCK. The direct ceruletide administration produces a more potent contraction of the gallbladder compared to contraction caused by a meal. The fluctuations in the rate, at which the stomach empties after eating, influenced by both individual factors and medical conditions, can result in the highest levels of BA in the blood occurring 1 to 8 h after consuming feed. Evidently, ceruletide serum BA stimulation is a viable and dependable method for diagnosing congenital PSS. The procedure is capable of being completed in 30 min as opposed to 120 min, and it is unaffected by the patient's meal intake or vomiting episodes [13,172–174].

As further effects of elevated CCK levels, the gastric motility and emptying are inhibited by constricting the pylorus and relaxing the stomach, while oppositely, motoric movements of the colon are stimulated. Furthermore, CCK has a slight stimulating effect on the bicarbonate secretion, supporting the impacts of the secretin hormone and also leading to an enzyme-rich pancreatic juice secretion, provided that the carbon chain of the fatty acids in chymus consists of more than 8–12 atoms [13]. Additionally, through a negative feedback mechanism of the two hormones in question, a duodenal peristaltic and secretory inhibition is performed as well [14,173,174].

An irregular contraction of the gallbladder during fasting can lead to a preprandial level exceeding the PP value. Clinically, it is common practice to measure fasting values and PP bile acid values in pairs to acquire two comparable values in the dynamically changeable factor in question (Pena-Ramos et al., 2021). For preventing lipemia, a minimal quantity of the meal is administered to induce a response, and blood collection is conducted 2 h following the meal [46,173].

As the liver is responsible for keeping TBA in the physiological range, even postprandially, an insufficient function of this organ can be suspected in case of a too high fasting level and/or PP TBA level [46]. It is not without reason that the determination of paired TBA, which is usually called a PP stimulation test, is the most relevant liver function parameter. Resting serum bile acids exhibit greater specificity, but poorer sensitivity, than PP BAs. Measuring PP serum BAs can serve as a more effective screening test for excluding liver disease compared to measuring resting serum BAs [171]. Prolonged fasting, unsuitable test meals, compromised intestinal absorption, and increased food transit in the intestinal segments, together or respectively, may decrease BA levels and impact on the accuracy of bile acid testing for hepatobiliary diseases. BA tests should not be performed in animals showing clinical jaundice or with elevated levels of direct (conjugated) bilirubin. This test does not provide any data on liver function or the existence of portosystemic shunting in the presence of cholestasis.

In the bibliographic sources, there are numerous studies in which extremely elevated PP (and even fasting) BA concentrations were measured, especially in the case of CPSS and cirrhosis [171,175–177].

5.2. Impact of Motility and Transit Times

A number of factors influence the enterohepatic circulation of BAs, which consequently affect the TBA level. The transit time (TT) of the intestines, which is determined by their motility, is one of these factors. Evidently, variations in digestive anatomy can significantly affect physicochemical parameters including pH, digestive secretions, TT, and gut microbiota [1,14].

Due to their more developed colon and caecum, large breed dogs produce feces which is softer and moister than those of small breed dogs, according to an analysis by Weber et al. based on four PhD theses published between 1998 and 2013. Large dogs' digestive mucosa exhibits increased permeability, which may cause an electrolyte backflow and account for the decreased consistency of their feces. This may account for the increased colonic fermentative activity and prolonged colonic TT. In addition, there was no significant correlation between the weight or size of the dog and the gastric emptying parameters nor a significant impact of the body size on small intestine TT, according to the review [48,178,179].

Gastric TTs, small intestinal transit periods, and maximal pressures during GI transit exhibit significant variations [180].

Having compared the gastrointestinal physiologies of cats and dogs, Tolbert et al. found that canines have shorter total TTs and quicker TTs. The increased potential capacity of the esophageal lumen in Beagle dogs and the reduced esophageal motility in cats account for this. In contrast to cats, dogs had a quicker gastric passage (with pre- and postprandial ranges of 6.8–15 h [1] and 31–829 min [181]) and a slower intestinal transit period (795–830 min). A reduction in gastrointestinal pH was observed in all species after feeding, with a substantially greater decrease in the first-hour small intestinal pH in dogs postprandially compared to cats [181]. The mean TTs in Beagle dogs pre- and postprandially were as follows: esophageal 3 min and 13 min; gastric 31 min and 829 min; and intestinal 795 and 830 min [181]. No correlation seems to occur between body weight (BW) and gastric emptying time, both in fed and fasted subjects. However, it is apparent that the consistency of feed has an impact on gastric emptying time. In addition, there is no consensus regarding the relationship between the dog size and the small and large intestinal TT. However, the available data indicate that the former is significantly shorter in the largest dog breeds. On the other hand, the data indicate a significant positive correlation between total TT and BW, highlighting the impact of breed as well as body size on this factor. The range is 22.9 to 31 h, 19.1 to 55 h, and 18.2 to 45 h for small dogs, medium dogs, and large dogs, respectively [1,181].

The pre-treatments had only a minor impact on the brief (0.57 ± 0.37 h) stomach TT in the fasting state. Gastric transit was evidently prolonged in the fed state (2.94 ± 0.91 h); compared to meat, a liquid meal resulted in a quicker gastric emptying time, with 90% emptying in 0.4 h and 50% in 1–3 h [1]. The study arms did not differ significantly in the maximum pressures detected in the canine GI tract; in the stomach, maximum pressures of up to 800 mbar were commonly reported [180].

The effects of biliary and pancreatic fluid diversion on canine upper gastrointestinal motility and hormone secretion were investigated by Sato et al. Upper gastrointestinal motility shows two types of patterns, being classified as either interdigestive or PP. The former pattern consists of periodically changing movements, the so-called migrating motor complex (MMC), while in the latter state, persistent contractions are performed. It is thought that BA pancreatic secretions regulate MMC. Although the MMC seems to decrease in canines undergoing total external biliary diversion, it reappears following BA administration. The research examined primarily the correlation between the MMC and bile or pancreatic juice; however, it is probable that the diversion of biliary or pancreatic juice modifies the properties of the MMC. Uncertainty surrounds the function of bile and pancreatic fluid in upper gastrointestinal motility after a meal [182].

Postprandial contractions in the upper gut were repressed. The duration of PP contractions in the proximal jejunum, which is correlated with gastric emptying, was prolonged. Additionally, the frequency and migration velocity of the MMC in the jejunum were reduced [182].

5.3. Impact of Breeds and Body Sizes

Due to decades of selective breeding, dog sizes have varied, which has been linked to differences in digestive physiology and susceptibility to disease. According to a review paper, dog sizes clearly show changes in interior anatomy. In dogs weighing 60 kg and 5 kg, the mature digestive tract length can account for 2.8% and 7% of the total body mass, respectively [178]. Colonic measurements, TT permeability, fiber degradation, fecal SCFA concentration, and fecal water content all rise with body size, but fecal bile acid concentration falls. Changes in microbiota composition based on dog BW may also be linked to modifications in BA concentrations, inversely correlated with BW, as certain bacteria are fully involved in BA deconjugation [178].

These findings imply that there are differences in microbiota activity between small and large breed sizes, particularly in the recycling of BA. Physicochemical characteristics including pH, TT, and digestive secretions can all be affected by variations in digestive anatomy, which in turn can have an impact on the microbes in the gut [1,178].

Canine body sizes also affect the fecal BAPs. As BW increases, there appears to be a decrease in both the primary to secondary BA ratios and TBAs. Deschamps et al. summarized the fecal BA concentration ranges of a few studies, according to which the values are most frequently between 5.1 and 7.5 μg of total BA per mg of dry feces. Data related to the ratio of primary and secondary BA content in feces were also collected. They indicate that fecal secondary BA (84.9%) has greater relative percentages than primary BA (15.5%). Furthermore, proportions of main BAs, such as CA and CDCA, appear to be connected with canine BW negatively, although secondary BA shows the opposite relationship (one study only). These findings imply that there are differences in microbial activity between small and large breed sizes, particularly in recycling BAs [1].

The former team created the CANIne Mucosal ARtificial COLon (CANIM-ARCOL), a new model of the canine colon, in order to overcome this limitation. This model replicates the key nutritional (ileal effluent composition), physiochemical (pH, TT, anaerobiosis), and microbial (lumen and mucus-associated microbiota) features of this ecosystem, and it is customized for three dog sizes: small (below 10 kg), medium (10 to 30 kg), and large (above 30 kg). In the application of this model, it was disclosed that body size had an obvious impact on daily BA concentrations, with small bioreactors exhibiting concentrations of about 200 μg/mL in the fermentation medium compared to 400 and 500 μg/mL in medium and large conditions, accordingly. The corresponding profiles showed a substantial difference also in relation to dog size, with a considerable rise in DCA and a drop in the LCA proportion in large dogs as opposed to small and medium dogs. Furthermore, the percentages of CDCA and a structural isomer of LCA tended to be larger and lower, respectively, in large bioreactors. Finally, there was a substantial rise in DCA and LCA concentrations with dog size as well as the luminal compartment's ammonia concentration [48].

Uniquely, it was observed in a recent study that TBA concentrations increased in vitro with dog size. However, contrasting trends were noticed in dog stools. Deschamps et al. were the first who described the efficient deconjugation of primary BAs into secondary BAs in an in vitro canine model using a nutritive medium. Furthermore, the proportions of secondary BAs increased proportionally with dog size, which aligns with the results obtained from fecal samples collected in living organisms [48].

It should also be taken into account that the focus of hepatic BA conjugation in dogs differs from that in other commonly investigated species, leading to further fundamental canine specificities in BA pools. Kook et al. noted that in accordance with the previously published results for six dogs, taurine-conjugated BAs comprised more than 99 percent of the total BA pool. Their own study confirmed these results using the gallbladder BAP for all 12 dogs examined (i.e., 99.04% for taurine-conjugated BAs, 0.15% for glycine-conjugated BAs, and 0.81% for unconjugated BAs) [50]. While rodents are capable of forming both taurine and glycine conjugates, dogs exhibit a predilection for taurine conjugation, while non-human primates and humans show a preference for glycine conjugation [50,59].

A small percentage of BAs can be converted by dogs via glycine conjugation, whereas cats appear to endure obligate conjugation with taurine. It appears that even taurine-depleted cats generate negligible quantities of glycine-conjugated BAs [46].

5.4. Impact of Age and Aging

Lee et al. described a method for profiling BAs in bile. Age-dependent alterations in rat BAs were characterized by applying a method for profiling BAs in bile. Unconjugated BAs and glycine-conjugated BAs were observed to decrease or remained unaffected; in turn, taurine-conjugated BAs exhibited a general increase. Consistent with the patterns observed in the targeted BAs, five of the taurine-conjugated BAs showed an increase, whereas a glycine-conjugated BA demonstrated a decline among the unknown BAs. The concentrations of 19 targeted BAs in rat bile obtained from subjects aged six weeks, six months, and fifteen months were investigated in this study. The predominant forms of BAs were taurine-conjugated, which constituted more than 80% of the entire pool of BAs. The levels of unconjugated BAs and glycine-conjugated secondary BAs decreased as the subjects aged. The levels of taurine-conjugated primary BAs increased. The research revealed that the concentration of primary taurine-conjugated BAs had increased in the absence of bacteroid activity; on the other hand, the concentration of glycine-conjugated primary BAs had been unaffected. Additionally, the research defined 22 unidentified BAs in rat bile, which may serve crucial functions and require additional investigations [163].

Secondary BAs are known to hinder the growth and spore germination of *C. difficile*. A recent study demonstrated that this hindrance was influenced by the dosage of secondary BAs. Blake et al. found that the levels of secondary BAs in the feces of puppies up to 5 to 6 weeks were much lower compared to adult dogs. A simultaneous rise in the levels of secondary BAs in the feces and a decline in the abundance of *C. difficile* provided evidence to support the concept of correlation in dogs [19].

5.5. Impact of Nutrient Content and Microbial Aspects

A study on C57BL/6J mice found that obesity and weight loss induced by diet had altered BA concentrations and BA-sensitive gene expression in insulin target tissues. La Farano et al. postulated that certain alterations in BA concentrations and associated gene expression would persist following weight loss in diet-induced obesity. They found that obesity induced by diet decreased hepatic BAs while increased numerous conjugated BAs in the plasma. The investigation on BAs in various tissues revealed that the levels of BAs in the hypothalamus, muscle, and epididymal white adipose tissue had been variable but largely stable. The observed changes in BA levels were substantially reversed, irrespective of tissue, when the feed was changed from high fat to low fat, and the subjects returned to a leaner physique and an insulin-sensitive state. The team established that further research is needed to examine the effects of BA deposition in metabolically significant tissues in relation to weight gain and loss [95].

According to a study on healthy canines, an excess of dietary lipids can impair gallbladder motility and BA metabolism [46]. For the duration of the two-week study, seven healthy Beagles were administered either a low-fat diet or a high-fat and high-cholesterol diet. The former increased total cholesterol and TCDCA in the plasma, while it decreased CCK-induced gallbladder motility. In contrast, the latter decreased TDCA in bile. Comparable alterations in the composition of BA and biliary hypomotility can be observed in canines with hyperlipidemia, according to the findings [12]. Kakimoto et al. referred to studies applying muscle strips or smooth muscle cells from the gallbladder to illustrate how hydrophobic BAs inhibit smooth muscle contraction through the induction of lipid peroxidation and oxidative stress (high H_2O_2 concentrations), in addition to their potent detergent effect on tissues [12].

It may be beneficial to consider an additional aspect in future perspectives, i.e., the associations between the intestinal microbiome of dogs fed with various feed forms and the BAP. Our concept is backed by the established strict correlation between gut microbial composition and BAP, as well as the observation by Schmidt et al. The research group found that the metabolome and microbial communities of BARF-fed dogs were significantly distinct from those of commercially fed ones [158].

Given that both BA compounds and microbes are influenced by the feed consumed, which in turn causes substantial changes in the diversity of the gut microbiota, it is indisputable that the quality of feed applied also impacts BAP [183,184].

Inconsistent serum BA results can be caused by variances in the type or amount of feed applied to the PP serum BA stimulation test, as well as differences in stomach emptying and gallbladder contraction [13].

To emphasize the relevant connection between the nutrient content of a feed and the state and species composition of the gut microbiota, Belchik et al. used a type of prebiotic to investigate the microbiome and the BAP in antibiotic-treated dogs. Supplementing with novel milk oligosaccharide biosimilars helped the dogs receiving antibiotics to recover from antibiotic treatment more quickly, to preserve microbial diversity, and to have more consistent fecal scores. The gut microbes and secondary BAPs are crucial for gut health. They revealed that dogs employ mainly *C. hiranonis* as a BA converter, whereas humans require primarily *C. scindens* in BA dehydroxylation reactions. It is known that in sick dogs, the chance of relapse is higher when secondary BA profiles and *C. hiranonis* cannot be recovered [184].

To conclude, under physiological conditions, the dynamics of bile acid circulation in canines is modulated by factors, such as the composition of the feedstuff, transit time, motility patterns, and the intestinal microbiome. Beyond these, the BA pool exhibits notable diversity with respect to canine breeds, sizes, and age. Despite advancements in our understanding, defining normative BAPs remains a challenge, leading to a lack of standardized data for canines.

6. Major Causes of Elevated TBA or Paired TBA

TBA or paired TBA is a useful liver function test. The TBA concentrations increase in the circulation when pathology alters the enterohepatic circulation. Decreased liver function, portosystemic vascular anomalies, and cholestasis are the most common reasons for its elevation [171,176].

6.1. Intrahepatic Diseases

When liver function is compromised, it decreases the removal of BAs from the portal circulation. Dysregulated BA metabolism affects lipid metabolism, immune environment, and intestinal bacteria, leading to inflammation, fibrosis, and hepatic steatosis. Gut microbes play a crucial role in regulating fat accumulation in hepatocytes, and changes in intestinal bacteria composition may affect immune function and inflammation [10,54,60,61]. The involvement of BAs and their receptors in liver repair and regeneration is critical. BA-induced proliferative and adaptive responses may play a role in both the biliary homeostasis maintenance and the regenerative response. During regeneration, BA receptors may collaborate to prevent toxic BA overload and restore a functional liver mass to its maximum capacity [185].

Yu et al. suggested that enhancing BA metabolism could be a promising therapeutic approach to prevent the onset and progression of non-alcoholic fatty liver disease (NAFLD) in human [60]. Also, Chow et al. published a report on NAFLD, which was characterized by the fat accumulation in the liver (macrosteatosis). This condition can evolve into a more severe version called nonalcoholic steatohepatitis, defined by inflammation and fibrosis. Steatohepatitis may occur as a result of an imbalance in BA pool. Presently, pharmacological therapy focuses on these routes [186].

The fecal BAPs of 28 adults with biopsy-confirmed NAFLD (non-alcoholic fatty liver: NAFL or non-alcoholic steatohepatitis: NASH), and healthy control subjects (HC) (n = 25) were analyzed in an investigation. In comparison to HC, patients with NAFLD had elevated concentrations of total fecal BA, CA, and CDCA and higher BA synthesis, according to the findings, similarly to previous publications on the serum/plasma BA content in individuals with NAFLD. In NASH, the proportion of primary to secondary BAs was greater than in the control group; however, there was no significant difference in the ratio of conjugated to unconjugated BAs between the groups. According to the study's hypothesis, patients with NAFLD have distinct BA homeostasis, associated with dysbiosis, which renders them more susceptible to liver impairment. To this date, research of Mouzaki et al. offers the most exhaustive examination of BA homeostasis in patients diagnosed with NAFLD [111].

Sugita et al. used LC-MS/MS to measure the serum concentrations of 16 distinct BAs and compared them across various liver disorders. Patients with hepatitis, patients with alcoholic liver disease, patients with biliary tract disorders, NAFLD patients, and other liver patients were among them. As per the results, the group of liver patients with alcoholism had much higher levels of UDCA and glycoursodeoxycholic acid (GUDCA), while the group with biliary tract illness had significantly lower levels of both DCA and UDCA than the group with viral hepatitis [126].

Staley et al. performed an analysis regarding hepatic injury induced by BAs. The research group clarified that BAs could damage the integrity of cell membranes, having toxic consequences, which can make the liver more susceptible [17]. The lipophilicity of individual BAs dictates their ability to dissolve in fats, and the movement of conjugated BAs through transporters may impact their profile of causing cell death in drug-induced liver injury (DILI). BAs play a crucial role in the development of DILI that can result in severe and irreversible liver failure. The research demonstrated notable increases in BA levels in the blood and urine of rats exposed to several hepatotoxins, as well as alterations in metabolites related to energy, urea, and BA metabolism in the urine and blood. Changes in BA profiles may derive from hepatotoxicity or the functional blockage of BA transporters. These variations in BA composition in the circulation can occur independently of liver cell injury [59].

According to Slopianka et al., BAs (particularly CA, CDCA, DCA, muricholic acid (b), UDCA in plasma) may serve as further markers for diagnosing DILI in plasma, potentially even distinguishing between damage to hepatocytes and the biliary system [97]. Qin et al. also specifically studied rats with DILI. Although it was determined similarly that BAs play a critical role in pre-diagnosing DILI, their utility as biomarkers remained uncertain. Bile acids found in urine are more appropriate as biomarkers for identifying cholestatic liver injury than those in serum, as indicated by a research group's findings [90].

According to the research by Rena-Ramos et al., dogs with cirrhosis and congenital circulatory abnormalities had the highest average resting serum bile acids (SBAs). Chronic hepatitis and congenital circulatory abnormalities were associated with the highest median PP concentrations [171]. The study included 341 dogs with various liver diseases. Liver biopsies were collected to evaluate the accuracy of resting and PP SBAs in detecting a variety of liver problems in dogs. The median resting SBAs were highest in dogs with cirrhosis and congenital circulatory abnormalities, while the conditions with the highest median PP levels were chronic hepatitis and circulatory anomalies. The resting SBA levels in dogs with liver illness across all groups ranged from less than 10 to more than 90 μmol/L [171].

Using ultra-performance liquid chromatography in tandem with mass spectrometry (UPLC-MS/MS), Zhang et al. assessed the plasma concentrations of 15 BAs in a total of 329 human subjects, including those with benign biliary diseases (BBD), CCA, gallbladder cancer (GC), hepatocellular carcinoma (HCC), and healthy individuals. Comparing HC and BBD groups to CCA patients, the majority of unconjugated BAs, such as CA, CDCA, UDCA, and LCA, were significantly lower in patients with CCA. On the other hand, from HC to BBD to CCA, there was a progressive stepwise rise in plasma-conjugated primary BAs, such as TCA, TCDCA, GCA, and glycochenodeoxycholic acid (GCDCA). Furthermore, compared to other groups, the absolute contents of total tauro-conjugated SBA and GUDCA were likewise significantly higher in CCA. Comparing the BBD groups to the malign groups, the multiple alterations of conjugated serum BAs in CCA were significantly smaller than those of conjugated primary BAs. Due to a decrease in total unconjugated BAs and an increase in conjugated BAs, the ratio of conjugated to unconjugated BAs was often significantly higher in CCA patients compared to the other study groups. Furthermore, CCA had considerably greater total primary BAs, but there was no difference in serum TBA across the three study groups. As a result, it was found that CCA patients had a lower SBA-to-primary-BA ratio. To sum up, distinct variations in the plasma concentration of BAs could function as diagnostic indicators to differentiate CCA from BBD and HC [141].

In a case report, the BAP of a dog with cholangiocarcinoma (CCA) and a healthy dog was compared by Thompson et al. When comparing the dog with neoplastic disorder to the healthy one, the former one's total serum BA content was over a hundred times greater. TCA accounted for 88.8% of this total. Cholestasis was characterized by an elevated level and a predominance of TCA in the BAP, and a tumor-induced obstruction of the biliary tract was the cause of this condition. Furthermore, in this case, the BAP was identical to that published for rats and humans with cholestatic conditions [128].

In beagle dogs with cholestasis brought on by rifampicin, BAs were investigated. UPCL-MS/MS technology based on enzyme digestion was used to determine BAs. For nine days, the subjects were given 300 mg of rifampicin twice a day. Their blood was drawn fifteen minutes before the medication administration, and daily urine and intestinal sludge samples were taken between 0 and 24 h of treatment. Moreover, 27 BAs, including primary, secondary, and tertiary BAs, were examined. Following rifampicin treatments, daily urination and defecation revealed a rise in BA diversity in serum or urine but a decrease in BA diversity in feces, suggesting that rifampicin therapies changed the intestinal microbe composition of BAs metabolism [115].

6.2. Portosystemic Vascular Anomalies (PSVAs)

The portosystemic vascular abnormalities bypass the hepatic sinusoids and directly connect the portal venous system, containing contaminated splanchnic blood, with the systemic circulation. The above impairment is classified as congenital portosystemic shunts (CPSSs) or acquired PSSs (APSSs) [175]. Improper portal flow to the liver can impact liver function tests such as fasting ammonia (FA) or ammonia tolerance test (ATT), serum fasting BAs, and paired TBA concentrations [175,176]. A comparative study demonstrated that ceruletide-stimulated SBA testing is a viable substitute for postprandial SBA stimulation when it comes to diagnosing hepatic dysfunction, in particular, in canines that are concurrently experiencing upper respiratory disease and PSS [13].

Especially when evaluated postprandially, the serum TBA concentration is a dependable marker for PSS diagnosis with extraordinarily high sensitivity and specificity [187]. However, it is important to consider further aspects. Straten et al. demonstrated that in dogs showing symptoms, an elevation in both fasting TBA and FA are adequate for the diagnosis of PSS and may hold similar significance. Compared to FA concentrations, ATT and fasting TBA exhibit greater sensitivity in identifying PSS; however, they do not possess the same level of specificity as increasing FA concentrations. Combining fasting TBA with FA or ATT and fasting TBA in serial testing provides the benefit of enhancing specificity.

The findings indicate that measuring fasting TBA concentrations is the most effective test for detecting CPSS in a healthy population [175,177].

Anglin et al. conducted an investigation involving 72 dogs diagnosed with PSS (cases collected between 2000 and 2008). The mean fasting BA values were as follows: 101.2 μmol/L at the initial presentation, 111.1 μmol/L at the first recheck examination, and 127.9 μmol/L at the second or final recheck examination (with a reference range of 0–10 μmol/L). The PP BA values were as follows: 145.6, 258.9 and 285.2 (with 0–30 μmol/L serving as the reference range) [188].

The majority of dogs with PSS had elevated concentrations of serum TBA according to a study involving 42 canines, where mean fasting values and mean PP values were 86.5 μmol/L and 165.5 μmol/L, respectively. This was the result of a median fasting period preceding the surgery [189].

Investigating the application of SBA as a prognostic indicator for congenital portosystemic shunt (PSS) in Maltese dogs, one study discovered that preoperative SBA concentrations were considerably greater (192 μmol/L) in this breed than those observed in other breeds (137 μmol/L) with portocaval shunt. However, postoperative SBA levels in Maltese and other dog breeds decreased following surgery. Regarding postoperative SBA levels, no significant difference was found between Maltese and other dog breeds. Measuring pre- and post-operative SBA levels may be a beneficial method for determining the prognosis of PSS in Maltese dogs, according to the study of Kim et al. [190].

The BA panel in dogs or other animals has not been examined for PSVA.

6.3. Cholestasis

A disrupted bile flow results in cholestasis, marked by increased levels of BA in the liver and bloodstream, leading to damage to hepatocytes and the bile epithelium [17]. Cholestasis may arise from either extrahepatic or intrahepatic causes. Li et al. concluded that despite extensive research into the etiology of cholestasis in humans, the precise molecular pathways, by which BAs induce liver damage, continue to be a subject of debate. As the research team described, in cholestatic liver injury, the ability to excrete BA is compromised due to either the direct inhibition or genetic defects of canalicular BA transporters in hepatocytes or due to a mechanical or immune-mediated blockage of the biliary ducts. Regardless of the underlying reason, there is an elevation in BA levels in both the liver and the bloodstream, which causes damage to hepatic cells and an increase in the growth of bile ducts. Untreated cholestatic liver damage frequently leads to the progression of liver fibrosis, cirrhosis, and ultimately liver failure. Li et al. conducted a comprehensive analysis of how BAs stimulate the formation of molecular agents that attract neutrophils, as well as the involvement of the inflammatory response in this pathogenic mechanism. These advancements indicate several new areas, where medications could potentially be considered useful treatments for cholestatic liver damage [79].

Pena-Ramos et al. suggested that the absence of an increase in SBAs in dogs could be caused possibly by bile being transported from the biliary tract into the duodenum ineffectively or slowly as a result of biliary epithelial inflammation. Due to the disrupted bile flow into the gut, BA absorption cannot occur properly. Consequently, two hours postprandially, the BA levels in the site of portal blood flow may not rise noticeably, which would restrict the rise in PP SBA concentrations in the systemic circulation in comparison with preprandial values [171].

Being consistent with research on humans, Jiang and Schnabl summarized that alterations in microbial composition were observed in an animal model of cholestasis (bile duct ligation) as early as three days following the procedure. In comparison to wild-type mice, the Mdr2/mouse, serving as an alternative mouse model of human primary sclerosing cholangitis (PSC), demonstrated an augmented translocation of bacteria and a modified composition of the intestinal microbiota. Significant intestinal dysbiosis and liver toxicity were induced when microbiota was transferred from Mdr2/mice to wild-type mice. This indicates a potential association between the gut microbiota and the progression of disease in Mdr2/mice. In summary, mounting evidence from clinical and preclinical investigations further substantiates the correlation between the microbiota and the cholestatic liver diseases. In a recent research study, alterations in the gastrointestinal microbiota were identified as a potential therapeutic target and biomarker for primary biliary cholangitis being a chronic, progressive consequence of cholestasis [61].

Increased damage to cholangiocytes may result from a protracted membrane exposure caused by the accumulation of hydrophobic BAs during cholestasis [17,51]. Bile salts accumulate and penetrate cholangiocytes extensively at low pH levels, in which the elevation in biliary acidity is a factor in the accelerated rates of apoptosis. The potential therapeutic applicability of UDCA in the prevention of cholestasis and the treatment of cholangiopathies is supported by its cytoprotective mechanisms. The activation of a BA nuclear receptor decreased systemic inflammation and improved cholangiocyte protection in mice but did not ameliorate primary sclerosing cholangitis, according to one study. Although it was observed to exacerbate pruritus induced by cholestasis, the protective properties of the nuclear receptor prompted research into its potential as a target for immune-mediated cholangiopathies [51].

Washizu et al. observed notable physiologic variations in the primary BAs, patterns of conjugation, and concentrations of various BAs in different species (dog, cow, horse, and human) [129].

Some years later, a biliary obstruction was induced in dogs via the surgical ligation of common bile ducts for the purpose of observing changes in the composition of SBAs. The results of the research suggest that the glycine/taurine ratio is not a practical indicator for detecting hepatopathy in canines. However, the CA:CDCA ratio may have a potential to fulfil this function [134].

The liver has a vital function in the systemic response to sepsis, and infectious bacteria can cause cholestasis and hyperbilirubinemia, even in the absence of direct hepatic invasion by the pathogen. Cholestasis occurs early in the sepsis process, and the SBA values were identified as an early indicator of short-term survival. Nevertheless, there is little evidence of the evaluation of SBAs in septic patients. Baptista et al. evaluated the significance of preprandial BA concentrations in a group of dogs diagnosed with sepsis. The hypothesis was that dogs with sepsis have elevated levels of resting SBAs. The data were retrospectively gathered over a 12-year period for analyzing data from 26 dogs, and they were contrasted in view of the SBAs of orthopedic patients with sepsis and healthy subjects. As a result, the septic group's SBA levels were considerably higher than those of the non-septic groups in a positive correlation with the bilirubin levels. Nonetheless, there was no discernible change in the SBA level between the orthopedic and septic groups. However, the research had several limitations, including a large number of cases with incomplete data, a small sample size, and the use of single BA measurements [191].

A milder form of intrahepatic cholestasis is observed when circumstances such as hypoxia arising from centrolobular necrosis or enlarged hepatocytes due to hepatic glycogen, fat, or copper accumulation are present. Consequently, measurable BA values will increase albeit not to the same extent as major causes [192–194].

The intrahepatic cholestasis of pregnancy is a hepatic disorder characterized by elevated levels of sulfated BAs in the urine. One of the research projects developed a method to detect and measure sulfated BAs in urine by means of successfully isolating and quantifying 16 different forms of SBAs over a span of 18 min. The research revealed a significant increase in the overall levels of urinary sulfated BAs in individuals diagnosed with intrahepatic cholestasis, especially in relation to sulfated glycine-amidated and taurine-amidated BAs. The research proposes that urine sulfated BAs may serve as an alternative to SBAs for non-invasive sample collections in the diagnosis of the intrahepatic cholestasis of pregnancy [146].

Humbert et al. examined BA secretion and circulation in patients with exocrine pancreatic insufficiency (EPI) and chronic pancreatitis (CP). The investigation was conducted by concurrently quantifying PP concentrations of specific BAs in the contents of the duodenum and blood plasma. CP is a medical disorder characterized by a decrease in the body's BAs, which leads to a diminished ability to absorb lipids. The levels and profiles of BAs in patients with exocrine pancreatic insufficiency did not undergo a comprehensive investigation. Humbert et al. revealed a significant decrease in the total BA levels in the duodenal contents of CP patients compared to those of healthy individuals. The study additionally found that patients with CP had decreased amounts of both glyco- and tauro-conjugated as well as unconjugated BA compared to the healthy group. Also, a contemporaneously elevated primary-to-secondary-BA ratio was experienced, which indicates an impaired transformation by the gut microbiota. The results indicated a significant 5-fold drop in the concentration of BA in the duodenal contents of patients with CP compared to healthy individuals. The study also discovered that patients with CP exhibited a significant rise in plasma BA concentration 30 min postprandially, which aligns with a diversion of BA circulation. These findings could provide an explanation for the 5-fold reduction in the BA pool observed in patients with CP compared to healthy individuals [102].

The physiological relationship between the composition of the intestinal microbiota and the metabolism of BAs suggests that changes in BA homeostasis may accompany dysbiosis, in this manner playing a role in the metabolic dysregulation observed in diabetes mellitus (DM). Alterations in both primary and secondary fecal unconjugated bile acids (fUBAs) were noted by Jergens et al. in dogs. These changes were characterized by elevated amounts of CA and the total primary fUBAs. On the other hand, LCA levels were lower, both as a proportion of all fUBAs and as an absolute number, when measured in dogs with diabetes. DM occurring naturally in dogs is insulin-dependent and may be accompanied by dysbiosis and an altered BA metabolism. Consequently, they may provide a clinically applicable model for analyzing human disease [154].

BAs perform endocrine functions, and they are implicated in glucose homeostasis. The interaction with the gut microbiome was demonstrated to be altered in humans and rodents with type 1 and type 2 diabetes. This connection further complicates the metabolic function of BAs [195]. By analyzing the serum metabolomic profiles of diabetic canines, O'Kell et al. identified parallels to type 1 diabetes in humans in their pilot study. They revealed that primary and secondary BAs (TCDCA, TDCA, and TUDCA), among other metabolites, were substantially lower in diabetic dogs compared to healthy dogs. Furthermore, disruptions in a variety of BAs were recognized in humans. The investigation displayed that metabolomics provides the potential of distinguishing between diabetic and non-diabetic dogs [78].

The concentrations of total taurine-conjugated BAs were found to be intermediate in impaired individuals and higher in type 2 diabetes according to Wewalka et al. upon comparing the SBA composition of individuals with normal glucose tolerance, type 2 diabetes, and impaired glucose tolerance. Positive associations with fasting and postload glucose levels were observed, whereas insulin resistance did not correlate with this association [196].

6.4. Other Causes of Elevated TBA Level

Microbiomes are critically involved in BA metabolism. Dysbiosis is characterized by the altered deconjugation of BA, particularly when *Clostridium hiranonis* populations are decreased. As a consequence, the efficiency of BA removal from portal circulation is compromised, leading to an accumulation of circulating total BAs. The research by Melgarejo et al. aimed to determine the diagnostic value of unconjugated BA concentrations in the serum of canines with bacterial proliferation. Fasting sera were collected from 23 dogs in all, from which 10 had dysbiosis (previously SIBO) confirmed by culture, 8 were diagnosed indirectly, and 5 were HC. Unconjugated BA concentrations were significantly elevated in dogs with SIBO and dogs with indirectly diagnosed SIBO. The unconjugated BA CA predominated in the serum of dogs afflicted with SIBO. The study emphasizes the importance of detecting and managing SIBO in dogs to prevent transmission and facilitate treatment. The high capacity of liver to conjugate BAs to amino acids taurine and glycine makes conjugated primary BAs constitute nearly all of normal hepatic bile. Serum unconjugated BA measurements provide a noninvasive, specific, and sensitive diagnostic tool for SIBO in canines. The effects of dietary taurine and glycine on canine BA conjugation were investigated. The results of the study in question, for the first time, evidenced that dogs with bacteriologically proven SIBO have a higher rate of intestinal BA deconjugation than normal healthy pet dogs [123]. However, the diagnostic value of these results was not verified by other researchers [123,197].

Comito et al. used high-performance liquid chromatography in tandem with mass spectrometry (HPLC-MS/MS) to conduct a thorough study of 31 compounds from among fecal BAs and related microbiota metabolites. The diseased participants had noticeably greater levels of CA in their stool. Additionally, pathological individuals had higher CDCA than the control participants. Conversely, the control subjects had greater DCA and LCA values for fecal sludge. Significantly increased β-muricholic acid, decreased 3,12-dioxo-DCA, and 3-oxo-LCA levels were found in relation to the other microbiota metabolites. For other BAs, no discernible variations were discovered. The number of distinct BA groups (primary, secondary, and oxo-BA), as well as the overall number of non-oxo-BAs and BAs were computed in order to comprehend the impact of CIE on the BA pool. The intestinal sludge of dogs with CIE had a notably greater primary BA concentration than for the control subjects [198].

Corticosteroids regulate fecal BAs in canines with CIE, according to a study by Guard et al. The patients had an elevated DI at baseline, which remained unchanged over time. In comparison to healthy dogs, secondary fUBA was reduced in dogs with CIE; nevertheless, the percentage of secondary fecal UBA increased after two to three months of treatment. Guard et al. observed BA malabsorption either as a primary condition or in combination with chronic inflammatory bowel disease (IBD) [83].

Untargeted metabolomics is an emerging discipline that investigates metabolic pathways in fecal samples. In a comparative analysis of metronidazole and fecal microbial transplantation (FMT) for canines with acute diarrhea, notable distinctions were observed in terms of fecal consistency, microbiota composition, and metabolome profiles. Fecal consistency improved with both FMT and the antibiotic agent, but FMT demonstrated superior consistency. In terms of bacterial taxa, both groups differed considerably. Although FMT treatment resulted in reduced percentages of primary BAs, metronidazole was unable to restore these profiles to normal. The hypothesis of the study, i.e., FMT could restore the composition of BAs in the feces, was confirmed. Thus, metronidazole failed to restore fecal metabolome, whereas FMT did. The investigation concluded that canines with acute diarrhea exhibited a normalization of microbiota and an increase in microbial diversity. On the other hand, fecal metabolomes continued to exhibit dysbiosis even after a period of 28 days [149].

Another study also confirmed that FMT could restore the initial balance of the BAP after antibiotic treatment-induced damage [17]. Belchik et al. also confirmed the observation that the administration of antibiotics led to BA dysmetabolism and intestinal dysbiosis in dogs [184].

Tylosin significantly altered the BAP and fecal microbiome of healthy canines in the investigation conducted by Marclay et al., which resulted in a decrease in secondary BA concentrations on day 7. However, on day 14, the majority of parameters returned to the baseline. BA dysmetabolism and intestinal dysbiosis were observed in canines as a result of this macrolide antibiotic, but the impact of FMT on the recovery of the microbiome and metabolome remained uncertain. After administering the drug, the values returned to the baseline [157]. In contrast, Manchester et al. observed that fecal dysbiosis and corresponding shifts in fecal unconjugated BAs did not resolve uniformly in healthy canines after tylosin cessation [156,157].

By raising secondary BA concentrations, *C. scindens* was shown, in a study by Marclay et al., to assist a mouse model fight infection from *C. difficile*. This suggests that manipulating bacteria metabolizing BAs may be a viable therapy for individuals susceptible to infections of this nature. In the investigation, antibiotic dosing influenced all particular BA concentrations and ratios but did not affect total BA concentrations [157].

Primary BAs significantly increased, while secondary BAs considerably decreased under the treatment of antibiotics. Rather than indicating alterations in re-absorption, the lack of influence on total BA points to a reduced transformation of primary to secondary BA by GI microorganisms [157].

Secondary BA concentrations were lower in dogs with GI disease compared to the healthy ones. In order to ascertain the clinical relevance of lactate and BA quantification in canine feces, additional research is required. While these metabolites may not serve as reliable indicators of dysbiosis on their own, they become promising targets for further investigations in the role of microbiota in health and disease when interpreted in conjunction with bacterial abundances [65].

The available studies, in which profiles of BAs were measured, are summarized by the conditions of dogs in the following table (Table 3).

Table 3. Studies on bile acid profiles in canines.

Bibliographic Source	Illness(es)
[115]	rifampicin-induced cholestasis (DILI)
[112]	healthy, hepatic disorders, clinical signs of hepatic disorder
[133]	gallbladder mucocele formation
[134]	biliary obstruction produced by surgical ligation of the common bile duct
[123]	culture-proven SIBO, indirectly diagnosed SIBO
[128]	metastatic cholangiocarcinoma
[129]	none (healthy dogs)
[130]	none (healthy dogs)
[144]	none (healthy dogs)
[143]	none (healthy dogs)
[3]	none (healthy dogs)
[65]	gastrointestinal diseases: chronic enteropathy (CE) and exocrine pancreatic insufficiency (EPI)
[19]	none (healthy dogs)
[149]	acute diarrhea (AD)
[83]	steroid-responsive chronic inflammatory enteropathy (CE)
[52]	none (healthy dogs)
[154]	insulin-dependent diabetes mellitus
[156]	none (healthy dogs)
[157]	none (healthy dogs)
[158]	none (healthy dogs)
[99]	none (healthy dogs)

In summary, the bibliographical evidence regarding the association between BAs and the disorders in question in canines is very far from solid. Furthermore, a sufficiently diverse profile of BAs is absent from the available studies to determine whether a well-defined pattern exists in BAPs for a particular canine disease. In cases where abnormalities are suspected, clinicians would also require clarification regarding the determination of whether feces or blood should be collected for the purpose of measuring BAPs. There is a lack of research that has compared the therapeutic benefits of concurrently taking blood and feces samples for BAP determination in identical conditions; nevertheless, such an analysis could reveal whether the profiles derived from the two matrices demonstrate parallel patterns or a correlation of any kind. Such research could potentially yield valuable insights and thereby provide therapeutic alternatives.

7. Conclusions

The examination of bile acids provides essential knowledge in various fields of health science, including physiology, microbiology, internal medicine, and pharmacology. They undergo a complex transformation in the gut, affecting their structure and biochemical properties, acquiring several types of physiologic roles. Classified based on their impact on organisms, bile acids are categorized as primary or secondary, and they are further differentiated by their hydroxylation status (mono-, di-, or trihydroxy) and conjugation (taurine- or glycine-conjugated, or unconjugated). Perturbations in gut microbiota composition can disrupt bile acid metabolism, which may lead to various pathological conditions in human as well as in veterinary cases.

In clinical diagnostic laboratories, the enzymatic reaction-based photometric method remains a fundamental tool for measuring total bile acids, owing to its simplicity and cost-effectiveness. On the other hand, there is an increasing demand for more comprehensive bile acid panels, which are likely to be acquired by the LC-MS/MS technique. While triple quadrupole (QqQ) mass spectrometers currently dominate routine targeted analyses, the expanding scope of non-targeted metabolomic studies is expected to elevate the significance of time-of-flight (TOF) and ion trap (IT) instruments in bile acid analysis. This evolution reflects a growing demand for detailed and nuanced assessments in clinical practice.

Bile acid levels in dogs are typically assessed following a 12 h fasting period, with measurements taken 2 h after the consumption of canned food to obtain paired values for comparison. Various factors influence bile acid circulation in canines, including meal composition, intestinal transit times, motility, and the gut microbiome, and obviously, the composition of the bile acid pool varies among dog breeds, sizes, and ages.

Multiple matrices, such as blood, feces, urine, liver tissue, and gallbladder bile, can be used to determine canine bile acid profiles. While many studies focus on analyzing a single matrix, achieving a precise diagnostic outcome may necessitate the examination of multiple matrices to obtain a comprehensive understanding of the bile acid pool. Although blood serum or plasma is often preferred for analysis, the choice of matrix depends on technical expertise and laboratory objectives. Despite feces being the second most commonly analyzed matrix in the literature, its interpretation remains challenging due to significant individual and breed-specific variations, as well as technical sampling difficulties (since sample collection conditions cannot be adapted from laboratory rodents or humans to dogs). Nonetheless, comparing bile acid levels between feces and biological fluids could provide valuable insights into physiological and pathological changes in the enterohepatic circulation of bile acids. Altogether, there is a significant shortage of research that thoroughly examines bile acid profiles in various matrices, especially taking into account breed- and age-specific differences and clinical situations, even in healthy dogs.

In clinical trials, the combined measurement of various bile acids is gaining increased attention; however, a consensus among experts regarding the specific matrix to examine remains precarious. While certain bile acids are measured individually and TBA is quantified, the range of compounds measured is frequently insufficiently defined and/or insufficiently broad to derive diagnostic and prognostic significance. Furthermore, the existence of a distinct pattern for each disease remains unknown. The bibliographical material regarding the association between BAs and a specific disorder in dogs is far from established. It is also noteworthy to highlight that based on the information presented in this study, the success of using the Beagle breed for experimental purposes to examine the bile acid profile is also highly doubtful.

Our review endeavored to navigate the ongoing debates concerning the diverse beneficial and detrimental properties of different bile acids while also elucidating the physiological effects associated with their various groups. Addressing a significant gap in the literature, our review shed light on species-specific variations observed in the measurement of bile acid parameters, with a particular emphasis on canines. We underscored the importance of utilizing appropriate matrices alongside the potential diagnostic implications derived from them, highlighting disparities between human and canine results and sampling methodologies. This differentiation is crucial given the distinct digestive system characteristics in dogs, which encompass disparate processes and microbiological composition.

Further investigations into alterations of the overall bile acid profile in canine pathological conditions would significantly contribute to a more comprehensive understanding of various pathomechanisms, interrelationships among companion animal illnesses, and the correlation between the gut microbiome and the extensive systemic effects of bile acids on the entire body. Additionally, a thorough exploration of the interactions between bile acids and the gastrointestinal microbiome holds promise for novel therapeutic avenues. Therefore, pursuing these research objectives presents an intriguing prospect from both a scientific and therapeutic standpoint.

Author Contributions: K.N.: original draft preparation and conceptualization, Á.S.: original draft preparation, conceptualization, and supervision, D.S.K.: original draft preparation, conceptualization, supervision, and review and editing, R.K.L.: original draft preparation, V.H.: original draft preparation, K.V.: original draft preparation, T.B.: conceptualization and supervision, A.B.: original draft preparation, K.L.: original draft preparation, conceptualization, supervision, and review and editing. All authors have read and agreed to the published version of the manuscript.

Funding: This research was funded by the University of Veterinary Medicine's Internal Research Grant No. KFKU106000_2023.

Institutional Review Board Statement: Not applicable.

Informed Consent Statement: Not applicable.

Data Availability Statement: Not applicable.

Acknowledgments: The authors would like to express their gratitude to István Lányi for the careful and deep technical language editing of the manuscript.

Conflicts of Interest: The authors declare no conflicts of interest.

References

1. Deschamps, C.; Humbert, D.; Zentek, J.; Denis, S.; Priymenko, N.; Apper, E.; Blanquet-Diot, S. From Chihuahua to Saint-Bernard: How did digestion and microbiota evolve with dog sizes. *Int. J. Biol. Sci.* **2022**, *18*, 5086–5102. [CrossRef]
2. Niu, X.; Xu, Y.; Yang, Q.; Tang, X.; Li, Y.; Wang, Z. Analytical methods for characterization of bile acids and its application in quality control of cow-bezoar and bear bile powder. *Am. J. Appl. Chem.* **2014**. [CrossRef]
3. Sangaraju, D.; Katavolos, P.; Liang, X.; Chou, C.; Zabka, T.S.; Dean, B.; Maher, J. Establishment of baseline profiles of 50 bile acids in preclinical toxicity species: A comprehensive assessment of translational differences and study design considerations for biomarker development. *Toxicol. Appl. Pharmacol.* **2022**, *443*, 116008. [CrossRef]
4. Molinero, N.; Ruiz, L.; Sánchez, B.; Margolles, A.; Delgado, S. Intestinal bacteria interplay with bile and cholesterol metabolism: Implications on host physiology. *Front. Physiol.* **2019**, *10*, 185. [CrossRef]
5. Russell, D.W. Fifty years of advances in bile acid synthesis and metabolism. *J. Lipid Res.* **2009**, *50*, S120–S125. [CrossRef]
6. Patel, S.K.; Billingsley, M.M.; Mukalel, A.J.; Thatte, A.S.; Hamilton, A.G.; Gong, N.; El-Mayta, R.; Safford, H.C.; Merolle, M.; Mitchell, M.J. Bile acid-containing lipid nanoparticles enhance extrahepatic mRNA delivery. *Theranostics* **2024**, *14*, 1–16. [CrossRef]
7. Ridlon, J.M.; Daniel, S.L.; Gaskins, H.R. The Hylemon-Björkhem pathway of bile acid 7-dehydroxylation: History, biochemistry, and microbiology. *J. Lipid Res.* **2023**, *64*, 100392. [CrossRef]
8. Ridlon, J.M.; Kang, D.J.; Hylemon, P.B. Bile salt biotransformations by human intestinal bacteria. *J. Lipid Res.* **2006**, *47*, 241–259. [CrossRef]
9. Larabi, A.B.; Masson, H.L.P.; Bäumler, A.J. Bile acids as modulators of gut microbiota composition and function. *Gut Microbes* **2023**, *15*, 2172671. [CrossRef]
10. Behr, C.; Slopianka, M.; Haake, V.; Strauss, V.; Sperber, S.; Kamp, H.; Walk, T.; Beekmann, K.; Rietjens, I.; van Ravenzwaay, B. Analysis of metabolome changes in the bile acid pool in feces and plasma of antibiotic-treated rats. *Toxicol. Appl. Pharmacol.* **2019**, *363*, 79–87. [CrossRef]
11. Di Ciaula, A.; Garruti, G.; Baccetto, R.L.; Molina-Molina, E.; Bonfrate, L.; Wang, D.Q.-H.; Portincasa, P. Bile Acid Physiology. *Ann. Hepatol.* **2017**, *16*, S4–S14. [CrossRef]
12. Kakimoto, T.; Kanemoto, H.; Fukushima, K.; Ohno, K.; Tsujimoto, H. Effect of a high-fat–high-cholesterol diet on gallbladder bile acid composition and gallbladder motility in dogs. *Am. J. Vet. Res.* **2017**, *78*, 1406–1413. [CrossRef]
13. Bridger, N.; Glanemann, B.; Neiger, R. Comparison of postprandial and ceruletide serum bile acid stimulation in dogs. *J. Vet. Intern. Med.* **2008**, *22*, 873–878. [CrossRef]
14. Center, S.A.; Thompson, M.; Guida, L. 3α-Hydroxylated bile acid profiles in clinically normal cats, cats with severe hepatic lipidosis, and cats with complete extrahepatic bile duct occlusion. *Am. J. Vet. Res.* **1993**, *54*, 681–688. [CrossRef]

15. Bhowmik, S.; Chiu, H.; Jones, D.H.; Chiu, H.; Miller, M.D.; Xu, Q.; Farr, C.L.; Ridlon, J.M.; Wells, J.E.; Elsliger, M.; et al. Structure and functional characterization of a bile acid 7α dehydratase BaiE in secondary bile acid synthesis. *Proteins Struct. Funct. Bioinform.* **2016**, *84*, 316–331. [CrossRef]
16. Jia, H.-M.; Yu, M.; Ma, L.-Y.; Zhang, H.-W.; Zou, Z.-M. Chaihu-Shu-Gan-San regulates phospholipids and bile acid metabolism against hepatic injury induced by chronic unpredictable stress in rat. *J. Chromatogr. B* **2017**, *1064*, 14–21. [CrossRef]
17. Staley, C.; Weingarden, A.R.; Khoruts, A.; Sadowsky, M.J. Interaction of gut microbiota with bile acid metabolism and its influence on disease states. *Appl. Microbiol. Biotechnol.* **2017**, *101*, 47–64. [CrossRef]
18. Doden, H.; Sallam, L.A.; Devendran, S.; Ly, L.; Doden, G.; Daniel, S.L.; Alves, J.M.P.; Ridlon, J.M. Metabolism of Oxo-Bile Acids and Characterization of Recombinant 12α-Hydroxysteroid Dehydrogenases from Bile Acid 7α-Dehydroxylating Human Gut Bacteria. *Appl. Environ. Microbiol.* **2018**, *84*, e00235-18. [CrossRef]
19. Blake, A.B.; Cigarroa, A.; Klein, H.L.; Khattab, M.R.; Keating, T.; Van De Coevering, P.; Lidbury, J.A.; Steiner, J.M.; Suchodolski, J.S. Developmental stages in microbiota, bile acids, and clostridial species in healthy puppies. *J. Vet. Intern. Med.* **2020**, *34*, 2345–2356. [CrossRef]
20. Zwicker, B.L.; Agellon, L.B. Transport and biological activities of bile acids. *Int. J. Biochem. Cell Biol.* **2013**, *45*, 1389–1398. [CrossRef]
21. Dawson, P.A.; Karpen, S.J. Intestinal transport and metabolism of bile acids. *J. Lipid Res.* **2015**, *56*, 1085–1099. [CrossRef]
22. Ridlon, J.M.; Kang, D.J.; Hylemon, P.B.; Bajaj, J.S. Bile acids and the gut microbiome. *Curr. Opin. Gastroenterol.* **2014**, *30*, 332–338. [CrossRef]
23. Sayin, S.I.; Wahlström, A.; Felin, J.; Jäntti, S.; Marschall, H.-U.; Bamberg, K.; Angelin, B.; Hyötyläinen, T.; Orešič, M.; Bäckhed, F. Gut microbiota regulates bile acid metabolism by reducing the levels of tauro-beta-muricholic acid, a naturally occurring FXR antagonist. *Cell Metab.* **2013**, *17*, 225–235. [CrossRef]
24. Wells, J.E.; Hylemon, P.B. Identification and characterization of a bile Acid 7α-dehydroxylation operon in *Clostridium* sp. strain TO-931, a highly active 7α-dehydroxylating strain isolated from human feces. *Appl. Environ. Microbiol.* **2000**, *66*, 1107–1113. [CrossRef]
25. Campbell, C.; McKenney, P.T.; Konstantinovsky, D.; Isaeva, O.I.; Schizas, M.; Verter, J.; Mai, C.; Jin, W.-B.; Guo, C.-J.; Violante, S.; et al. Bacterial metabolism of bile acids promotes generation of peripheral regulatory T cells. *Nature* **2020**, *581*, 475–479. [CrossRef]
26. Winston, J.A.; Theriot, C.M. Diversification of host bile acids by members of the gut microbiota. *Gut Microbes* **2020**, *11*, 158–171. [CrossRef]
27. Poland, J.C.; Flynn, C.R. Bile acids, their receptors, and the gut microbiota. *Physiology* **2021**, *36*, 235–245. [CrossRef]
28. Tang, Y.; Zhang, J.; Li, J.; Lei, X.; Xu, D.; Wang, Y.; Li, C.; Li, X.; Mao, Y. Turnover of bile acids in liver, serum and caecal content by high-fat diet feeding affects hepatic steatosis in rats. *Biochim. et Biophys. Acta (BBA) Mol. Cell Biol. Lipids* **2019**, *1864*, 1293–1304. [CrossRef]
29. Ridlon, J.M.; Harris, S.C.; Bhowmik, S.; Kang, D.-J.; Hylemon, P.B. Consequences of bile salt biotransformations by intestinal bacteria. *Gut Microbes* **2016**, *7*, 22–39. [CrossRef]
30. Zheng, L. New insights into the interplay between intestinal flora and bile acids in inflammatory bowel disease. *World J. Clin. Cases* **2022**, *10*, 10823–10839. [CrossRef]
31. Kim, K.H.; Park, D.; Jia, B.; Baek, J.H.; Hahn, Y.; Jeon, C.O. Identification and Characterization of Major Bile Acid 7α-Dehydroxylating Bacteria in the Human Gut. *mSystems* **2022**, *7*, e0045522. [CrossRef]
32. Griffiths, W.J.; Sjovall, J. Bile acids: Analysis in biological fluids and tissues. *J. Lipid Res.* **2010**, *51*, 23–41. [CrossRef]
33. James, S.B.; Raphael, B.L.; Clippinger, T. Diagnosis and Treatment of Hepatic Lipidosis in a Barred Owl (*Strix varia*). *J. Avian Med. Surg.* **2000**, *14*, 268–272. [CrossRef]
34. Harris, S.C.; Devendran, S.; García, C.M.; Mythen, S.M.; Wright, C.L.; Fields, C.J.; Hernandez, A.G.; Cann, I.; Hylemon, P.B.; Ridlon, J.M. Bile acid oxidation by *Eggerthella lenta* strains C592 and DSM 2243T. *Gut Microbes* **2018**, *9*, 523–539. [CrossRef]
35. Anwer, M.S.; Stieger, B. Sodium-dependent bile salt transporters of the SLC10A transporter family: More than solute transporters. *Pflügers Arch. Eur. J. Physiol.* **2014**, *466*, 77–89. [CrossRef]
36. Dawson, P.A. Role of the intestinal bile acid transporters in bile acid and drug disposition. *Handb. Exp. Pharmacol.* **2011**, *201*, 169–203. [CrossRef]
37. Zhang, Y.; Jiang, R.; Zheng, X.; Lei, S.; Huang, F.; Xie, G.; Kwee, S.; Yu, H.; Farrar, C.; Sun, B.; et al. Ursodeoxycholic acid accelerates bile acid enterohepatic circulation. *Br. J. Pharmacol.* **2019**, *176*, 2848–2863. [CrossRef]
38. Kliewer, S.A.; Mangelsdorf, D.J. Bile Acids as Hormones: The FXR-FGF15/19 Pathway. *Dig. Dis.* **2015**, *33*, 327–331. [CrossRef]
39. Dawson, P.A.; Lan, T.; Rao, A. Bile acid transporters. *J. Lipid Res.* **2009**, *50*, 2340–2357. [CrossRef]
40. Copple, B.L.; Li, T. Pharmacology of bile acid receptors: Evolution of bile acids from simple detergents to complex signaling molecules. *Pharmacol. Res.* **2016**, *104*, 9–21. [CrossRef]
41. Fang, C.; Filipp, F.V.; Smith, J.W. Unusual binding of ursodeoxycholic acid to ileal bile acid binding protein: Role in activation of FXRα. *J. Lipid Res.* **2012**, *53*, 664–673. [CrossRef]
42. Chiang, J.Y.L. Bile Acid Metabolism and Signaling. *Compr. Physiol.* **2013**, *3*, 1191–1212.
43. Yeo, X.Y.; Tan, L.Y.; Chae, W.R.; Lee, D.-Y.; Lee, Y.-A.; Wuestefeld, T.; Jung, S. Liver's influence on the brain through the action of bile acids. *Front. Neurosci.* **2023**, *17*, 1123967. [CrossRef]

44. Ticho, A.L.; Malhotra, P.; Dudeja, P.K.; Gill, R.K.; Alrefai, W.A. Intestinal absorption of bile acids in health and disease. *Compr. Physiol.* **2019**, *10*, 21–56. [CrossRef]
45. Ridlon, J.M.; Devendran, S.; Alves, J.M.; Doden, H.; Wolf, P.G.; Pereira, G.V.; Ly, L.; Volland, A.; Takei, H.; Nittono, H.; et al. The 'in vivo lifestyle' of bile acid 7α-dehydroxylating bacteria: Comparative genomics, metatranscriptomic, and bile acid metabolomics analysis of a defined microbial community in gnotobiotic mice. *Gut Microbes* **2020**, *11*, 381–404. [CrossRef]
46. Center, S.A. Serum Bile Acids in Companion Animal—Medicine. *Vet. Clin. N. Am. Small Anim. Pr.* **1993**, *23*, 625–657. [CrossRef]
47. Dawson, P.A.; Haywood, J.; Craddock, A.L.; Wilson, M.; Tietjen, M.; Kluckman, K.; Maeda, N.; Parks, J.S. Targeted deletion of the ileal bile acid transporter eliminates enterohepatic cycling of bile acids in mice. *J. Biol. Chem.* **2003**, *278*, 33920–33927. [CrossRef]
48. Deschamps, C.; Denis, S.; Humbert, D.; Priymenko, N.; Chalancon, S.; De Bodt, J.; Van de Wiele, T.; Ipharraguerre, I.; Alvarez-Acero, I.; Achard, C.; et al. Canine Mucosal Artificial Colon: Development of a new colonic in vitro model adapted to dog sizes. *Appl. Microbiol. Biotechnol.* **2024**, *108*, 166. [CrossRef]
49. Praslickova, D.; Torchia, E.C.; Sugiyama, M.G.; Magrane, E.J.; Zwicker, B.L.; Kolodzieyski, L.; Agellon, L.B. The ileal lipid binding protein is required for efficient absorption and transport of bile acids in the distal portion of the murine small intestine. *PLoS ONE* **2012**, *7*, e50810. [CrossRef]
50. Kook, P.H.; Schellenberg, S.; Rentsch, K.M.; Reusch, C.E.; Glaus, T.M. Effect of twice-daily oral administration of hydrocortisone on the bile acids composition of gallbladder bile in dogs. *Am. J. Vet. Res.* **2011**, *72*, 1607–1612. [CrossRef]
51. Jones, H.; Alpini, G.; Francis, H. Bile acid signaling and biliary functions. *Acta Pharm. Sin. B* **2015**, *5*, 123–128. [CrossRef]
52. Herstad, K.M.V.; Rønning, H.T.; Bakke, A.M.; Moe, L.; Skancke, E. Changes in the faecal bile acid profile in dogs fed dry food vs high content of beef: A pilot study. *Acta Vet. Scand.* **2018**, *60*, 29. [CrossRef]
53. Sitkin, S.; Pokrotnieks, J. Bad "good" bile acids and gut microbiota dysbiosis in inflammatory bowel disease: Mice and humans are not the same. *Dig. Dis. Sci.* **2021**, *66*, 925–927. [CrossRef]
54. Ju, J.; Zhang, C.; Yang, J.; Yang, Q.; Yin, P.; Sun, X. Deoxycholic acid exacerbates intestinal inflammation by modulating interleukin-1β expression and tuft cell proportion in dextran sulfate sodium-induced murine colitis. *PeerJ* **2023**, *11*, e14842. [CrossRef]
55. Lenci, I.; Milana, M.; Signorello, A.; Grassi, G.; Baiocchi, L. Secondary bile acids and the biliary epithelia: The good and the bad. *World J. Gastroenterol.* **2023**, *29*, 357–366. [CrossRef]
56. Kang, J.D.; Myers, C.J.; Harris, S.C.; Kakiyama, G.; Lee, I.-K.; Yun, B.-S.; Matsuzaki, K.; Furukawa, M.; Min, H.-K.; Bajaj, J.S.; et al. Bile Acid 7α-Dehydroxylating Gut Bacteria Secrete Antibiotics that Inhibit Clostridium difficile: Role of Secondary Bile Acids. *Cell Chem. Biol.* **2019**, *26*, 27–34.e4. [CrossRef]
57. Watanabe, M.; Fukiya, S.; Yokota, A. Comprehensive evaluation of the bactericidal activities of free bile acids in the large intestine of humans and rodents. *J. Lipid Res.* **2017**, *58*, 1143–1152. [CrossRef]
58. Kurdi, P.; Kawanishi, K.; Mizutani, K.; Yokota, A. Mechanism of growth inhibition by free bile acids in lactobacilli and bifidobacteria. *J. Bacteriol.* **2006**, *188*, 1979–1986. [CrossRef]
59. Schadt, H.S.; Wolf, A.; Pognan, F.; Chibout, S.-D.; Merz, M.; Kullak-Ublick, G.A. Bile acids in drug induced liver injury: Key players and surrogate markers. *Clin. Res. Hepatol. Gastroenterol.* **2016**, *40*, 257–266. [CrossRef]
60. Yu, Q.; Jiang, Z.; Zhang, L. Bile acid regulation: A novel therapeutic strategy in non-alcoholic fatty liver disease. *Pharmacol. Ther.* **2018**, *190*, 81–90. [CrossRef]
61. Jiang, L.; Schnabl, B. Gut microbiota in liver disease: What do we know and what do we not know? *Physiology* **2020**, *35*, 261–274. [CrossRef]
62. Slattery, S.A.; Niaz, O.; Aziz, Q.; Ford, A.C.; Farmer, A.D. Systematic review with meta-analysis: The prevalence of bile acid malabsorption in the irritable bowel syndrome with diarrhoea. *Aliment. Pharmacol. Ther.* **2015**, *42*, 3–11. [CrossRef]
63. Li, L.; Liu, T.; Gu, Y.; Wang, X.; Xie, R.; Sun, Y.; Wang, B.; Cao, H. Regulation of gut microbiota-bile acids axis by probiotics in inflammatory bowel disease. *Front. Immunol.* **2022**, *13*, 974305. [CrossRef]
64. Wang, X.; Chen, L.; Wang, H.; Cai, W.; Xie, Q. Modulation of bile acid profile by gut microbiota in chronic hepatitis B. *J. Cell Mol. Med.* **2020**, *24*, 2573–2581. [CrossRef]
65. Blake, A.B.; Guard, B.C.; Honneffer, J.B.; Lidbury, J.A.; Steiner, J.M.; Suchodolski, J.S. Altered microbiota, fecal lactate, and fecal bile acids in dogs with gastrointestinal disease. *PLoS ONE* **2019**, *14*, e0224454. [CrossRef]
66. Luo, L.; Aubrecht, J.; Li, D.; Warner, R.L.; Johnson, K.J.; Kenny, J.; Colangelo, J.L. Assessment of serum bile acid profiles as biomarkers of liver injury and liver disease in humans. *PLoS ONE* **2018**, *13*, e0193824. [CrossRef]
67. Kakiyama, G.; Pandak, W.M.; Gillevet, P.M.; Hylemon, P.B.; Heuman, D.M.; Daita, K.; Takei, H.; Muto, A.; Nittono, H.; Ridlon, J.M.; et al. Modulation of the fecal bile acid profile by gut microbiota in cirrhosis. *J. Hepatol.* **2013**, *58*, 949–955. [CrossRef]
68. Swann, J.R.; Want, E.J.; Geier, F.M.; Spagou, K.; Wilson, I.D.; Sidaway, J.E.; Nicholson, J.K.; Holmes, E. Systemic gut microbial modulation of bile acid metabolism in host tissue compartments. *Proc. Natl. Acad. Sci. USA* **2011**, *108* (Suppl. S1), 4523–4530. [CrossRef]
69. van Best, N.; Rolle-Kampczyk, U.; Schaap, F.G.; Basic, M.; Damink, S.W.M.O.; Bleich, A.; Savelkoul, P.H.M.; von Bergen, M.; Penders, J.; Hornef, M.W. Bile acids drive the newborn's gut microbiota maturation. *Nat. Commun.* **2020**, *11*, 3692. [CrossRef]
70. Kelly, O.B.; Mroz, M.S.; Ward, J.B.J.; Colliva, C.; Scharl, M.; Pellicciari, R.; Gilmer, J.F.; Fallon, P.G.; Hofmann, A.F.; Roda, A.; et al. Ursodeoxycholic acid attenuates colonic epithelial secretory function. *J. Physiol.* **2013**, *591*, 2307–2318. [CrossRef]

71. da Silva, T.C.; Polli, J.E.; Swaan, P.W. The solute carrier family 10 (SLC10): Beyond bile acid transport. *Mol. Asp. Med.* **2013**, *34*, 252–269. [CrossRef]
72. Ward, J.B.J.; Lajczak, N.K.; Kelly, O.B.; O'dwyer, A.M.; Giddam, A.K.; Gabhann, J.N.; Franco, P.; Tambuwala, M.M.; Jefferies, C.A.; Keely, S.; et al. Ursodeoxycholic acid and lithocholic acid exert anti-inflammatory actions in the colon. *Am. J. Physiol. Liver Physiol.* **2017**, *312*, G550–G558. [CrossRef]
73. MahmoudianDehkordi, S.; Arnold, M.; Nho, K.; Ahmad, S.; Jia, W.; Xie, G.; Louie, G.; Kueider-Paisley, A.; Moseley, M.A.; Thompson, J.W.; et al. Altered bile acid profile associates with cognitive impairment in Alzheimer's disease—An emerging role for gut microbiome. *Alzheimer's Dement.* **2019**, *15*, 76–92. [CrossRef]
74. Chiang, J.Y.L. Bile acids: Regulation of synthesis. *J. Lipid Res.* **2009**, *50*, 1955–1966. [CrossRef]
75. Winston, J.A.; Rivera, A.; Cai, J.; Patterson, A.D.; Theriot, C.M. Secondary bile acid ursodeoxycholic acid alters weight, the gut microbiota, and the bile acid pool in conventional mice. *PLoS ONE* **2021**, *16*, e0246161. [CrossRef]
76. Yang, Z.; Danzeng, A.; Liu, Q.; Zeng, C.; Xu, L.; Mo, J.; Pingcuo, C.; Wang, X.; Wang, C.; Zhang, B.; et al. The Role of Nuclear Receptors in the Pathogenesis and Treatment of Non-alcoholic Fatty Liver Disease. *Int. J. Biol. Sci.* **2024**, *20*, 113–126. [CrossRef]
77. Calzadilla, N.; Comiskey, S.M.; Dudeja, P.K.; Saksena, S.; Gill, R.K.; Alrefai, W.A. Bile acids as inflammatory mediators and modulators of intestinal permeability. *Front. Immunol.* **2022**, *13*, 1021924. [CrossRef]
78. O'kell, A.L.; Garrett, T.J.; Wasserfall, C.; Atkinson, M.A. Untargeted metabolomic analysis in naturally occurring canine diabetes mellitus identifies similarities to human Type 1 Diabetes. *Sci. Rep.* **2017**, *7*, 9467. [CrossRef]
79. Li, M.; Cai, S.-Y.; Boyer, J.L. Mechanisms of bile acid mediated inflammation in the liver. *Mol. Asp. Med.* **2017**, *56*, 45–53. [CrossRef]
80. Mackei, M.; Talabér, R.; Müller, L.; Sterczer, Á.; Fébel, H.; Neogrády, Z.; Mátis, G. Altered intestinal production of volatile fatty acids in dogs triggered by lactulose and psyllium treatment. *Vet. Sci.* **2022**, *9*, 206. [CrossRef]
81. Li, B.; Zhang, J.; Chen, Y.; Wang, Q.; Yan, L.; Wang, R.; Wei, Y.; You, Z.; Li, Y.; Miao, Q.; et al. Alterations in microbiota and their metabolites are associated with beneficial effects of bile acid sequestrant on icteric primary biliary Cholangitis. *Gut Microbes* **2021**, *13*, 1946366. [CrossRef]
82. Alshawaqfeh, M.K.; Wajid, B.; Minamoto, Y.; Markel, M.; Lidbury, J.A.; Steiner, J.M.; Serpedin, E.; Suchodolski, J.S. A dysbiosis index to assess microbial changes in fecal samples of dogs with chronic inflammatory enteropathy. *FEMS Microbiol. Ecol.* **2017**, *93*, fix136. [CrossRef] [PubMed]
83. Guard, B.C.; Honneffer, J.B.; Jergens, A.E.; Jonika, M.M.; Toresson, L.; Lawrence, Y.A.; Webb, C.B.; Hill, S.; Lidbury, J.A.; Steiner, J.M.; et al. Longitudinal assessment of microbial dysbiosis, fecal unconjugated bile acid concentrations, and disease activity in dogs with steroid-responsive chronic inflammatory enteropathy. *J. Vet. Intern. Med.* **2019**, *33*, 1295–1305. [CrossRef] [PubMed]
84. Liu, Y.; Rong, Z.; Xiang, D.; Zhang, C.; Liu, D. Detection technologies and metabolic profiling of bile acids: A comprehensive review. *Lipids Health Dis.* **2018**, *17*, 121. [CrossRef] [PubMed]
85. Vaz, F.M.; Ferdinandusse, S. Bile acid analysis in human disorders of bile acid biosynthesis. *Mol. Asp. Med.* **2017**, *56*, 10–24. [CrossRef] [PubMed]
86. Sarafian, M.H.; Lewis, M.R.; Pechlivanis, A.; Ralphs, S.; McPhail, M.J.W.; Patel, V.C.; Dumas, M.-E.; Holmes, E.; Nicholson, J.K. Bile acid profiling and quantification in biofluids using ultra-performance liquid chromatography tandem mass spectrometry. *Anal. Chem.* **2015**, *87*, 9662–9670. [CrossRef] [PubMed]
87. Dosedělová, V.; Itterheimová, P.; Kubáň, P. Analysis of bile acids in human biological samples by microcolumn separation techniques: A review. *Electrophoresis* **2021**, *42*, 68–85. [CrossRef] [PubMed]
88. Łuczykowski, K.; Warmuzińska, N.; Bojko, B. Current approaches to the analysis of bile and the determination of bile acids in various biological matrices as supportive tools to traditional diagnostic testing for liver dysfunction and biliary diseases. *TrAC Trends Anal. Chem.* **2021**, *142*, 116307. [CrossRef]
89. Gómez, C.; Stücheli, S.; Kratschmar, D.V.; Bouitbir, J.; Odermatt, A. Development and Validation of a Highly Sensitive LC-MS/MS Method for the Analysis of Bile Acids in Serum, Plasma, and Liver Tissue Samples. *Metabolites* **2020**, *10*, 282. [CrossRef]
90. Qin, S.; Tian, J.; Wang, L.; Zhao, Y.; Wang, D.; Wang, F.; Meng, J.; Liu, M.; Liang, A. Ultra-performance chromatography-electrospray tandem mass spectrometry analysis of bile acid profiles in the enterohepatic circulation following geniposide and acetaminophen-induced liver injury. *J. Chromatogr. A* **2022**, *1680*, 463417. [CrossRef]
91. Tian, M.; Yan, J.; Zhang, H.; Wei, Y.; Zhang, M.; Rao, Z.; Zhang, M.; Wang, H.; Wang, Y.; Li, X. Screening and validation of biomarkers for cadmium-induced liver injury based on targeted bile acid metabolomics. *Environ. Pollut.* **2022**, *300*, 118837. [CrossRef]
92. García-Cañaveras, J.C.; Donato, M.T.; Castell, J.V.; Lahoz, A. Targeted profiling of circulating and hepatic bile acids in human, mouse, and rat using a UPLC-MRM-MS-validated method. *J. Lipid Res.* **2012**, *53*, 2231–2241. [CrossRef]
93. Suzuki, Y.; Kaneko, R.; Nomura, M.; Naito, H.; Kitamori, K.; Nakajima, T.; Ogawa, T.; Hattori, H.; Seno, H.; Ishii, A. Simple and rapid quantitation of 21 bile acids in rat serum and liver by UPLC-MS-MS: Effect of high fat diet on glycine conjugates of rat bile acids. *Nagoya J. Med. Sci.* **2013**, *75*, 57–71. [PubMed]
94. Hagio, M.; Matsumoto, M.; Fukushima, M.; Hara, H.; Ishizuka, S. Improved analysis of bile acids in tissues and intestinal contents of rats using LC/ESI-MS. *J. Lipid Res.* **2009**, *50*, 173–180. [CrossRef]
95. La Frano, M.R.; Hernandez-Carretero, A.; Weber, N.; Borkowski, K.; Pedersen, T.L.; Osborn, O.; Newman, J.W. Diet-induced obesity and weight loss alter bile acid concentrations and bile acid–sensitive gene expression in insulin target tissues of C57BL/6J mice. *Nutr. Res.* **2017**, *46*, 11–21. [CrossRef]

96. Liu, Z.; Zhang, Y.; Zhang, R.; Gu, L.; Chen, X. Promotion of classic neutral bile acids synthesis pathway is responsible for cholesterol-lowing effect of Si-miao-yong-an decoction: Application of LC–MS/MS method to determine 6 major bile acids in rat liver and plasma. *J. Pharm. Biomed. Anal.* **2017**, *135*, 167–175. [CrossRef] [PubMed]
97. Slopianka, M.; Herrmann, A.; Pavkovic, M.; Ellinger-Ziegelbauer, H.; Ernst, R.; Mally, A.; Keck, M.; Riefke, B. Quantitative targeted bile acid profiling as new markers for DILI in a model of methapyrilene-induced liver injury in rats. *Toxicology* **2017**, *386*, 1–10. [CrossRef]
98. Wu, W.; Liu, X.; Peng, X.; Xue, R.; Ji, L.; Shen, X.; Chen, S.; Gu, J.; Zhang, S. Bile acids override steatosis in farnesoid X receptor deficient mice in a model of non-alcoholic steatohepatitis. *Biochem. Biophys. Res. Commun.* **2014**, *448*, 50–55. [CrossRef] [PubMed]
99. Thakare, R.; Alamoudi, J.A.; Gautam, N.; Rodrigues, A.D.; Alnouti, Y. Species differences in bile acids II. Bile acid metabolism. *J. Appl. Toxicol.* **2018**, *38*, 1336–1352. [CrossRef]
100. Wang, W.; Shi, Z.; Zhang, R.; Yu, J.; Wang, C.; Hou, J.; Sun, J.; Liu, Y.; Qin, K.; Liu, Y.; et al. Liver proteomics analysis reveals abnormal metabolism of bile acid and arachidonic acid in Chinese hamsters with type 2 diabetes mellitus. *J. Proteom.* **2021**, *239*, 104186. [CrossRef]
101. Wei, Y.; Cheng, J.; Luo, M.; Yang, S.; Xing, Q.; Cheng, J.; Lv, J.; Yu, C.; Sun, L.; Shi, D.; et al. Targeted metabolomics analysis of bile acids and cell biology studies reveal the critical role of glycodeoxycholic acid in buffalo follicular atresia. *J. Steroid Biochem. Mol. Biol.* **2022**, *221*, 106115. [CrossRef]
102. Humbert, L.; Rainteau, D.; Tuvignon, N.; Wolf, C.; Seksik, P.; Laugier, R.; Carrière, F. Postprandial bile acid levels in intestine and plasma reveal altered biliary circulation in chronic pancreatitis patients. *J. Lipid Res.* **2018**, *59*, 2202–2213. [CrossRef] [PubMed]
103. Naritaka, N.; Suzuki, M.; Sato, H.; Takei, H.; Murai, T.; Kurosawa, T.; Iida, T.; Nittono, H.; Shimizu, T. Profile of bile acids in fetal gallbladder and meconium using liquid chromatography-tandem mass spectrometry. *Clin. Chim. Acta* **2015**, *446*, 76–81. [CrossRef] [PubMed]
104. Ďurč, P.; Dosedělová, V.; Foret, F.; Dolina, J.; Konečný, Š.; Himmelsbach, M.; Buchberger, W.; Kubáň, P. Analysis of major bile acids in saliva samples of patients with Barrett's esophagus using high-performance liquid chromatography-electrospray ionization-mass spectrometry. *J. Chromatogr. A* **2020**, *1625*, 461278. [CrossRef] [PubMed]
105. Han, J.; Liu, Y.; Wang, R.; Yang, J.; Ling, V.; Borchers, C.H. Metabolic profiling of bile acids in human and mouse blood by lc–ms/ms in combination with phospholipid-depletion solid-phase extraction. *Anal. Chem.* **2015**, *87*, 1127–1136. [CrossRef] [PubMed]
106. Scherer, M.; Gnewuch, C.; Schmitz, G.; Liebisch, G. Rapid quantification of bile acids and their conjugates in serum by liquid chromatography–tandem mass spectrometry. *J. Chromatogr. B* **2009**, *877*, 3920–3925. [CrossRef]
107. Ulaszewska, M.M.; Mancini, A.; Garcia-Aloy, M.; Del Bubba, M.; Tuohy, K.M.; Vrhovsek, U. Isotopic dilution method for bile acid profiling reveals new sulfate glycine-conjugated dihydroxy bile acids and glucuronide bile acids in serum. *J. Pharm. Biomed. Anal.* **2019**, *173*, 1–17. [CrossRef] [PubMed]
108. Zheng, J.J.; Shields, E.E.; Snow, K.J.; Nelson, D.M.; Olah, T.V.; Reily, M.D.; Robertson, D.G.; Shipkova, P.A.; Stryker, S.A.; Xin, B.; et al. The utility of stable isotope labeled (SIL) analogues in the bioanalysis of endogenous compounds by LC-MS applied to the study of bile acids in a metabolomics assay. *Anal. Biochem.* **2016**, *503*, 71–78. [CrossRef] [PubMed]
109. Amplatz, B.; Zöhrer, E.; Haas, C.; Schäffer, M.; Stojakovic, T.; Jahnel, J.; Fauler, G. Bile acid preparation and comprehensive analysis by high performance liquid chromatography–high-resolution mass spectrometry. *Clin. Chim. Acta* **2017**, *464*, 85–92. [CrossRef]
110. Peng, Z.; Zhang, Q.; Mao, Z.; Wang, J.; Liu, C.; Lin, X.; Li, X.; Ji, W.; Fan, J.; Wang, M.; et al. A rapid quantitative analysis of bile acids, lysophosphatidylcholines and polyunsaturated fatty acids in biofluids based on ultraperformance liquid chromatography coupled with triple quadrupole tandem massspectrometry. *J. Chromatogr. B* **2017**, *1068–1069*, 343–351. [CrossRef]
111. Mouzaki, M.; Wang, A.Y.; Bandsma, R.; Comelli, E.M.; Arendt, B.M.; Zhang, L.; Fung, S.; Fischer, S.E.; McGilvray, I.G.; Allard, J.P. Bile Acids and Dysbiosis in Non-Alcoholic Fatty Liver Disease. *PLoS ONE* **2016**, *11*, e0151829. [CrossRef]
112. Balkman, C.E.; Center, S.A.; Randolph, J.F.; Trainor, D.; Warner, K.L.; Crawford, M.A.; Adachi, K.; Erb, H.N. Evaluation of urine sulfated and nonsulfated bile acids as a diagnostic test for liver disease in dogs. *J. Am. Vet. Med Assoc.* **2003**, *222*, 1368–1375. [CrossRef]
113. Ferslew, B.C.; Xie, G.; Johnston, C.K.; Su, M.; Stewart, P.W.; Jia, W.; Brouwer, K.L.; Barritt, A.S. Altered Bile Acid Metabolome in Patients with Nonalcoholic Steatohepatitis. *Dig. Dis. Sci.* **2015**, *60*, 3318–3328. [CrossRef]
114. Zhu, P.; Zhang, J.; Chen, Y.; Yin, S.; Su, M.; Xie, G.; Brouwer, K.L.R.; Liu, C.; Lan, K.; Jia, W. Analysis of human C24 bile acids metabolome in serum and urine based on enzyme digestion of conjugated bile acids and LC-MS determination of unconjugated bile acids. *Anal. Bioanal. Chem.* **2018**, *410*, 5287–5300. [CrossRef] [PubMed]
115. Gui, L.; Wu, Q.; Hu, Y.; Zeng, W.; Tan, X.; Zhu, P.; Li, X.; Yang, L.; Jia, W.; Liu, C.; et al. Compensatory Transition of Bile Acid Metabolism from Fecal Disposition of Secondary Bile Acids to Urinary Excretion of Primary Bile Acids Underlies Rifampicin-Induced Cholestasis in Beagle Dogs. *ACS Pharmacol. Transl. Sci.* **2021**, *4*, 1001–1013. [CrossRef] [PubMed]
116. Humbert, L.; Maubert, M.A.; Wolf, C.; Duboc, H.; Mahé, M.; Farabos, D.; Seksik, P.; Mallet, J.M.; Trugnan, G.; Masliah, J.; et al. Bile acid profiling in human biological samples: Comparison of extraction procedures and application to normal and cholestatic patients. *J. Chromatogr. B* **2012**, *899*, 135–145. [CrossRef]

117. Leibovitzh, H.; Nayeri, S.; Borowski, K.; Hernandez-Rocha, C.; Lee, S.H.; Turpin, W.; Stempak, J.; Sandhu, I.; Milgrom, R.; I Smith, M.; et al. P066 Lower serum concentration of sulfated bile acids is associated with Primary Sclerosing Cholangitis Inflammatory Bowel Disease comorbidity and advanced liver fibrosis. *J. Crohn's Colitis* **2022**, *16*, i173–i174. [CrossRef]
118. Vijayvargiya, P.; Busciglio, I.; Burton, D.; Donato, L.; Lueke, A.; Camilleri, M. Bile acid deficiency in a subgroup of patients with irritable bowel syndrome with constipation based on biomarkers in serum and fecal samples. *Clin. Gastroenterol. Hepatol.* **2018**, *16*, 522–527. [CrossRef] [PubMed]
119. Cui, Y.; Xu, B.; Zhang, X.; He, Y.; Shao, Y.; Ding, M. Diagnostic and therapeutic profiles of serum bile acids in women with intrahepatic cholestasis of pregnancy-a pseudo-targeted metabolomics study. *Clin. Chim. Acta* **2018**, *483*, 135–141. [CrossRef] [PubMed]
120. Gao, J.; Xu, B.; Zhang, X.; Cui, Y.; Deng, L.; Shi, Z.; Shao, Y.; Ding, M. Association between serum bile acid profiles and gestational diabetes mellitus: A targeted metabolomics study. *Clin. Chim. Acta* **2016**, *459*, 63–72. [CrossRef] [PubMed]
121. Lawrance, D.; Williamson, C.; Boutelle, M.G.; Cass, A.E.G. Development of a disposable bile acid biosensor for use in the management of cholestasis. *Anal. Methods* **2015**, *7*, 3714–3719. [CrossRef]
122. Luo, C.; Zhang, X.; He, Y.; Chen, H.; Liu, M.; Wang, H.; Tang, L.; Tu, G.; Ding, M. A pseudo-targeted metabolomics study based on serum bile acids profiling for the differential diagnosis of benign and malignant breast lesions. *Steroids* **2021**, *175*, 108914. [CrossRef]
123. Melgarejo, T.; Williams, D.A.; O'connell, N.C.; Setchell, K.D. Serum unconjugated bile acids as a test for intestinal bacterial overgrowth in dogs. *Dig. Dis. Sci.* **2000**, *45*, 407–414. [CrossRef] [PubMed]
124. Qi, Y.; Jiang, C.; Cheng, J.; Krausz, K.W.; Li, T.; Ferrell, J.M.; Gonzalez, F.J.; Chiang, J.Y. Bile acid signaling in lipid metabolism: Metabolomic and lipidomic analysis of lipid and bile acid markers linked to anti-obesity and anti-diabetes in mice. *Biochim. et Biophys. Acta (BBA) Mol. Cell Biol. Lipids* **2015**, *1851*, 19–29. [CrossRef]
125. Steiner, C.; von Eckardstein, A.; Rentsch, K.M. Quantification of the 15 major human bile acids and their precursor 7α-hydroxy-4-cholesten-3-one in serum by liquid chromatography–tandem mass spectrometry. *J. Chromatogr. B* **2010**, *878*, 2870–2880. [CrossRef]
126. Sugita, T.; Amano, K.; Nakano, M.; Masubuchi, N.; Sugihara, M.; Matsuura, T. Analysis of the serum bile acid composition for differential diagnosis in patients with liver disease. *Gastroenterol. Res. Pract.* **2015**, *2015*, 717431. [CrossRef]
127. Taguchi, K.; Fukusaki, E.; Bamba, T. Simultaneous and rapid analysis of bile acids including conjugates by supercritical fluid chromatography coupled to tandem mass spectrometry. *J. Chromatogr. A* **2013**, *1299*, 103–109. [CrossRef]
128. Thompson, M.B.; Chappell, J.D.; Kunze, D.J.; Blair, P.C. Bile acid profile in a dog with cholangiocarcinoma. *Vet. Pathol.* **1989**, *26*, 75–78. [CrossRef] [PubMed]
129. Washizu, T.; Tomoda, I.; Kaneko, J.J. Serum bile acid composition of the dog, cow, horse and human. *J. Vet. Med Sci.* **1991**, *53*, 81–86. [CrossRef]
130. Weiler, K.; Kleber, K.; Zielinsky, S.; Moritz, A.; Bauer, N. Analytical performance and method comparison of a quantitative point-of-care immunoassay for measurement of bile acids in cats and dogs. *J. Vet. Diagn. Investig.* **2020**, *33*, 35–46. [CrossRef]
131. Xie, G.; Wang, Y.; Wang, X.; Zhao, A.; Chen, T.; Ni, Y.; Wong, L.; Zhang, H.; Zhang, J.; Liu, C.; et al. Profiling of serum bile acids in a healthy chinese population using UPLC–MS/MS. *J. Proteome Res.* **2015**, *14*, 850–859. [CrossRef]
132. Xiong, Y.; Shi, C.; Zhong, F.; Liu, X.; Yang, P. LC-MS/MS and SWATH based serum metabolomics enables biomarker discovery in pancreatic cancer. *Clin. Chim. Acta* **2020**, *506*, 214–221. [CrossRef]
133. Gookin, J.L.; Mathews, K.G.; Cullen, J.; Seiler, G. Qualitative metabolomics profiling of serum and bile from dogs with gallbladder mucocele formation. *PLoS ONE* **2018**, *13*, e0191076. [CrossRef] [PubMed]
134. Washizu, T.; Ishida, T.; Washizu, M.; Tomoda, I.; Kaneko, J.J. Changes in bile acid composition of serum and gallbladder bile in bile duct ligated dogs. *J. Vet. Med Sci.* **1994**, *56*, 299–303. [CrossRef] [PubMed]
135. Woolbright, B.L.; Dorko, K.; Antoine, D.J.; Clarke, J.I.; Gholami, P.; Li, F.; Kumer, S.C.; Schmitt, T.M.; Forster, J.; Fan, F.; et al. Bile acid-induced necrosis in primary human hepatocytes and in patients with obstructive cholestasis. *Toxicol. Appl. Pharmacol.* **2015**, *283*, 168–177. [CrossRef] [PubMed]
136. Ashritha, A.; Lal, B.B.; Sood, V.; Khanna, R.; Tripathi, G.; Maras, J.; Alam, S. Analysis of plasma bile acid composition in Indian children with progressive familial intrahepatic cholestasis. *J. Clin. Exp. Hepatol.* **2023**, *13*, S2–S3. [CrossRef]
137. De Giorgi, S.; Campos, V.; Egli, L.; Toepel, U.; Carrel, G.; Cariou, B.; Rainteau, D.; Schneiter, P.; Tappy, L.; Giusti, V. Long-term effects of Roux-en-Y gastric bypass on postprandial plasma lipid and bile acids kinetics in female non diabetic subjects: A cross-sectional pilot study. *Clin. Nutr.* **2015**, *34*, 911–917. [CrossRef] [PubMed]
138. Gu, P.; Liu, R.-J.; Cheng, M.-L.; Wu, Y.; Zheng, L.; Liu, Y.-J.; Ma, P.-C.; Ding, L. Simultaneous quantification of chlorogenic acid and taurocholic acid in human plasma by LC-MS/MS and its application to a pharmacokinetic study after oral administration of Shuanghua Baihe tablets. *Chin. J. Nat. Med.* **2016**, *14*, 313–320. [CrossRef] [PubMed]
139. de Paiva, M.J.N.; Menezes, H.C.; da Silva, J.C.C.; Resende, R.R.; Cardeal, Z.d.L. New method for the determination of bile acids in human plasma by liquid-phase microextraction using liquid chromatography-ion-trap-time-of-flight mass spectrometry. *J. Chromatogr. A* **2015**, *1388*, 102–109. [CrossRef] [PubMed]
140. Wang, Z.; Li, H.; Yun, Y.; Wang, H.; Meng, B.; Mu, Y.; Gao, S.; Tao, X.; Chen, W. A dynamic multiple reaction monitoring strategy to develop and optimize targeted metabolomics methods: Analyzing bile acids in capecitabine-induced diarrhea. *J. Pharm. Biomed. Anal.* **2022**, *219*, 114938. [CrossRef]

141. Zhang, X.; Yang, Z.; Shi, Z.; Zhu, Z.; Li, C.; Du, Z.; Zhang, Y.; Wang, Z.; Jiao, Z.; Tian, X.; et al. Analysis of bile acid profile in plasma to differentiate cholangiocarcinoma from benign biliary diseases and healthy controls. *J. Steroid Biochem. Mol. Biol.* **2021**, *205*, 105775. [CrossRef] [PubMed]
142. Cao, Y.; Li, W.; Gong, X.; Niu, X.; Zheng, J.; Yu, J.; Li, J.; Tu, P.; Song, Y. Widely quasi-quantitative analysis enables temporal bile acids-targeted metabolomics in rat after oral administration of ursodeoxycholic acid. *Anal. Chim. Acta* **2022**, *1212*, 339885. [CrossRef] [PubMed]
143. Thakare, R.; Alamoudi, J.A.; Gautam, N.; Rodrigues, A.D.; Alnouti, Y. Species differences in bile acids I. Plasma and urine bile acid composition. *J. Appl. Toxicol.* **2018**, *38*, 1323–1335. [CrossRef] [PubMed]
144. Nakagaki, M.; Danzinger, R.G.; Hofmann, A.F.; DiPietro, R.A. Biliary secretion and hepatic metabolism of taurine-conjugated 7 alpha-hydroxy and 7 beta-hydroxy bile acids in the dog. Defective hepatic transport and bile hyposecretion. *Gastroenterology* **1984**, *87*, 647–659. [CrossRef] [PubMed]
145. Chen, X.; Zhang, X.; Xu, B.; Cui, Y.; He, Y.; Yang, T.; Shao, Y.; Ding, M. The urinary bile acid profiling analysis of asymptomatic hypercholanemia of pregnancy: A pseudo-targeted metabolomics study. *Clin. Chim. Acta* **2019**, *497*, 67–75. [CrossRef] [PubMed]
146. Li, Y.; Zhang, X.; Chen, J.; Feng, C.; He, Y.; Shao, Y.; Ding, M. Targeted metabolomics of sulfated bile acids in urine for the diagnosis and grading of intrahepatic cholestasis of pregnancy. *Genes Dis.* **2018**, *5*, 358–366. [CrossRef] [PubMed]
147. Lyu, J.; Li, H.; Yin, D.; Zhao, M.; Sun, Q.; Guo, M. Analysis of eight bile acids in urine of gastric cancer patients based on covalent organic framework enrichment coupled with liquid chromatography-tandem mass spectrometry. *J. Chromatogr. A* **2021**, *1653*, 462422. [CrossRef]
148. Muto, A.; Takei, H.; Unno, A.; Murai, T.; Kurosawa, T.; Ogawa, S.; Iida, T.; Ikegawa, S.; Mori, J.; Ohtake, A.; et al. Detection of Δ4-3-oxo-steroid 5β-reductase deficiency by LC–ESI-MS/MS measurement of urinary bile acids. *J. Chromatogr. B* **2012**, *900*, 24–31. [CrossRef]
149. Chaitman, J.; Ziese, A.-L.; Pilla, R.; Minamoto, Y.; Blake, A.B.; Guard, B.C.; Isaiah, A.; Lidbury, J.A.; Steiner, J.M.; Unterer, S.; et al. Fecal microbial and metabolic profiles in dogs with acute diarrhea receiving either fecal microbiota transplantation or oral metronidazole. *Front. Vet. Sci.* **2020**, *7*, 192. [CrossRef]
150. Grundy, S.M.; Ahrens, E.; Miettinen, T.A. Quantitative isolation and gas–liquid chromatographic analysis of total fecal bile acids. *J. Lipid Res.* **1965**, *6*, 397–410. [CrossRef]
151. He, Z.; Ma, Y.; Yang, S.; Zhang, S.; Liu, S.; Xiao, J.; Wang, Y.; Wang, W.; Yang, H.; Li, S.; et al. Gut microbiota-derived ursodeoxycholic acid from neonatal dairy calves improves intestinal homeostasis and colitis to attenuate extended-spectrum β-lactamase-producing enteroaggregative Escherichia coli infection. *Microbiome* **2022**, *10*, 79. [CrossRef]
152. Horvath, T.D.; Haidacher, S.J.; Hoch, K.M.; Auchtung, J.M.; Haag, A.M. A high-throughput LC-MS/MS method for the measurement of the bile acid/salt content in microbiome-derived sample sets. *MethodsX* **2020**, *7*, 100951. [CrossRef]
153. Hu, C.; Wang, W.; Garey, K.W. Heterogeneity and lyophilization comparison of stool processing for gastrointestinal bile acid measurement by LC-MS/MS. *J. Chromatogr. B* **2023**, *1214*, 123569. [CrossRef]
154. Jergens, A.E.; Guard, B.C.; Redfern, A.; Rossi, G.; Mochel, J.P.; Pilla, R.; Chandra, L.; Seo, Y.-J.; Steiner, J.M.; Lidbury, J.; et al. Microbiota-Related Changes in Unconjugated Fecal Bile Acids Are Associated with Naturally Occurring, Insulin-Dependent Diabetes Mellitus in Dogs. *Front. Vet. Sci.* **2019**, *6*, 199. [CrossRef]
155. Kakiyama, G.; Muto, A.; Takei, H.; Nittono, H.; Murai, T.; Kurosawa, T.; Hofmann, A.F.; Pandak, W.M.; Bajaj, J.S. A simple and accurate HPLC method for fecal bile acid profile in healthy and cirrhotic subjects: Validation by GC-MS and LC-MS. *J. Lipid Res.* **2014**, *55*, 978–990. [CrossRef]
156. Manchester, A.C.; Webb, C.B.; Blake, A.B.; Sarwar, F.; Lidbury, J.A.; Steiner, J.M.; Suchodolski, J.S. Long-term impact of tylosin on fecal microbiota and fecal bile acids of healthy dogs. *J. Vet. Intern. Med.* **2019**, *33*, 2605–2617. [CrossRef]
157. Marclay, M.; Dwyer, E.; Suchodolski, J.S.; Lidbury, J.A.; Steiner, J.M.; Gaschen, F.P. Recovery of Fecal Microbiome and Bile Acids in Healthy Dogs after Tylosin Administration with and without Fecal Microbiota Transplantation. *Vet. Sci.* **2022**, *9*, 324. [CrossRef] [PubMed]
158. Schmidt, M.; Unterer, S.; Suchodolski, J.S.; Honneffer, J.B.; Guard, B.C.; Lidbury, J.A.; Steiner, J.M.; Fritz, J.; Kölle, P. The fecal microbiome and metabolome differs between dogs fed Bones and Raw Food (BARF) diets and dogs fed commercial diets. *PLoS ONE* **2018**, *13*, e0201279. [CrossRef] [PubMed]
159. Shin, A.; Camilleri, M.; Vijayvargiya, P.; Busciglio, I.; Burton, D.; Ryks, M.; Rhoten, D.; Lueke, A.; Saenger, A.; Girtman, A.; et al. Bowel functions, fecal unconjugated primary and secondary bile acids, and colonic transit in patients with irritable bowel syndrome. *Clin. Gastroenterol. Hepatol.* **2013**, *11*, 1270–1275.e1. [CrossRef] [PubMed]
160. Cardinelli, C.d.S.; Torrinhas, R.S.; Sala, P.; Pudenzi, M.A.; Angolini, C.F.F.; da Silva, M.M.; Machado, N.M.; Ravacci, G.; Eberlin, M.N.; Waitzberg, D.L. Fecal bile acid profile after Roux-en-Y gastric bypass and its association with the remission of type 2 diabetes in obese women: A preliminary study. *Clin. Nutr.* **2019**, *38*, 2906–2912. [CrossRef]
161. Vijayvargiya, P.; Camilleri, M.; Chedid, V.; Carlson, P.; Busciglio, I.; Burton, D.; Donato, L.J. Analysis of fecal primary bile acids detects increased stool weight and colonic transit in patients with chronic functional diarrhea. *Clin. Gastroenterol. Hepatol.* **2019**, *17*, 922–929.e2. [CrossRef]
162. Weingarden, A.R.; Chen, C.; Zhang, N.; Graiziger, C.T.B.; Dosa, P.I.; Steer, C.J.; Shaughnessy, M.K.; Johnson, J.R.; Sadowsky, M.J.; Khoruts, A. Ursodeoxycholic Acid Inhibits Clostridium difficile Spore Germination and Vegetative Growth, and Prevents the Recurrence of Ileal Pouchitis Associated With the Infection. *J. Clin. Gastroenterol.* **2016**, *50*, 624–630. [CrossRef]

163. Lee, G.; Lee, H.; Hong, J.; Lee, S.H.; Jung, B.H. Quantitative profiling of bile acids in rat bile using ultrahigh-performance liquid chromatography–orbitrap mass spectrometry: Alteration of the bile acid composition with aging. *J. Chromatogr. B* **2016**, *1031*, 37–49. [CrossRef]
164. Rees, D.O.; Crick, P.J.; Jenkins, G.J.; Wang, Y.; Griffiths, W.J.; Brown, T.H.; Al-Sarireh, B. Comparison of the composition of bile acids in bile of patients with adenocarcinoma of the pancreas and benign disease. *J. Steroid Biochem. Mol. Biol.* **2017**, *174*, 290–295. [CrossRef]
165. Mooranian, A.; Ionescu, C.M.; Walker, D.; Jones, M.; Wagle, S.R.; Kovacevic, B.; Chester, J.; Foster, T.; Johnston, E.; Kuthubutheen, J.; et al. Single-Cellular Biological Effects of Cholesterol-Catabolic Bile Acid-Based Nano/Micro Capsules as Anti-Inflammatory Cell Protective Systems. *Biomolecules* **2022**, *12*, 73. [CrossRef] [PubMed]
166. Qiao, X.; Ye, M.; Liu, C.-F.; Yang, W.-Z.; Miao, W.-J.; Dong, J.; Guo, D.-A. A tandem mass spectrometric study of bile acids: Interpretation of fragmentation pathways and differentiation of steroid isomers. *Steroids* **2012**, *77*, 204–211. [CrossRef] [PubMed]
167. Trifunović, J.; Borčić, V.; Vukmirović, S.; Kon, S.G.; Mikov, M. Retention data of bile acids and their oxo derivatives in characterization of pharmacokinetic properties and in silico ADME modeling. *Eur. J. Pharm. Sci.* **2016**, *92*, 194–202. [CrossRef] [PubMed]
168. Dutta, M.; Cai, J.; Gui, W.; Patterson, A.D. A review of analytical platforms for accurate bile acid measurement. *Anal. Bioanal. Chem.* **2019**, *411*, 4541–4549. [CrossRef] [PubMed]
169. Zhao, X.; Liu, Z.; Sun, F.; Yao, L.; Yang, G.; Wang, K. Bile Acid Detection Techniques and Bile Acid-Related Diseases. *Front. Physiol.* **2022**, *13*, 826740. [CrossRef] [PubMed]
170. Schermerhorn, T.; Center, S.A.; Dykes, N.L.; Rowland, P.H.; Yeager, A.; Erb, H.; Oberhansley, K.; Bonda, M. Characterization of hepatoportal microvascular dysplasia in a kindred of cairn terriers. *J. Vet. Intern. Med.* **1996**, *10*, 219–230. [CrossRef] [PubMed]
171. Pena-Ramos, J.; Barker, L.; Saiz, R.; Walker, D.J.; Tappin, S.; Hare, C.H.Z.; Roberts, M.L.; Williams, T.L.; Bexfield, N. Resting and postprandial serum bile acid concentrations in dogs with liver disease. *J. Vet. Intern. Med.* **2021**, *35*, 1333–1341. [CrossRef]
172. McLaughlin, J. Fatty Acid-Induced Cholecystokinin Secretion and Effects of Human Gastric Motility. Ph.D. Thesis, Hope Hospital, Department of Medicine, Manchester, UK, 1998.
173. Pasman, W.J.; Heimerikx, J.; Rubingh, C.M.; Berg, R.v.D.; O'Shea, M.; Gambelli, L.; Hendriks, H.F.; Einerhand, A.W.; Scott, C.; Keizer, H.G.; et al. The effect of Korean pine nut oil on in vitro CCK release, on appetite sensations and on gut hormones in post-menopausal overweight women. *Lipids Health Dis.* **2008**, *7*, 10. [CrossRef] [PubMed]
174. Wang, B.J.; Cui, Z.J. How does cholecystokinin stimulate exocrine pancreatic secretion? From birds, rodents, to humans. *Am. J. Physiol. Integr. Comp. Physiol.* **2007**, *292*, R666–R678. [CrossRef] [PubMed]
175. van Straten, G.; Spee, B.; Rothuizen, J.; Favier, R. Diagnostic value of the rectal ammonia tolerance test, fasting plasma ammonia and fasting plasma bile acids for canine portosystemic shunting. *Vet. J.* **2015**, *204*, 282–286. [CrossRef] [PubMed]
176. Konstantinidis, A.O.; Patsikas, M.N.; Papazoglou, L.G.; Adamama-Moraitou, K.K. Congenital portosystemic shunts in dogs and cats: Classification, pathophysiology, clinical presentation and diagnosis. *Vet. Sci.* **2023**, *10*, 160. [CrossRef] [PubMed]
177. Gerritzen-Bruning, M.J.; van den Ingh, T.S.G.A.M.; Rothuizen, J. Diagnostic value of fasting plasma ammonia and bile acid concentrations in the identification of portosystemic shunting in dogs. *J. Vet. Intern. Med.* **2006**, *20*, 13–19. [CrossRef] [PubMed]
178. Weber, M.; Canin, A.R.; de France, P.A.V. Influence of size on the dog's digestive function. *Bull. L'académie Vétérinaire Fr.* **2006**, *159*, 327–332. [CrossRef]
179. Weber, M.P.; Biourge, V.C.; Nguyen, P.G. Digestive sensitivity varies according to size of dogs: A review. *J. Anim. Physiol. Anim. Nutr.* **2017**, *101*, 1–9. [CrossRef] [PubMed]
180. Koziolek, M.; Grimm, M.; Bollmann, T.; Schäfer, K.J.; Blattner, S.M.; Lotz, R.; Boeck, G.; Weitschies, W. Characterization of the GI transit conditions in Beagle dogs with a telemetric motility capsule. *Eur. J. Pharm. Biopharm.* **2019**, *136*, 221–230. [CrossRef]
181. Tolbert, M.K.; Telles, N.J.; Simon, B.T.; Scallan, E.M.; Price, J.M.; Gould, E.N.; Papich, M.G.; Lidbury, J.A.; Steiner, J.M.; Kathrani, A. Gastrointestinal transit time is faster in Beagle dogs compared to cats. *J. Am. Vet. Med Assoc.* **2022**, *260*, S8–S14. [CrossRef]
182. Sato, M.; Shibata, C.; Kikuchi, D.; Ikezawa, F.; Imoto, H.; Sasaki, I. Effects of biliary and pancreatic juice diversion into the ileum on gastrointestinal motility and gut hormone secretion in conscious dogs. *Surgery* **2010**, *148*, 1012–1019. [CrossRef] [PubMed]
183. Tanprasertsuk, J.; Shmalberg, J.; Maughan, H.; Tate, D.E.; Perry, L.M.; Jha, A.R.; Honaker, R.W. Heterogeneity of gut microbial responses in healthy household dogs transitioning from an extruded to a mildly cooked diet. *PeerJ* **2021**, *9*, e11648. [CrossRef] [PubMed]
184. Belchik, S.E.; Oba, P.M.; Wyss, R.; Asare, P.T.; Vidal, S.; Miao, Y.; Adesokan, Y.; Suchodolski, J.S.; Swanson, K.S. Effects of a milk oligosaccharide biosimilar on fecal characteristics, microbiota, and bile acid, calprotectin, and immunoglobulin concentrations of healthy adult dogs treated with metronidazole. *J. Anim. Sci.* **2023**, *101*, skad011. [CrossRef] [PubMed]
185. Merlen, G.; Ursic-Bedoya, J.; Jourdainne, V.; Kahale, N.; Glenisson, M.; Doignon, I.; Rainteau, D.; Tordjmann, T. Bile acids and their receptors during liver regeneration: "Dangerous protectors". *Mol. Asp. Med.* **2017**, *56*, 25–33. [CrossRef] [PubMed]
186. Chow, M.D.; Lee, Y.-H.; Guo, G.L. The role of bile acids in nonalcoholic fatty liver disease and nonalcoholic steatohepatitis. *Mol. Asp. Med.* **2017**, *56*, 34–44. [CrossRef] [PubMed]
187. Park, J.; Jang, K.; Jo, H.M.; Kim, S.E. Laparoscopic attenuation of a congenital extrahepatic portosystemic shunt in a dog—A thin-film banding for splenophrenic shunt: A case report. *Front. Vet. Sci.* **2022**, *9*, 918153. [CrossRef]
188. Anglin, E.V.; Lux, C.N.; Sun, X.; Folk, C.A.; Fazio, C. Clinical characteristics of, prognostic factors for, and long-term outcome of dogs with multiple acquired portosystemic shunts: 72 cases (2000–2018). *J. Am. Vet. Med Assoc.* **2021**, *260*, S30–S39. [CrossRef]

189. Ishigaki, K.; Asano, K.; Tamura, K.; Sakurai, N.; Terai, K.; Heishima, T.; Yoshida, O. Percutaneous transvenous coil embolization (PTCE) for treatment of single extrahepatic portosystemic shunt in dogs. *BMC Vet. Res.* **2023**, *19*, 215. [CrossRef]
190. Kim, D.; Oh, H.; Ahn, H.; An, B.; Park, D.; NA, K.-J.; Kim, G. Comparison of serum bile acid concentrations between maltese and other breeds of dogs with portosystemic shunt. *Vivo* **2023**, *37*, 709–713. [CrossRef] [PubMed]
191. Baptista, L.; Pollard, D.; Di Bella, A. Evaluation of Resting Serum Bile Acid Concentrations in Dogs with Sepsis. *Vet. Sci.* **2022**, *9*, 627. [CrossRef] [PubMed]
192. Ullal, T.V.; Lakin, S.; Gallagher, B.; Sbardellati, N.; Abdo, Z.; Twedt, D.C. Demographic and histopathologic features of dogs with abnormally high concentrations of hepatic copper. *J. Vet. Intern. Med.* **2022**, *36*, 2016–2027. [CrossRef] [PubMed]
193. Alvarenga, I.C.; Aldrich, C.G.; Jewell, D.E. Influence of Liver Condition and Copper on Selective Parameters of Post-Mortem Dog Tissue Samples. *Animals* **2018**, *8*, 237. [CrossRef] [PubMed]
194. Spee, B.; Arends, B.; van den Ingh, T.S.G.A.M.; Penning, L.C.; Rothuizen, J. Copper metabolism and oxidative stress in chronic inflammatory and cholestatic liver diseases in dogs. *J. Vet. Intern. Med.* **2006**, *20*, 1085–1092. [CrossRef] [PubMed]
195. Tai, N.; Wong, F.S.; Wen, L. The role of gut microbiota in the development of type 1, type 2 diabetes mellitus and obesity. *Rev. Endocr. Metab. Disord.* **2015**, *16*, 55–65. [CrossRef] [PubMed]
196. Wewalka, M.; Patti, M.-E.; Barbato, C.; Houten, S.M.; Goldfine, A.B. Fasting serum taurine-conjugated bile acids are elevated in type 2 diabetes and do not change with intensification of insulin. *J. Clin. Endocrinol. Metab.* **2014**, *99*, 1442–1451. [CrossRef]
197. German, A.; Day, M.; Ruaux, C.; Steiner, J.; Williams, D.; Hall, E. Comparison of Direct and Indirect Tests for Small Intestinal Bacterial Overgrowth and Antibiotic-Responsive Diarrhea in Dogs. *J. Vet. Intern. Med.* **2003**, *17*, 33–43. [CrossRef]
198. Comito, R.; Porru, E.; Interino, N.; Conti, M.; Terragni, R.; Gotti, R.; Candela, M.; Simoni, P.; Roda, A.; Fiori, J. Metabolic bile acid profile impairments in dogs affected by chronic inflammatory enteropathy. *Metabolites* **2023**, *13*, 980. [CrossRef] [PubMed]

MDPI AG
Grosspeteranlage 5
4052 Basel
Switzerland
Tel.: +41 61 683 77 34

Metabolites Editorial Office
E-mail: metabolites@mdpi.com
www.mdpi.com/journal/metabolites

www.ingramcontent.com/pod-product-compliance
Lightning Source LLC
Chambersburg PA
CBHW041346210926
51850CB00013B/3005